T0264551

HOW TO ORDER THIS BOOK

BY PHONE: 800-233-9936 or 717-291-5609, 8AM–5PM Eastern Time

BY FAX: 717-295-4538

BY MAIL: Order Department
Technomic Publishing Company, Inc.
851 New Holland Avenue, Box 3535
Lancaster, PA 17604, U.S.A.

BY CREDIT CARD: American Express, VISA, MasterCard

PERMISSION TO PHOTOCOPY–POLICY STATEMENT

Authorization to photocopy items for internal or personal use, or the internal or personal use of specific clients, is granted by Technomic Publishing Co., Inc. provided that the base fee of US $3.00 per copy, plus US $ .25 per page is paid directly to Copyright Clearance Center, 222 Rosewood Drive, Danvers, MA 01923, USA. For those organizations that have been granted a photocopy license by CCC, a separate system of payment has been arranged. The fee code for users of the Transactional Reporting Service is 1-56676/95 $5.00 + $ .25.

POLYMER FILMS IN
SENSOR APPLICATIONS

POLYMER FILMS in SENSOR APPLICATIONS

TECHNOLOGY, MATERIALS, DEVICES AND THEIR CHARACTERISTICS

Gábor Harsányi, Ph.D.

Technical University of Budapest
Budapest, Hungary

CRC Press
Taylor & Francis Group
Boca Raton London New York

CRC Press is an imprint of the
Taylor & Francis Group, an **informa** business

Polymer Films in Sensor Applications

a TECHNOMIC publication

CRC Press
Taylor & Francis Group
6000 Broken Sound Parkway NW, Suite 300
Boca Raton, FL 33487-2742

First issued in hardback 2017

© 1995 by Taylor & Francis Group, LLC
CRC Press is an imprint of Taylor & Francis Group, an Informa business

No claim to original U.S. Government works

This book contains information obtained from authentic and highly regarded sources. Reasonable efforts have been made to publish reliable data and information, but the author and publisher cannot assume responsibility for the validity of all materials or the consequences of their use. The authors and publishers have attempted to trace the copyright holders of all material reproduced in this publication and apologize to copyright holders if permission to publish in this form has not been obtained. If any copyright material has not been acknowledged please write and let us know so we may rectify in any future reprint.

Except as permitted under U.S. Copyright Law, no part of this book may be reprinted, reproduced, transmitted, or utilized in any form by any electronic, mechanical, or other means, now known or hereafter invented, including photocopying, microfilming, and recording, or in any information storage or retrieval system, without written permission from the publishers.

For permission to photocopy or use material electronically from this work, please access www. copyright.com (http://www.copyright.com/) or contact the Copyright Clearance Center, Inc. (CCC), 222 Rosewood Drive, Danvers, MA 01923, 978-750-8400. CCC is a not-for-profit organization that provides licenses and registration for a variety of users. For organizations that have been granted a photocopy license by the CCC, a separate system of payment has been arranged.

Trademark Notice: Product or corporate names may be trademarks or registered trademarks, and are used only for identification and explanation without intent to infringe.

Main entry under title:
  Polymer Films in Sensor Applications: Technology, Materials, Devices
    and Their Characteristics

A Technomic Publishing Company book
Bibliography: p.
Includes index p. 425

Library of Congress Catalog Card No. 94-61559
ISBN-13: 978-1-56676-201-4 (pbk)
ISBN-13: 978-1-138-45960-1 (hbk)

Visit the Taylor & Francis Web site at
http://www.taylorandfrancis.com

and the CRC Press Web site at
http://www.crcpress.com

To my Professor and mentor
Teréz Kormány
(She taught me not only science
but humanity as well)
and
To my wife, Éva
(Her impatience forced me to finish it in time)
and
To her (and my!) ambitions
(to give us a little relaxation)

Foreword ix

Preface xi

Acknowledgements xiii

Introduction xv

In the second half of the 20th century, the interest in the development of electric sensing devices, called sensors, has considerably increased. The scientific research was followed by an emerging demand from various fields of life.

In the 1960s, the technique of chemical sensors has grown rapidly and gave the possibility for direct detection of various ion- and molecule-types with certain selectivity limits. The research and development of conventional macrosensors soon turned in the direction of microsensors as a result of the miniaturization in microelectronics and expanding applications in biology, medicine, safety, and environmental protection. Meanwhile, automation in industry and, especially, automotive electronics required the development of various low-cost and reliable physical and chemical sensors.

Various material structures are used in sensors, depending on the nature of the quantity to be measured. Their spectrum ranges from single crystals to amorphous polymers. The sphere of various sensor types also means a broad spectrum from the mechanical and acoustic sensors to measure electrochemical, chemical, and biological quantities, including the conventional electric sensing transducers and even the new type fibre-optic sensors.

Although the practical application of sensors has been developing rapidly, the theoretical background of their operation is clarified only partly or not at all. There are debates about the signal excitation mechanisms, signal conditioning methods, and the interpretation of practically measurable and theoretically accepted results.

This book gives a broad survey of sensor types that are based on polymer films. The sensor technologies and the application possibilities of sensitive polymer films are described in detail. It also deals with various physical,

ix

chemical sensors and biosensors for different measurement purposes and signal ranges, pointing out their applications as well.

I recommend this book to anyone wishing to be informed about the possibilities for polymers in sensors (including technologies, sensor structures, sensing polymers, and applications) or wanting to start research and development in this area.

PROF. DR. H.C. ERNÖ PUNGOR
Member of Hungarian Academy of Sciences
Minister, President of the National Committee
for Technological Development
Budapest, Hungary

In recent years, polymeric materials have conquered a large part of sensorics. During my discussion with the publisher in the last months of 1992, I had not realized yet how enormous this area was. I have been working and teaching sensorics for more than a decade (including polymer-based sensors!); however, I was also surprised about the various application possibilities of polymers in this field. When the search of abstracting journal data bases was done for the purpose of this book, much information remained hidden because the authors of the articles often do not emphasize the application of polymers in their problems. Only detailed analysis of the papers revealed enough information for a general overview.

One surprising result was discovered: almost half of the promising chemical sensors are nowadays based on polymers. The great variety of properties that can be altered and controlled with different additives in polymers is the basis of their application for various chemical purposes. Physical sensors are based on only a few types of material, but the great choice of the different sensor types that can be operated with them makes the situation just as complicated.

One important reason for the increase in sensor applications is the remarkable progress that has been made in the field of disordered polymeric materials in both theoretical and experimental aspects and also the great advantages made in the polymer film preparation technologies. From another point of view, the progress is also accelerated by microelectronics, its devices, and its techniques, which tend to force sensorics in a direction to consume low-cost materials compatible with electronic technologies.

As mentioned above, there have been a number of articles published on the subject of sensing polymer films; however, there is still no comprehensive survey on this topic. This book gives a broad summary about polymer films used in sensors.

The reader will find a strong emphasis not only on "what" and "how," but also on "why." Both specialists and newcomers will find this compact survey easy to use. Moreover, researchers and development engineers can access the book according to physical, chemical, and technical fundamentals, and since it contains comparisons and assessments of the various types of sensors with respect to their practical applications, users will also benefit from the volume. Newcomers coming both from the area of engineering and from chemistry can use it easily.

The most important definitions in connection with sensorics and polymer materials are summarized in the Introduction. However, a little knowledge from the following areas is helpful (but not necessary) for a better understanding: Solid State Materials, Devices and Technologies, Networks and Systems, Polymer Science, Analytical Chemistry, and Biochemistry.

The first chapter is devoted to the technology of sensors, including both inorganic and polymeric films. It introduces the reader to the most important technologies to see how up-to-date sensor structures can be fabricated and to clarify which processes are compatible with the polymer film deposition techniques.

The second chapter describes the sensor structures in which sensing polymers can be applied, also giving their principles of operation. It is also a good general survey about microsensors.

The third chapter completes the presentation of operation principles by describing the most important physical and chemical sensing mechanisms that can be achieved in polymer materials. This is a very broad area that needs a lot of theoretical and experimental experience to be able to write a summary; therefore, I decided to collaborate with the best scientists from each particular field. Thus, this chapter is partly a contributed one.

The discussion then turns to applications in Chapter 4, with a comprehensive description of the sensor types operating with sensing polymer films.

The Appendix gives a survey of the common polymer structures and a few examples of realized sensor structures. A Glossary of Abbreviations is also given to make reading easier.

As mentioned, many sources, in addition to the author's personal knowledge and experience, have been used to compile the information presented here: experts as contributors, trade literature, and papers that contain relevant up-to-date material. The sources are referred to and listed in the "References and Supplementary Reading" section at the end of each chapter. A number of copyrighted figures have been borrowed from other sources with permission of the copyright owners.

The author is deeply grateful to the many individuals and institutions who made their time and knowledge available to help put this material together. First of all, special thanks to the contributors of special sections, who accepted the invitation to improve the quality of this book: Prof. R. D. Armstrong, University of Newcastle upon Tyne, UK; Dr. M. H. Abraham, University College London, UK; Dr. G. Bidan, Grenoble Nuclear Research Center, France; Prof. W. Göpel, University of Tübingen, Germany; Dr. J. W. Grate, Pacific Northwest Laboratory, USA; Prof. G. Horvai, Technical University of Budapest, Hungary; Dr. R. A. McGill, Naval Research Laboratory, USA; Prof. Y. Sakai, Ehime University, Japan; and Dr. K. D. Schierbaum, University of Tübingen, Germany.

I would also like to express my thanks to Prof. Ernő Pungor for writing the Foreword for this book. He is not only the leader of the present scientific life in Hungary (as minister), but is an outstanding scientist in the field of ion-selective polymeric membranes.

I am also grateful to all the publishers and authors who have given their consent to publish borrowed material. Their names and credit are given in the appropriate places.

Many thanks to my friends and colleagues for reviewing the book and for helpful advice: to Dr. András Bezegh, who has personally conducted research in connection with polymer based ISFET (at the University of Utah), for his criticizing comments in connection with chemical sensors and biosensors; to Prof. Rajendra P. Agarwal (University of Roorkee, India) for the good ideas in connection with silicon processing and for general review; and to Prof. Emil Hahn who taught me Sensors and Actuators at the Technical University of Budapest, and for helping me in preparing the Introduction.

And last but not least, I have to mention the great help and technical assistance provided by Éva Antal for the word processing, by Anikó Gyürki in the preparation of the drawings and figures, and by Erika Tóth in searching for literature sources.

Finally, much of the work described here was accomplished with the help of a number of outstanding graduate and postgraduate students, and I would like to acknowledge their enthusiasm by giving their names: Zsolt Keresztes-Nagy, Csaba Császár, Levente Pércsi, Róbert Sali, and János Skorutyák.

I am also grateful to the publisher (namely, Joseph Eckenrode, Ph.D., and Ms. Susan G. Farmer) for their willingness to give prompt assistance during the whole period of work.

And I would be remiss if I did not acknowledge those who have financially supported my work through the past few years: the National Scientific Research Fund (OTKA, project No. F007365) and The National Committee for Technological Development (OMFB).

GÁBOR HARSÁNYI, PH.D.

Synthetic polymers seem to be the most widely used materials of the 20th century. Their ancestors, naturally occurring macromolecular materials such as natural rubber, cotton, and hardened oils, have been known and used much longer, but synthetic polymers were first produced in the 19th century. These materials now dominate so many applications that it is hardly possible to estimate their importance in this book.

There is, at present, considerable interest in the use of polymers as components of microelectronic systems. The particular features of polymers that make them attractive for these applications are their mouldability, conformability, and extreme ease of deposition in thin- or thick-film form. Thin polymeric films are excellent interlayer dielectrics in multilayer metallization structures, which means an application as insulating layers. However, it has become clear in the last fifteen years that some polymers can be processed to have semiconducting or even metallic behaviour. A relatively new application is the field of electric (or optical) sensing devices, which is exactly the subject of present studies.

Integrated circuits have also changed many aspects of our life. Moreover, while the microelectronic revolution has been spreading, the IC technologies have been gradually transforming the way of engineers' and scientists' thinking about sensing devices.

The microprocessor is deaf, dumb, and blind without suitable sensors to provide input from the surrounding world of physical and chemical variables, such as pressure, acceleration, flowrate, temperature, humidity, concentration of chemical compounds, etc. Without actuators, it is powerless to carry out control functions. Widespread application of microprocessors and related memory units, which are able to handle and store signals (the carriers of information), has created the need for low-cost, high-performance sensors and actuators in large quantities. These devices that can be created using IC batch processing techniques similar to the one used to manufacture microprocessors, memories, and/or interconnecting systems are finding their way into a myriad of new applications.

Conventional application fields of sensors are measuring and testing different physical and chemical quantities and the industrial process control and/or automation. The new areas are mainly from consumer fields (e.g., domestic and automotive electronics, household, safety, comfort, and pleasure) and from the high advanced scientific areas (e.g., medicine, environment protection, research, etc.). The former needs low cost and the latter high reliability sensors.

The great technical possibilities and the relatively low cost of microprocessor-based systems have also turned the most important requirements into other directions. A few years ago, the expected properties were as follows:

- good linearity
- small hysteresis
- small offset
- low temperature drift
- low interference effects
- good interchangeability
- good long-term stability and reliability

From these items, only the last one should be kept. The others can be compensated or calculated using microprocessor-based systems and additional sensors (e.g. temperature compensation).

On the other hand, the most important new requirements are low cost, physical compatibility with IC and interconnection technologies in order to be able to integrate sensors into arrays and together with signal processing units, and electrical compatibility with microprocessors [it is desirable to provide digital or pseudo-digital (e.g., frequency) output], and the demand on multifunction sensors and integrated sensor arrays is increasing because of the necessity for compensation of interferences. One group of sensors of current interest is therefore silicon sensors.

Silicon is a very favourable material for sensors because it shows many physical and chemical effects for sensing purposes. The application of silicon technology offers the following advantages:

- There is the possibility of batch fabrication, thereby reducing the costs.
- The dimensions of the sensors are very small, the power consumption and time constant is also small, and the sensor does not modify the quantity to be measured.

- The sensor and the processor may be integrated on the same chip. All the steps of IC technology such as photo- or electron-lithography, ion implantation, wet and dry etching, metallization, etc., can be used for sensor fabrication.
- There is the possibility of micromachining, i.e., the formation of three-dimensional structures by isotrope and anisotrope etching.

However, the use of silicon has some drawbacks as well:

- The electric parameters have a strong temperature dependence.
- The semiconductor processing needs very high-cost machinery and investment.
- The processing of many sensor materials, such as ceramics, glasses, metals, etc., is not compatible with that of the silicon.

Another group of sensors is based on ceramics and glasses. These materials are generally produced at high-temperature processes. The batch production also has some difficulties. The latter can be eliminated combining them with thin and thick films, which is a usual technique to produce high-density interconnection systems. Combining silicon or other semiconductor chips with thin- and thick-film interconnection and/or passive networks, hybrid devices and sensors can be fabricated. Polymer thick-film technology and surface mounting means an extension of hybrids on different polymeric substrates.

The hybrid technology in sensors also has several advantages:

- Batch fabrication is possible with moderate investment; hence, low-cost sensors can be produced.
- It means an easy way of integration with multilayer interconnection systems.
- A wide choice of materials is applicable.
- There is a flexibility in the production.
- High stability and low temperature drift of the passive network elements exist.

Recently, a great competition has been started between electric and fibre-optic sensing devices. Optical fibres offer the same advantages to transducer systems that they have in telecommunication:

- low signal attenuation
- high information transfer capacity
- elimination of electromagnetic interference problems
- flexibility

Further advantages, especially for sensors, are explosion safety and biocompatibility of most materials.

During recent years, polymeric materials have gained a wide theoretical interest and practical application in sensorics. They can be used for very different purposes and may offer unique possibilities.

Polymers offer many advantages for sensor technologies: they are relatively low-cost materials; their fabrication techniques are quite simple (there is no need for special clean-room and/or high-temperature processes); they can be deposited on various types of substrates; and the wide choice of their molecular structure and the possibility to build in side chains, charged or neutral particles, and even grains of specific behaviour into the bulk material or on its surface region enables the production films with various physical and chemical properties, including also sensing behaviour.

The active sensing polymers are used in sheet or film form built into inorganic solid-state devices as an integral part. The latter is fabricated using monolithic semiconductor processing, processing of ceramics and glasses, or thin- and thick-film technologies. The deposition and patterning of microsensor polymer films include the spinning and/or casting and photolithography of photosensitive polymers, the same of nonphotosensitive polymers, printing and subsequent cross-linking (UV, IR, or heat) of polymers, electrochemical polymerization, vacuum deposition of polymer films (evaporation, sputtering, plasma polymerization), and other techniques such as the Langmuir-Blodgett method, gamma-irradiation, etc.

The typical device structures that can be used to measure the changes in the properties of the polymer films can be categorized into the following groups: impedance-type sensors, semiconductor-based ones (e.g., FETs), resonant sensors (BAW, SAW, FPW, etc.), electrochemical cells, calorimetric sensors, and fibre-optic sensors.

The sensing effects in polymers can be described using the following grouping of materials: dielectrics, conductive composites, electrolytes, electroconducting conjugated polymers, sorbents, ion-exchange membranes, permselective membranes, membranes with specific recognition sites, and optically sensitive polymers.

The various sensors applying sensing polymers can be grouped according to the quantity to be measured: temperature sensors, mechanical sensors (touch switch devices, deformation sensors, tactile sensors, pressure sensors, accelerometers, vibrometers, etc.), acoustic sensors (microphones, ultrasonic sensors, hydrophones, etc.), radiation sensors, humidity sensors, gas sensors, ion-selective sensors, special sensors in medicine and biology, and others (liquid component sensors, material identification, etc.). More recently, the application of polymers in actuators has also gained considerable attention.

As a summary, it can be recognized that

- Sensors and actuators mean the extension of intelligent electronic systems to be able to do a self-controlled communication with their environment; thus, they might be the key devices of the next century.
- Polymers and semiconductors are the materials, and their processing techniques are the technologies of our century. Semiconductor devices are the basis of intelligent systems and polymers mean the substrates for "intelligent materials."
- If these objects happen to find each other, really great horizons might be opened.

I.1 SENSORS AND THEIR CHARACTERISTICS

The most difficult problem is to define the basic ideas that we are dealing with. The author of this book has been teaching "Sensors and Actuators" and conducting research in connection with this topic for more than a decade. However, it has been very difficult to find a good definition for sensors. Even the authors of the books and articles about this topic try to avoid dealing with this problem.

Norton [1] defines sensors in connection with instrumentation systems that can be categorized into *measurement* (especially analysis) or *control* systems:

> In *measurement systems,* a quantity or property is measured and the measured value is displayed. *Analyzers* are measuring systems whose purpose is to display the nature and proportion of the constituents of a substance or quantity. In *control systems* the information about a quantity or property that is being measured is used to control the quantity or property so that its measured value should equal a desired value.

The simplest measuring system is a measuring device that also displays the measured value, e.g., a mercury-in-glass thermometer. However, nobody thinks that it is a "sensor" or that it contains one. Moreover, a bi-metal switching unit cannot be handled as an integrated "sensor-actuator," although it is a controlling device.

Janata and Bezegh [2] reported in 1988 in a review paper about chemical sensors: "It seems that so far nobody has been able to give an unambiguous and universally accepted definition." Then they gave the following definition: "A chemical sensor is a transducer which provides direct information about the chemical composition of its environment." However, the problem is still there to exclude analytical instruments that are not considered sensors. The "direct information" is also too much: a pH indicator dye alone is not a sensor, although it provides direct information about pH.

According to Norton's book, the sensor "converts the measurand into a usable electrical output." This approach does not contain the fibre-optic sensors that give optical output. If the word *electrical* in this definition is replaced by *electrical or optical,* the definition will be more general, but it is only a simplification of the problem.

The nature of the output (electrical or optical) does not seem to be important. The importance is that there is a signal or a change of a signal, which holds the information about the measurand and can be directly transmitted by one of the information transmission channels. Thus, the following definition can be suggested: *the sensor is a transducer that converts the measurand into a signal.* Similarly, *the actuator converts a signal into an action.* The schematic structure of such a system is shown in Figure I.1.

There are several possible levels when integrating signal conditioning units with sensors. All of them rely on technologies developed for monolithic and hybrid integrated circuits. The first level of signal conditioning that can be integrated on a chip is balancing of the offset, compensating the temperature drift, etc. A higher level of integration may include amplification and signal conversion. The third level, which is the visible technology trend of *smart sensors,* is the incorporation of a microprocessor for performing different functions at the sensor level, such as

- analog/digital conversion and digital processing
- compensation of static errors
- self-calibration and testing
- automatic switching of the measurement ranges
- recording of the measured data
- calculation of mean values, tolerances, etc.
- possibility of multisensor signal processing

The smart sensors represent the "New Wave" of sensors in our age and in the future.

Sensor arrays are integrated sensors consisting of the same or similar sensor structures with the same or similar function.

Multisensors consist of several sensors having different functions.

A *multifunction sensor* is a single device that can realize several different sensor functions under different conditions.

Generator-type sensors operate without external excitation, while *modulator-type sensors* need an external source.

Sensors can be classified according to the *nature of interaction,* which means the basis of operation into the groups of physical sensors, chemical sensors, and biosensors.

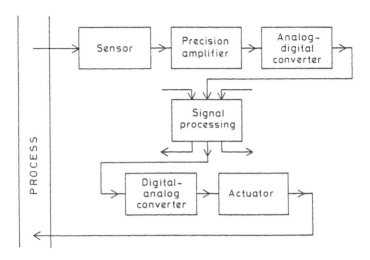

FIGURE I.1. Schematic structure of a process-controlling unit using sensors and actuators.

Another classification is possible according to the *quantity to be measured* into the groups of sensors for the measurement of

- geometrical quantities (position, displacement, etc.)
- mechanical quantities (force, acceleration, pressure, flowrate, rpm, etc.)
- thermal quantities (temperature, heat flow, etc.)
- electrostatic and magnetic fields
- radiation intensity (electromagnetic, nuclear, sound, etc.)
- chemical quantities (concentration of humidity, gas components, ions, etc.)
- biological quantities (concentration of enzymes, antibodies, etc.)

Static characteristics give the most important behaviour of sensors: the relationship between the output signal and the measurand.

Sensitivity is defined as the slope of the former function. In linear ranges, it is a constant [see Figure I.2(a)].

Full-scale output (FSO) of the sensor is the maximum (or nominal) output signal.

Linearity is the closeness of the sensor's calibration curve to a specified straight line. It is expressed as a percent of FSO and means the maximum deviation of any calibration point from the corresponding point on the specified straight line.

Hysteresis is the maximum difference in output, at any measurand values within the specified range, when the value is approached first with increasing and then with decreasing measurand [see Figure I.2(b)]. It is also given in percent FSO.

The *limit of detection* is the lowest value of measurand that can be detected by the sensor.

Resolution is the smallest increment in the output given in percent FSO.

Repeatability (or sometimes called reproducibility) is the ability of a sensor to reproduce output readings when the same measurand value is applied to it consecutively, under the same conditions, and in the same direction.

The *zero-measurand output* ("the zero") is the output when zero measurand is applied.

Zero shift (or drift) is a change in the zero-measurand output under specified conditions (e.g., temperature change, long-term storage, aging, etc.). Similarly, *sensitivity shift* is a change in the slope of the calibration curve.

The *response time* is the length of time required for the output to rise to a specified percentage of its final value (as a result of a step change in measurand).

Selectivity means the suppression of the environmental interferences. The ideal sensor will only respond to changes in the measurand. However, in practice, sensors can also respond to changes in other quantities, e.g., temperature. This cross-effect can be eliminated by compensation, multi-sensor operation, etc.

Lifetime is also an important property of sensors: it is the length of time that they remain sensitive under normal operational conditions.

I.2 POLYMERS AND THEIR PROPERTIES [3–6]

It can hardly be imagined that someone starts to study polymer sensing films in sensors without any knowledge in connection with polymer sciences. However, it is also difficult to write this book without a short survey about polymers.

Polymers are very long molecules, and while a few of commercial interest are formed in nature, such as cellulose, most are synthesized using a chemical process called polymerization. In this process, small molecules, called monomers, are joined together, end to end, in a growing chain. Eventually, the length of the polymer chain reaches a point where the macroscopic properties are altered significantly.

A *monomer* is a small molecule that is reacted to produce a polymer. Only the portion that is actually used in the chemical reaction is considered

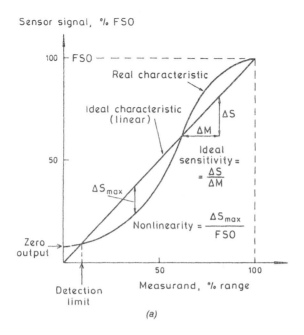

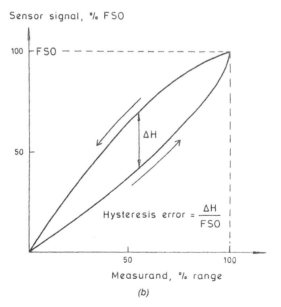

FIGURE I.2 Calibration curves for sensors indicating the most important properties: (a) sensitivity, nonlinearity, detection limit, zero output and (b) hysteresis error.

to be the monomer. The monomeric units of the common polymers that are often mentioned in this book are given in Appendix 1. The special sensing polymers are generally described within the chapters.

The most common interaction between atoms within a polymer molecule is the *covalent bond*. This type of chemical bonding not only produces a stable bond of high strength, but also allows the entire molecule to remain uniformly strong with each additional bond added to it. The electrons that form the covalent bond are in the outer shell of the bonding atom and are either *s*- or *p*-type. Almost all covalent bonds have some element of ionic bonding. The bent bond confiruation does not increase the amount of ionic character normally present, but the resonating bond and multiple bonds do increase the ionic portion of the bonding. Additionally, there are often major components of ionic bonding existing in a polymer bond either due to pedant groups that are ionic in nature or due to a backbone that contains dissimilar atoms. An example of the former is the OH radical; of the latter is the Si-O backbone bonds of silicates.

Backbone is the series of atoms chemically bonded in a linear array that characterizes the structure of the polymer.

A *carbon ring* is a ring of six carbon atoms with three resonating double bonds.

A *conjugated double bond* is the alteration of a single and double bond along a backbone of a polymer. A ring of atoms, such as a carbon ring, may be considered as a unit along a backbone.

Resonating multiple bonds are present when an atom is bonded with a multiple bond on one side and a lesser number of bonds on the other side. One or more on the extra bonds may be able to switch positions.

Pedant group is a cluster of atoms chemically bonded to the backbone of a polymer.

A *radical* is a chemically bonded group of atoms with a net electrical charge.

The so-called *van der Waals bond* is a form of secondary bonding that may be considered to include hydrogen bonding. It acts between separate polymeric chains.

Addition polymerization results in a product incapable of further reaction.

Condensation polymerization is a reaction that results in a product, which itself is further capable of reaction. Usually in condensation reactions, a small molecule such as water is split off, and the remainder of the molecule reacts with other similar molecules.

A *linear polymer* has chain molecules without branching and cross-linking.

Cross-linking is a chemical bonding, often with the addition of another atom, of one polymeric chain to another.

A *copolymer* is a polymeric material made up of two or more different types of monomers.

Graft polymerization is the addition of a second monomer along a polymer chain other than at the chain's end.

Photopolymerization is the selective polymerization (cross-linking) of light-sensitive monomers upon exposure to a light source.

An *initiator* is a reactive specimen, such as a free radical, which can be added to an unsaturated bond.

A *plasticizer* is a liquid or solid additive used to make a polymer more fluid.

A *filler* is an additive used to increase the volume of the polymer or to enter new properties.

A *reinforcement* additive serves to increase the tensile or strength of the polymer.

Foam is a two-phase mixture of polymer and gas in which the gas is the foaming agent.

Below its melting point, a *crystalline polymer* has a crystal-like structure: it arranges itself into a regular three-dimensional array with a small repeating unit.

Amorphous polymers possess no regular crystalline structure.

A *glassy state* is a metastable phase of polymers. It is characterized by the short-range order and short-time stiffness.

A *rubber state* is exhibited by the most amorphous polymers. The material is capable of recovering from large deformations quickly and forcibly.

A *visco-fluid state* exists in addition to the rubber and glassy state for amorphous polymers. It does not have the fluidity of a true liquid, but the molecules do possess sufficient thermal mobility to move past one another.

Swelling means the dissolving of a solvent into a cross-linked polymer, which will elastically stretch to accommodate the solvent.

Thermoplastic polymers can melt without decomposing.

Thermosets are polymers that become hard when heated and will not thereafter melt without decomposing.

Elastomers are polymers that are able to withstand large deformation and return to their original shape when the force is removed. The most effective elastomers are cross-linked polymers.

Extrusion is the process in which a plastic material is forced through a die.

Injection moulding is a process in which a viscous polymer is injected at high temperature and under pressure into a closed, relatively cold cavity where it cools and hardens.

I.3 REFERENCES AND SUPPLEMENTARY READING

1. Norton, H. N., *Sensor and Analyzer Handbook,* Prentice Hall, Inc., Englewood Cliffs, NJ (1982).
2. Janata, J. and Bezegh, A., "Chemical Sensors," *Anal. Chem.,* 60 (1988), pp. 62R–74R.
3. Hawkins, W. L., "Polymer Chemistry," in *Physical Design of Electronic Systems,* Vol. II., (ed. Everitt, W. L.) Prentice-Hall, Inc., Englewood Cliffs, NJ (1970).
4. Osborn, K. R. and Jenkins, W. A., *Plastic Films, Technology and Packaging Applications,* Technomic Publishing Co., Inc., Lancaster, PA (1992).
5. Daniels, C. A., *Polymers: Structure and Properties,* Technomic Publishing Co., Inc., Lancaster, PA (1989).
6. Leevers, P. S., "Plastics and Polymers," in *Materials Science* (eds., Anderson, J. C., Leaver, K. D., Rawlings, R. D. and Alexander, J. M.), Chapman and Hall, London, New York (1990).

CHAPTER 1

Sensor Technologies

Over the years, a number of materials and technologies have been developed in the fabrication sensors. Some of these are relatively new, while others are well developed and have been mastered to a sufficiently advanced level, like silicon technology.

Both large categories of materials, viz., inorganic and organic types, are nowadays used in the fabrication of sensors. Inorganic materials include single crystals such as quartz, silicon, compound semiconductors; polycrystalline and amorphous materials such as ceramics, glasses, and their composites; and metals. Organic materials applied in sensors are mainly polymers; however, lipids, enzymes, and biochemical compounds have also found an expanding use recently.

In recent years, polymers have found an increasing role in sensorics due to their unique characteristics, and a number of new sensors have been developed. Their application is relatively new in this field, but they seem to be very promising for sensorics to build up cheap and reliable sensors. The great variety of properties gives a lot of application possibilities. Materials and technologies are mostly in the state of development.

The active sensing polymers are used in sheet or film form built into inorganic solid-state devices as an integral part. The interaction flowchart of sensor operation is summarized in Figure 1.1. It demonstrates that the design of polymer sensor structures requires the understanding of physical and chemical behaviour, principles of operation, and fabrication processes of both polymer and inorganic device parts.

In this chapter, first a survey about the technologies of inorganic device parts commonly used in sensorics will be given, and then the special techniques for forming and shaping polymer foils and films will be described.

Because of the great demand for sensors that are compatible with

1

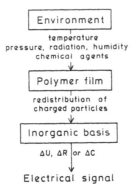

FIGURE 1.1. Interaction flowchart of sensors applying sensing polymer films.

microelectronics, the fabrication processes of microsensors originate from the different integrated circuit and interconnection technologies. These techniques can be categorized into the following groups:

- Monolithic semiconductor processing has become a unique method for structuring single crystals in order to fabricate sophisticated microcircuits and, recently, even components for micromechanics.
- Ceramics and glasses are widely used as substrates for fabricating multilayer interconnection systems and for the integration of passive and/or sensing devices.
- Thin- and thick-film technologies give possibilities to realize passive networks and/or interconnection systems on the surface of semiconductor chips and ceramic or glass substrates.

Several film deposition and related processes that had been used for the fabrication of devices based on inorganic materials could successfully be adopted for making polymer films, sheets, and fibres. Therefore, we start with the discussion of the former ones.

1.1 MONOLITHIC SEMICONDUCTOR PROCESSING [1,2]

Nowadays, microelectronic circuit fabrication is based on silicon and on compound semiconductors, mainly GaAs. However, semiconductor micro-sensorics is dominated by the former material. Therefore, the silicon processing will be discussed in detail, and special processes used in the compound semiconductor technology will also be mentioned.

1.1.1 Silicon Single Crystal

Single crystal silicon has a diamond structure with cubic symmetry, as shown in Figure 1.2(a). The various directions in the lattice are denoted by three indices. The electrical, chemical, and mechanical properties of silicon depend on the orientation. For example, certain planes of the lattice will etch much more rapidly than others when the crystal is subjected to chemical etching.

Silicon in single crystal form is a unique material due to its electrical and mechanical properties. Its electrical resistivity can be varied many orders of magnitude by adding minute quantities of impurities. These impurities are of two types: n-type and p-type, and they are called dopants. Boron is an example of a p-type dopant, while phosphorus and arsenic are examples of n-type dopants.

In single crystal form, every silicon atom is bonded to four other silicon atoms. The dopant atoms are substituted for the silicon in the crystal lattice and these try to form bonds with four other silicon atoms. However, p-type dopants have one electron too few, while n-type dopants have one electron too many to make a perfect crystal fit. These "extra" and "missing" electrons impart unique electrical properties to the doped silicon regions.

If a boundary exists between a p-type region and an n-type region, a p-n junction is formed. Such a junction will allow an electric current to pass only in one direction. Semiconductor devices (diodes, transistors, integrated circuits) can be fabricated by forming such junctions.

Bulk form single crystal silicon is grown from a very pure melt. A schematic of the often used vertical-pull method (generally known as the Czochralski method) is shown in Figure 1.2(b). A tiny single crystal seed is dipped into the melt and slowly withdrawn under conditions of tight temperature and motion control. By controlling the rate of withdrawal, a sausage-shaped "boule" of silicon is produced with a given diameter. The orientation of the crystal, with respect to the axis of the boule, is determined by the orientation of the seed. The electrical properties of the crystal are controlled by small amounts of dopants introduced into the melt.

After crystal growth and purification by zone refining, the boule is sliced into wafers about 0.5-mm thick. These wafers are then chemically-mechanically ground and polished to a very high surface quality in preparation for use in producing integrated circuits or silicon sensors. Although, for integrated circuits, one side is typically polished, for sensors, it is more common to begin with wafers that are polished to a mirror surface on both sides. The surface orientation of the silicon wafers is determined by the boule. The most common types of wafers produced have a surface oriented

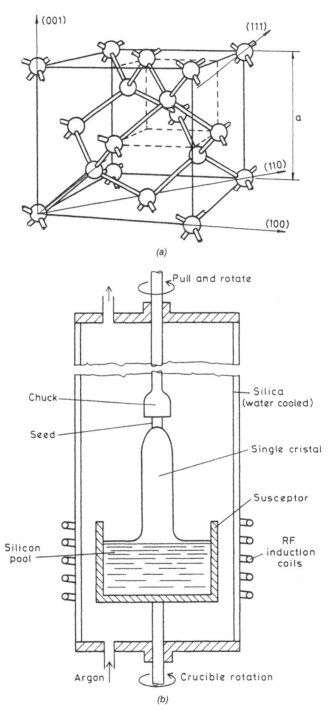

(a)

(b)

FIGURE 1.2. Silicon single crystal: (a) lattice structure and (b) fabrication technology.

along the (100) or (111) direction. Silicon sensors almost always employ (100) wafers because of their unique anisotropic etching characteristics.

1.1.2 Epitaxial Growth

Single crystal wafers fabricated by the processes described previously still contain surface crystal defects. It is possible to improve the crystal surface structure by growing a thin single crystal layer onto the surface of the wafer by the process known as epitaxy. The expression originates from the Greek word meaning "arranging upon." In this, atoms are deposited onto the crystal surface under conditions such that, under the influence of thermal agitation, each atom can run around until it finds its correct crystal lattice position where it forms bonds to the atoms in the solid surface. Thus, a new crystal is formed layer by layer as long as the supply of atoms continues. The source of silicon is usually a vapour, which may be produced by evaporating silicon from a molten pool in vacuum but, more usually, comes from a chemical reaction, as described later.

There are three main types of epitaxial layer growing: liquid phase epitaxy (LPE), vapour phase epitaxy (VPE), and molecular beam epitaxy (MBE). In VPE growth, the materials to be deposited are brought to the surface of the slice in vapour phase. At the appropriate temperature, chemical reactions take place, which result in the deposition of atoms onto the surface where they replicate the underlaying crystal structure.

The generally used method to carry out VPE is chemical vapour deposition (CVD). In the case of silicon, the often used source of atoms is silicon tetrachloride, which is reduced by hydrogen on a hot silicon surface to deposit silicon atoms and gaseous hydrochloric acid. Doping can be achieved by mixing an appropriate gaseous source of dopant atoms diluted in a stream of hydrogen with the $SiCl_4$. For *n*-type doping the most usual dopant gas is phosphine (PH_3), which decomposes to phosphorus and hydrogen on the hot substrate surface. If a *p*-type epilayer is required, diborane (B_2H_6) is used.

The conditions for epitaxial growth of good single crystal layer are quite critical. The substrate temperature must be sufficiently high so that the arriving atoms can migrate rapidly enough across the crystal surface to find their required crystal lattice positions. Their rate of arrival must be slow enough to allow this process to occur for each atom in turn. Above an upper limit, a polycrystalline layer will be produced.

Liquid phase epitaxy (LPE) is the simplest crystal growth technique for depositing semiconducting epitaxial films. It is based on the crystallization of semiconducting materials from a solution saturated or supersaturated with the material to be grown. The saturated solution is prepared at a high temperature (1000°C), then gradually cooled. The solution becomes supersaturated, and a crystalline phase begins to grow over a given substrate. The solution is then removed by chemical or mechanical means from the substrate. By this process, monocyrstalline films with a low number of crystal defects can easily be prepared.

MBE is the most recent method for epitaxial layer growing. It is a sophisticated evaporation of materials in ultrahigh vacuum. The slices are placed in a vacuum chamber, and element species are evaporated from heated sources, impinging upon the heated surface, and condense on the substrate. With proper control of sources, almost any material composition may be deposited within a resolution of a few atomic layers. The equipment can also contain the analyzers to obtain a good control of crystal structure and composition.

MBE offers a number of advantages over the CVD process, the most important of which is that substrate temperatures can be as low as 400°C, as opposed to up to 1250°C for the CVD process. However, it is relatively slow, requires ultrahigh vacuum, and is therefore expensive.

1.1.3 Structuring Silicon

Silicon sensors and microstructures are formed on slicon wafer by modifying both their mechanical structure and electrical characteristics. To form a mechanical structure, silicon can be etched away from the starting wafer using various etching techniques. Alternatively, silicon can be added to the structure by depositing amorphous or polycrystalline silicon (polysilicon), by growing an "epitaxial" layer of single crystal silicon onto a single crystal silicon surface, or by bonding an additional silicon wafer to a partially processed first wafer. When removing silicon by etching, those areas where etching is not required must be protected by suitable masking layers.

Similarly, the electrical properties of silicon can be modified by selective doping technques. Solid-state diffusion and ion implantation are two common techniques for increasing the dopant concentration in the silicon. It is desirable to modify the characteristics only in selected areas of the wafer. Suitable masking layers must therefore also be available for selective doping.

There are a number of techniques for producing masking layers on the silicon surface. The most common is the thermal oxidation process. By heating the wafer to a temperature near or about 1000°C in an oxidizing atmosphere, a thin layer of silicon is converted to the insulator silicon dioxide (SiO_2). This dielectric layer is an excellent electrical insulator. In addition, it acts as a mask to the diffusion of dopants used to vary the conductivity and to the etchants used to shape the mechanical structure of the silicon devices and sensors.

Another common insulator and masking layer is silicon nitride (Si_3N_4). This insulator is deposited by CVD on the wafers, rather than being grown. Again, the wafers are heated in a furnace, and gases are introduced, which react with each other on the hot surface of the wafer, forming silicon nitride. Silicon nitride has the unique property of being almost impervious to most impurities and to all of the silicon etches used to produce sensors and microstructures.

Phosphosilicate glass deposited by CVD is also an often used insulating and passivation layer on the top of the chips.

As can be seen, CVD is a commonly used technique for film deposition: single crystal silicon epitaxial layers, polysilicon, insulating layers, and even conducting films such as $MoSi_2$ can be deposited. The conventional CVD is performed at atmospheric pressure, but the more advanced methods are the low-pressure CVD (LPCVD) and the plasma-enhanced CVD (PECVD), which use lower temperature ranges.

Metal layers and interconnections on the silicon surface are deposited using thin film techniques described in more detail in Section 1.3.

1.1.4 Lithography and Etching

The insulators described above, as well as any other etchable film (such as metal) or soluble layers (such as polymers) deposited on a wafer, are patterned using a process called lithography. Lithography is a technique by which patterns are replicated onto the substrate using auxiliary layers, for example, etch resistant materials. The most widely used conventional photolithography is illustrated in Figure 1.3.

Using photolithography, features of the pattern as small as 1 μm can be defined. In addition, layer upon layer of such features can be successively aligned to one another with the same high accuracy. Photolithography also has a very important role in the shaping and patterning of thin films and even polymers. Thus, special attention must be paid to this process.

It begins when a photosensitive film called a photoresist (or simply resist) is spun onto the wafer. After a heat cycle, the wafer is brought into close proximity with a glass mask plate having a UV-nontransparent layer in the desired pattern. The wafer is then precisely positioned so that any pattern on the wafer is aligned with the additional pattern on the mask. The mask/wafer combination is then locked in place and exposed to ultraviolet light. The light passes through the openings in the mask and produces a chemical change in the photoresist, which is often also a polymer-based material. The resist is then developed in a chemical bath. This removes the resist in either the exposed areas (positive resist) or the unexposed areas (negative resist). In this way, the pattern is transferred from the mask to the resist film.

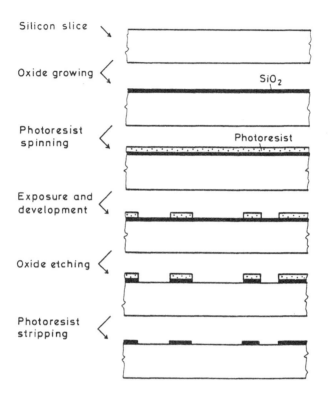

FIGURE 1.3. Silicon photolithography process.

For the realization of patterns with submicron sizes, techniques other than photolithography must be used. The exposure may also be accomplished using electron beams, X-rays, or ion beams. These are the newer techniques in lithography.

Once the resist pattern is formed and hardened by a heat treatment, the wafer is put through various etching processes to transfer the pattern from the resist into the oxide, nitride, or other film (such as metal) to be patterned on the wafer.

There are both wet chemical etchants and dry etching techniques using plasma reactors. The latter ones are denoted as plasma etching. In dry etching, the plasma serves as a source of ionized species that produces or in some manner catalyses etching. Noble gas ions perform a simple sputtering, while ionized reactive gas molecules can take part in chemical reactions with the particles of the film to be etched.

After etching, the photoresist is removed, and the wafer is ready for further processing steps. These would include doping cycles or further etching steps.

1.1.5 Anisotropic Etching

A fundamental process for the production of microstructures in sensors is the anisotropic etching of silicon. Certain chemical etchants such as hydrazine, "EDP," and KOH attack the (100) and (110) planes of silicon much faster than the (111) planes. This fact is used to produce a number of accurately defined shapes in a silicon wafer with (100) surface orientation. Typically, the wafer is first oxidized, and then this oxide is patterned using photolithography. When such a wafer is immersed in an anisotropic etchant, silicon will be removed only from the areas where no oxide is present. The etch proceeds downward in the (100) direction very rapidly, but when (111) planes are encountered, the etching effectively stops. In a (100) wafer, the (111) planes are oriented at an angle of 54.7° with respect to the surface. Thus "V"-shaped grooves and cavities can precisely be formed in the wafers, as shown in Figure 1.4.

1.1.6 Doping Techniques

There are two major techniques to introduce dopants into the silicon crystal structure, viz., diffusion and ion implantation. Diffusion is a process in which impurity atoms penetrate into the crystal lattice due to a concentration gradient at the surface. It is carried out in a high-temperature furnace by depositing a large amount of the dopant on the surface of the silicon. The dopant on the surface is at a very high concentration compared to the silicon, and at high temperatures (usually above 800°C), it will diffuse into the silicon to try to equalize the concentration difference.

Ion implantation is a process in which the ionized dopant atoms are accelerated through an electromagnetic field and physically blasted into the silicon. The acceleration energy determines how far the dopant ion will penetrate into the silicon. Later on, the dopant needs to see a high-temperature annealing (at about 500°C in order to become incorporated into the crystal structure and become electrically activated). This high-

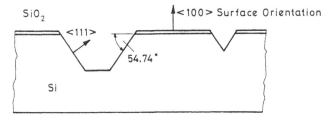

FIGURE 1.4. Cross section of V-groove and cavity-type etch profiles in silicon.

temperature operation may also be used to diffuse the dopant deeper into the wafer from the initially implanted depth.

Although ion implantation is usually more expensive than diffusion, it has the important advantage that the amount of dopant can be precisely monitored and controlled. As a result, resistors made by ion implantation, for example, have very tight tolerances over resistance values and their temperature coefficients. Equivalent tolerances are usually impossible to obtain with diffusion techniques.

1.1.7 Integrated Processing

In the previous sections, the basic technologies used in silicon devices have been outlined. In practice, a technology process flow should be built up connecting the individual process steps in series. The optimum technology is the one that minimizes the number of processing steps while at the same time guaranteeing a specified device performance. In microsensor structures, "MOSFET" devices are of great importance; therefore, in this chapter, a survey will be given about the MOSFET devices and their technology.

One of the broad classes of semiconductor devices is known as Field Effect Transistor, or FET. The FET makes use of an electric field to control the flow of current between the "source" and "drain" electrodes. The various types of FETs can be categorized according to the method of developing this electric field. When the field is generated with a voltage applied on a capacitor composed of a Metal Oxide Semiconductor (called a gate), the resulting transistors are called MOSFETs, or more simply, MOS transistors.

Numerous variations of MOS transistors are in use today. These variations include starting material, dielectric material, gate material, diffusion techniques, and etching techniques. One major category refers to the conducting state under zero bias, with normally off devices termed *enhancement,* and normally on devices termed *depletion.* A second major category is according to the polarity of current carriers: NMOS conducts with negative charges and PMOS employs positive charges.

Figure 1.5 shows a cross section of a PMOS transistor. The substrate is *n*-type, and the source and drain are *p*-type. A negative voltage applied to the gate will tend to attract free positive charges (holes) to form a shallow conducting channel between the source and drain.

The gate of an MOS transistor is electrically isolated from the remainder of the transistor, and no DC gate current flows, regardless of the conducting state of the source and drain electrodes.

"CMOS" means "Complementary MOS," where both NMOS and PMOS transistors are produced on the same chip.

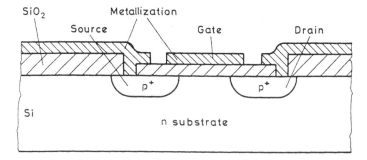

FIGURE 1.5. Cross section of a P-channel MOS (PMOS) transistor.

We will illustrate the application of individual technology steps in integrated processing through the example of conventional CMOS technology, which is often used in microsensor fabrication. It should be stressed that VLSI ICs need more sophisticated and much more complicated processes; however, the given example is a good demonstration for integrated processing.

The starting material for the CMOS process is a wafer of lightly doped *n*-type silicon. The manufacturing process is summarized in Figure 1.6. The steps are as follows:

(1) Oxidation, photolithography, window etching for the first diffusion
(2) *P*-well diffusion, oxidation
(3) Photolithography, window etching for the second diffusion
(4) *P*-diffusion and oxidation to define source and drain areas for PMOS transistors and, at the same time, to define a boundary or "channel stop" around the areas where NMOS transistors will be fabricated
(5) Photolithography and window etching for the third diffusion
(6) *N*-diffusion and oxidation to define NMOS transistors and "channel stop" around PMOS transistors
(7) Photolithography and window etching for the gates
(8) Thin gate oxide growing for MOS gate dielectric layers
(9) Photolithography and window etching for contacts
(10) Metal deposition, patterning, and passivation

1.2 PROCESSING OF CERAMICS AND GLASSES [3,4]

Ceramics and glasses play an important role in microsensorics and are used in the following areas:

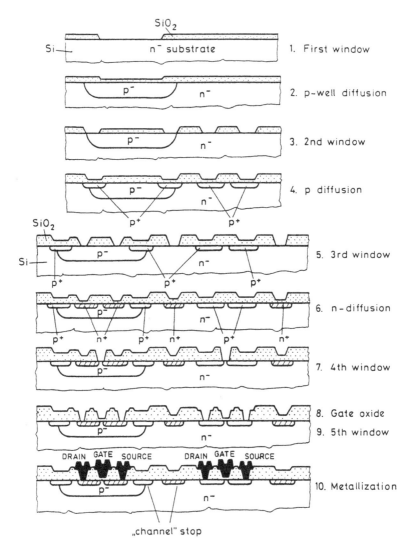

FIGURE 1.6. Processing of CMOS structures.

- ceramic and glass substrates for sensor structures
- active sensing ceramics and glass membranes
- piezoelectric ceramic resonators
- glass fibres for fibre-optic sensors

1.2.1 Microstructure of Ceramics

The microstructure of polycrystalline ceramics is usually complex, as shown by Figure 1.7. It can be distinguished by the existence of grain boundaries, which are not seen in single crystals. Furthermore, the existence of pores, imperfections, and multiphase composition enables the production of a great variety of properties.

Up to now, grain boundaries and additional phases were thought to be undesirable, and the goal was to eliminate them to obtain a structure as close to single crystal structures as possible. However, new processes have been found that make positive use of these surfaces and grain boundaries; thus, functional ceramics in which these properties are important are developing rapidly.

Figure 1.7 explains the role of grain boundaries. In the grain boundary region, energy is increased, so impurities tend to gather there. The impurities exist as a second or third phase among the constituent particles or segregate into the grain boundaries.

In general, ceramics are produced from powdered raw material by a sintering process. Ceramics obtained in this way are polycrystalline, an aggregation of fine crystalline grains, and grain boundaries inevitably exist. They play an important role in the sintering process and have a large influence on chemical and physical properties.

1.2.2 Manufacturing of Ceramics

Figure 1.8 shows the typical manufacturing schematic for ceramics. The process begins with the mixing of powdered raw materials and additives by

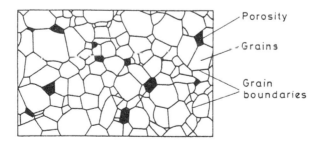

FIGURE 1.7. Typical microstructure of ceramics.

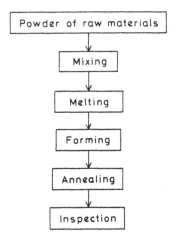

FIGURE 1.8. Processing of ceramics.

weight. When the starting material is a powder, the synthesis, or refining, of the powder is one of the most important processes affecting the quality of the final product. Ceramics are different from ordinary metals or plastics in that the powder forming varies diversely with regard to the desired product.

After the mixing, an organic resin — a temporary binder material — is added to the raw material powder. Then the powder is compacted in a definite form. Depending on shape and required characteristics, numerous powder compacting processes have been developed. The five main ones are as follows:

(1) Die pressing
(2) Rubber mould pressing
(3) Extrusion moulding
(4) Slip casting
(5) Injection moulding

However, powder itself consists of solid brittle particles; therefore, it is difficult to fill a die by pressure alone. As the pressure is increased, there is more strain on the compact, and cracks that can cause the compact to fracture are formed. The role of the binder is to enhance the fluidity of the powder.

The binders used vary with the different compacting processes (1) through (5), and the kind and quantity of binder required depend on the powder used and the product desired. Both hot pressing and HIP (hot isostatic press) are occasionally included in these forming and/or compacting processes, but these are used as aids for increasing the external pressure during sintering. Under (4), one can also place film forming, which is widely used in the production of alumina IC substrates (see Figure 1.9).

Prior to sintering, it is necessary to get rid of the organic resin used for forming by heating the compact at a low temperature. The formed product is sintered at a high temperature and becomes dense through contraction. Figure 1.10 shows the changes in structure associated with the sintering process. It is an aggregation of particles ranging in size from several micrometers to several tens of micrometers, which are connected by grain boundaries.

Sintering is the consolidation of a powder by means of prolonged use of elevated temperature, which is, however, below the melting point of any major constituents of the ceramic. Figure 1.11 shows that sintering involves the replacement of high-energy solid-gas interfaces by lower energy solid-solid interfaces (grain boundaries). It is this reduction in total interface energy that is the driving force for the sintering process. Clearly, sintering requires the movement of atoms or molecules through the component, and it has been found that the mechanism of mass transport, e.g., lattice diffusion, surface diffusion, and evaporation-condensation varies from ceramic to ceramic.

Vitrification is the densification in the presence of a viscous liquid and is the process that takes place in the majority of ceramics produced on a large scale, often called "glass ceramics." Upon cooling from the firing temperature, the liquid phase solidifies as a glass, thereby further increasing the bonding between particles and so forming a solid material, albeit with some porosity.

A solid compact of a few specialized ceramics is produced not by either sintering or vitrification, but by a chemical reaction, the process being termed reaction bonding. This technique is used particularly for silicon-based ceramics and in thick-film technology (see Section 1.3).

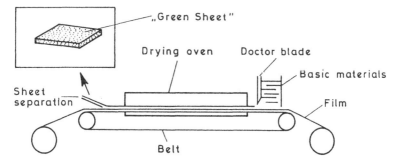

FIGURE 1.9. Green sheet formation in the production of alumina substrates.

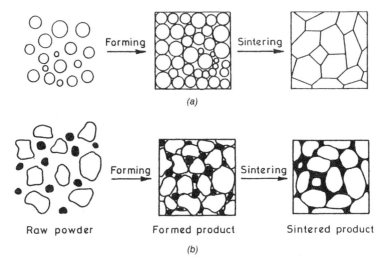

(a)

Raw powder   Formed product   Sintered product

(b)

FIGURE 1.10. Structural changes during sintering: (a) solid phase sintering and (b) liquid phase sintering.

It should be mentioned here that special technologies have been developed for high-quality alumina and glass ceramic multilayer substrates commonly used in IC and sensor laminations [5]. These technologies originate from the IC packaging, where the goal is to make an inexpensive hermetic package with metallization for output-lead connections without electronic leakage.

The manufacturing of an alumina multilayer system also begins with the mixing of powdered raw materials. A sheet of the powder mixed with organic resin, called a "green sheet," has to be prepared in advance by screen-printing a conductive line network on its surface and in its through holes for vias, using a paste composed mostly of tungsten or molybdenum powder (see Section 1.3). The manufacturing flowchart is given in Figure 1.12. Sev-

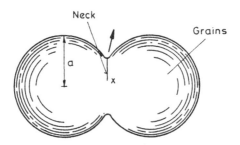

FIGURE 1.11. Neck growing during sintering.

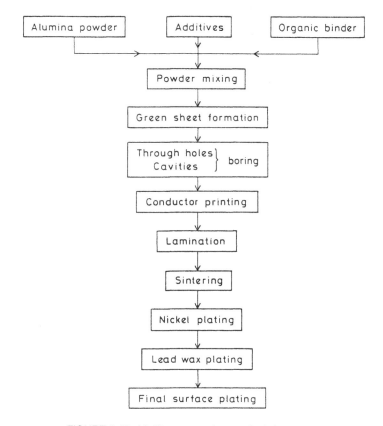

FIGURE 1.12. Multilayer ceramic manufacturing process.

eral of these sheets are then laminated and sintered over 1550°C in a hydrogen-containing atmosphere to avoid the oxidation of the metallization system.

Tungsten is widely used as a conductor since it can resist the high temperatures of sintering; its sintering characteristics are similar to those of alumina, and it has great bonding strength while retaining high conductivity. Therefore, elaborate integrated circuit substrates can be formed using tungsten conductor networks. In such cases, the outer surface of the tungsten is nickel (Ni) plated, and finally, a fine gold plating is added after the leads have been silver coated.

The technology of multilayer glass ceramic systems is almost the same, but the sintering temperature is much lower; it is about 850°C, which is also the typical firing temperature of thick films (see Section 1.3). Thus, precious metal thick-film conductive layers can be used for a metallization network.

The former method is often called HTCC (High Temperature Cofired Ceramic) and the latter LTCC (Low Temperature Cofired Ceramic) technology. The term "MLC" (Multilayer Ceramic) technology is used for both cases.

1.2.3 Microstructure and Behaviour of Glasses

Glasses have amorphous microstructures without long-range ordered crystalline form and are also produced from different metal-oxide components. Silica (SiO_2) is an example of a material that may be produced in either the crystalline or the glassy state, and the following discussion of this material will show the relationships between these states.

If silica is melted and cooled very slowly, it will crystallize at a particular temperature T_m, called the freezing or melting point, in an identical manner to that of a metal. The specific volume as a function of temperature exhibits a discontinuity at the melting point, as shown in Figure 1.13. Silica crystallizes in a number of forms, an example of which is shown schematically in two dimensions in Figure 1.14. If the silica is cooled rapidly from the molten state, it is unable to attain the long-range order of the crystalline state. The temperature dependence of the specific volume is given in Figure 1.13. The temperature T_g on this curve is called the glass transition tempera-

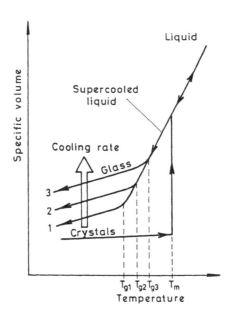

FIGURE 1.13. Effect of cooling rate on transition temperature at glasses.

Crystalline   Glassy silica   Glasses
silica

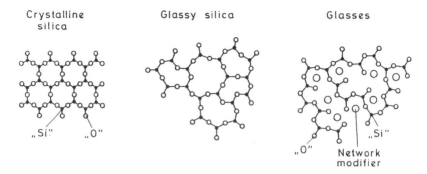

FIGURE 1.14. Microstructure of silica and glasses.

ture, which is not a well defined temperature and depends on the cooling rate. The slope of the curve between T_g and T_m is the same as that above T_m, indicating that there is no change in structure at T_m. This means that between T_g and T_m, the material is a supercooled liquid.

The state below T_g, termed the glassy state, is very closely akin to the liquid state. This is indeed the case: the glassy state consists of a short-range ordered network, as shown in Figure 1.14. This is a metastable structure and will very slowly tend to change to the lower free energy crystalline form. A glass is called a vitreous solid, and if the material transforms to the crystalline state, it is said to have been devitrified. At room temperature the rate at which devitrification to the crystalline state occurs is infinitely slow.

One of the important features of the structure of glass is that it is a very open network and can easily accommodate atoms of different species, such as sodium, potassium, calcium, and boron atoms, called network modifiers or glass formers, depending on their positions in the network of microstructure (see Figure 1.14). In microelectronics and sensorics, the sodium- and potassium-free glasses with P_2O_5, B_2O_3, PbO, CdO, etc., additives have great importance. Their glass transition temperature and physical properties can be modified in wide ranges by composition changes. Phosphosilicate glasses are used, for instance, in silicon chips as protective layers (see Section 1.1.2); borosilicate glasses are substrates for thin film sensors and main binder materials of thick-film pastes (see Section 1.3).

Devitrification of glasses can also be enhanced by the presence of foreign particles in the glass, which act as nucleation sites for crystallization. If this crystallization is well controlled, new types of material, called "glass ceramics" or, better, "crystallizable glasses," can be produced. A two-stage heat treatment is generally used to transform the glass to a glass ceramic. First, the glass is held at a low temperature to produce a large number of well-dispersed nuclei. The glass is then heated to a higher temperature at which the crystal growth rate reaches its maximum. On holding at this temperature, the crystalline phase, which may be of different structure and composition to the nuclei, grows upon the nuclei until crystallization is almost complete, with only a small amount of residual glass remaining. A glass ceramic, like a glass, has negligible porosity, and its properties are intermediate between those of a glass and the ceramics.

1.2.4 Manufacturing of Glasses

Figure 1.15 shows the classical manufacturing process for glasses. It begins with mixing powdered raw materials, which is similar to the ceramic processing. The melting of glass consists of three distinct stages. The first involves the evaporation of water and various chemical reactions, such as the decomposition of carbonates, sulphates, and nitrates, which results in a highly viscous liquid full of bubbles. The second stage, known as refining, is concerned with the removal of the bubbles by raising the temperature and the addition of refining agents that liberate large bubbles that sweep the smaller bubbles to the surface. At this stage the glass is in fluid state. The final stage in the melting process is to cool until the glass has a higher viscosity suitable for forming.

Blow moulding is a common technique for forming glass vessels; however, it is rarely used in sensor applications. Glass for sensor substrates has to be of constant thickness and free from surface blemishes.

A *float process* does not involve mechanical finishing of the glass. In this process the molten glass leaves the furnace in a continuous ribbon and then

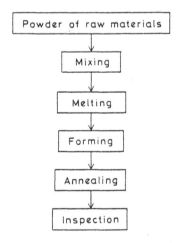

FIGURE 1.15. Processing flowchart of glasses.

floats on the surface of a bath of molten tin. The bath of molten tin is surrounded by an atmosphere of nitrogen to prevent oxidation of the tin, since tin oxide imparts a bloom to glass. With good temperature control, it is possible to produce parallel-sided flat glass with an excellent surface finish because the top surface of the glass is protected by the nitrogen atmosphere and the bottom surface by the liquid tin. After forming and cooling, glass is annealed at a temperature near T_g to eliminate the mechanical stresses.

Quenching of molten glass between cooled cylinders in motion is the production of glass powder row materials for thick-film frits (see Section 1.3).

CVD. Thin glass layers on the surface of silicon wafers or thin-film structures can be synthesized directly by Chemical Vapour Deposition (CVD) as could be seen at silicon processing (see Section 1.1).

Fibre-drawing. A specially developed technique is applied for forming optical fibres that are widely used in sensorics. The fibre can be drawn directly from the core and cladding glasses, which results in a continuous process, making the fibre cheap to produce. Such a process is the double crucible method of fibre manufacturing (also known as the direct melt technique).

Figure 1.16 shows a schematic of a double crucible pulling tower consisting of two concentric funnels. The outer funnel contains the cladding material, while the inner funnel contains the core glass. The crucibles are usually made of platinum to reduce contamination. The crucibles are heated to melt the glasses, and the fibre can then be drawn from the tip and attached to a take-up drum at the base of the tower. As the drum rotates, it pulls the fibre. The rate of drum rotation determines the thickness of the fibre. A noncontact thickness gauge regulates the drum speed by means of a feedback loop. Below the gauge, the fibre passes through a funnel containing a plastic coating to protect the fibre from impurities and structural damage. A curing lamp ensures that the coating is a solid before the fibre reaches the take-up drum.

1.3 THIN- AND THICK-FILM TECHNOLOGIES

Originally thin- and thick-film technologies were developed as metallization techniques of silicon wafers, glasses, and ceramics. Later, resistor and insulating layer materials also appeared; thus, these technologies began their own life as integrated circuit and microsensor processing methods. The most important differences between the two technologies are in the layer thickness, in the applied materials and processing steps. A short comparison is demonstrated in Table 1.1.

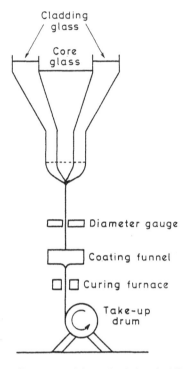

FIGURE 1.16. Double crucible method of optical fibre production.

1.3.1 Thin-Film Processing [6,7]

Fabrication technologies of preparing thin films may be divided into three main groups, namely, chemical deposition (including electrochemical methods), vacuum evaporation, and cathode sputtering. In the practice, they are applicable to a lot of substrate types, such as single crystals (sili-

TABLE 1.1. Typical Properties of Thick and Thin Films.

	Thick Films	Thin Films
Raw materials	Pastes: colloid compounds	High purity metals, alloys, compounds
Typical processing	Screen printing and curing	Vacuum deposition processes, CVD, etc.
Layer thickness	10–50 μm	10–200 nm
Layer structure	Active particles in a binder matrix	Polycrystalline, discontinuous, physical thin films

con, sapphire), borosilicate glasses, ceramics, glass-coated ceramics, polymers (polyimid, PTFE, epoxy), and to a great range of thickness.

Among chemical and electrochemical methods, the most important are: electrochemical deposition, electroless deposition, anodic and thermal oxidation, and chemical vapour deposition.

In *cathodic electrochemical deposition,* the substance (metal) to be deposited is present in a solution or melt in the form of ions. If two electrodes are inserted into the solution (or melt), the positive ions of the metal will be attracted to the cathode where the metal will be deposited. The mass of the substance deposited is proportional to the electrical charge necessary to be transferred. The proportionality constant is the electrochemical equivalent of the given substance. The properties of the deposited films, for example, its adhesion to the substrate and its crystal structure (the size of microcrystals), may be influenced by the composition of the electrolyte. By this method, it is, of course, possible to deposit films only on conductive substrates, and the films may be contaminated by substances that are present in the electrolyte. This method of deposition is used, for example, in the formation of nickel layers on the surface of other metals (see Section 1.2.2).

Electroless deposition is based upon a similar principle, but in this case the metal is deposited from the solution by electrochemical processes without the presence of an externally applied field. The rate of deposition depends on the temperature and pH value of the bath, and in some cases, the deposition needs to be stimulated by a catalyst. This is often used for the through hole metallization of insulating substrates.

Anodic oxidation is used mainly in the formation of films of the oxides of certain metals such as Al, Ta, Nb, Ti, and Zr. The metal to be oxidized is an anode dipped in the electrolyte from which it attracts the oxygen ions. The ions pass through the already formed oxide film by diffusion forced by a strong electric field and combine with the metallic atoms to form molecules of the oxide. For anodic oxidation it is possible to use either the constant-current or constant-voltage method. Solutions or melts of various salts, or in some cases acids, are used as electrolytes. The oxide ions needed for oxidation can be obtained not only from an electrolyte, but also from a discharge. Recently, anode oxidation in a glow discharge has been tried, and it has been found that it is governed by laws similar to those of oxidation in an electrolyte and results in an oxide of a very high quality.

Thermal oxidation was already explained in connection with the silicon processing (see Section 1.1.3).

Chemical Vapour Deposition (CVD) is a widely used method in semiconductor technology for the preparation of thin monocrystalline films of high purity called epitaxial layers (see Section 1.1.2). The method can also be used for the deposition of polysilicon layers, insulating films such as Si_3N_4 or phosphosilicate glass, or metallization ($MoSi_2$) layers. Chemical transport reactions are also used for the preparation of III-V compound semiconductor films, for instance, GaAs. Recently, new low-temperature versions of CVD have been developed: the LPCVD (Low Pressure CVD) and the PECVD (Plasma Enhanced CVD) processes.

The preparation method of thin films of some high molecular weight compounds has been elaborated by Langmuir and Blodgett and is often called the *LB-method* and the films LB films. A small amount of a high molecular weight substance, which has polar molecules possessing hydrophobic tails and hydrophilic heads (e.g., fatty acids or higher alcohols), is dissolved in a volatile solvent, and one drop of the solution is sprinkled on the surface of the water. The solvent evaporates, and the molecules of the substance diffuse over the surface of the water, all orientated in the same manner due to their polarity. According to their concentration, either a "two-dimensional gas" is formed or a monomolecular film of liquid or solid. Such a film can be lifted up and put upon a plate or directly deposited onto the surface of a plate-shaped substrate using dipping. Several such films can be applied on top of each other, and in this way even films of several hundreds of nanometer thickness can be built up gradually (see Figure 1.17).

The most important characteristic of the Langmuir-Blodgett (LB) technology is the ability to produce ordered mono- and multilayers of organic molecules such that the number of layers is precisely known and a high degree of order normal to the plane within the organic film is achieved. Recently, a wide field of application has been prophesied for LB-films for chemical and biosensors. The method has a great importance also when preparing polymer films and membranes on sensor surfaces.

From this brief survey, it is already clear that the majority of chemical methods of preparations of thin films are only applicable to a small group of materials. Some of the methods are excellent and are frequently used in sensor technologies. Nevertheless, the most widely adopted methods of

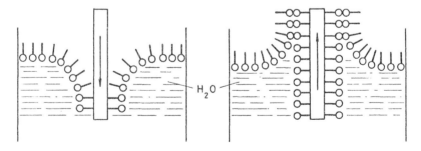

FIGURE 1.17. Langmuir-Blodgett deposition process of organic thin films.

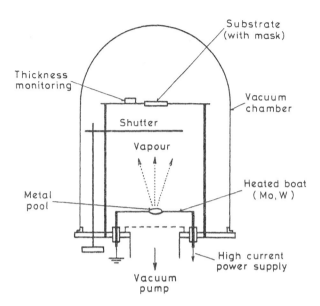

FIGURE 1.18. Schematic diagram of a vacuum evaporation unit.

preparation of thin films remain the physical methods: sputtering and vacuum evaporation. Both methods require low pressure in the working space and, therefore, make use of vacuum techniques.

The process of film formation by *vacuum evaporation* consists of several physical stages:

- transformation of the material to be deposited by evaporation or sublimation into the gaseous state
- transfer of particles to the substrate
- deposition onto the substrate
- rearrangement on the surface

In the vacuum evaporation technique, the material to be deposited is heated, usually by putting small lumps of it in an electrically heated "boat" made of tungsten or molybdenum, in an ultra-high vacuum (i.e., 10^{-4} Pa). The schematic diagram of a vacuum deposition unit is shown in Figure 1.18. In the case of metals, the deposition starts when the material has formed a liquid pool in the boat and the shutter is opened. The difference between the various evaporation methods is in the heating method and in the type of evaporation sources.

Electron-beam evaporation is now the most commonly used evaporation technique. An electron beam of sufficient intensity is ejected from a cathode, accelerated and focused onto an evaporant material. Impinging electrons heat its surface to the temperature required for evaporation. The method enables to attain a very high temperature and to evaporate materials which would otherwise be evaporated with difficulty or cannot be evaporated at all. An additional advantage of the method is the prevention of contamination by the evaporation source holder material. The beam heats only the evaporant whereas the support holder is usually cooled.

The simplest arrangement for *cathode sputtering* is shown in Figure 1.19. The material to be sputtered is used as a cathode target in a system in which a glow discharge is established in an inert gas (e.g., argon or xenon) at a pressure of 1 to 10 Pa and a voltage of several kilovolts. The substrate on which the film is to be deposited is placed on the anode of the system. The positive ions of the gas created by the discharge are accelerated towards the cathode (target). Under the bombardment of the ions, the material is removed from the cathode (mostly in the form of neutral atoms and, in part, also in the form of ions). The liberated components condense on surrounding areas and consequently on the substrate placed on the anode.

In RF (radio frequency) sputtering, AC discharges of 5–30 MHz are employed. In the case of the DC process described in the previous paragraph, problems are encountered in initiating the process and with sputtering insulating materials. In the RF process, one electrode is coupled capacitively to an RF generator. It then develops a negative DC bias with respect

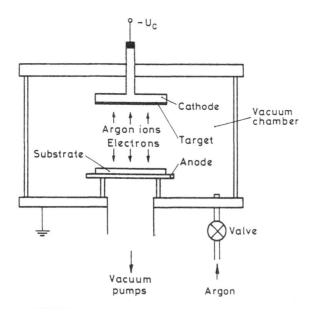

FIGURE 1.19. Apparatus for diode cathode sputtering.

to the other electrode. This is related to the easier response of electrons than ions to an applied RF field.

A special form of deposition by sputtering is called reactive sputtering. It is performed in the atmosphere of a pure reactive gas, which is eventually to be a component of the film, or a mixture of inert and reactive gases.

Magnetron sputtering uses magnetic fields to deflect electrons from striking the substrate in order to increase the ionization efficiency. Various magnetically confined discharge configurations have evolved.

For applications in microelectronics and sensorics, it is necessary to form the films in specific, sometimes complicated patterns. For this purpose a suitable mask is used to prevent condensation on the areas that we desire to keep clean. The mask is usually formed from a metal and is placed in the vapour stream as close as possible to the substrate on which the film is to be deposited. This is the so-called noncontact masking method.

There are several technologies for preparation of "in contact" masks, which are deposited directly on the substrate. The most widely used type is the photolithographic one combined with selective etching of different layers, similarly to the silicon photolithography process shown in Figure 1.3. Figure 1.20(a) illustrates the process in the case of a double-layer system for formation of a nichrome resistor with gold contacts.

Figure 1.20(b) shows the often used, so-called lift-off technique. In this method, the deposition of the film is carried out on a substrate that is covered previously with a photolithographically patterned mask. After the deposition, the latter will be removed by etching or stripping from the unnecessary areas, together with the deposited film.

1.3.2 Thick-Film Processing [8,9]

The most commonly used technology for making thick films is the screen printing and firing/curing at elevated temperature. The classical so-called "CERMET" version of the technology uses inorganic-based glasses, glass ceramics and ceramic-glass-metal composite raw materials for films, and relatively high-temperature firing processes. The emerging polymer thick-film (PTF) technology is based on polymers and conductive polymer-composites with low curing temperatures.

The CERMET thick-film technology generally uses ceramic-type substrates: alumina (Al_2O_3-SiO_2-MgO) is the commonly used substrate in thick-film ICs and sensors. It has good electrical, mechanical, and thermal properties; beryllia (BeO) offers high thermal conductivity for high-power applications; aluminium-nitride (AlN) is a recently developed type that offers high thermal conductivity with lower prices than beryllia. Enamelled steel substrates for thick-film circuits give a new type of flexible solution.

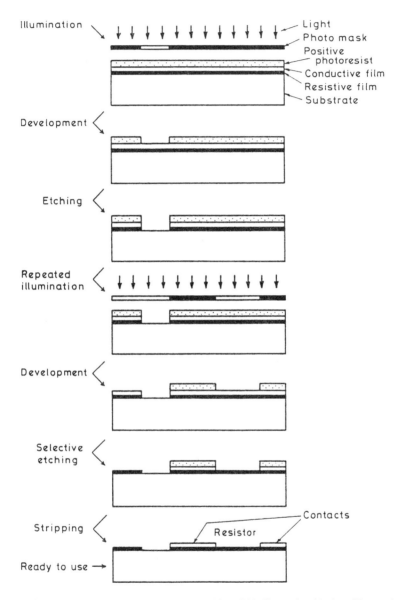

FIGURE 1.20. Processing steps of contact masking of thin films using (a) photolithography and selective etching.

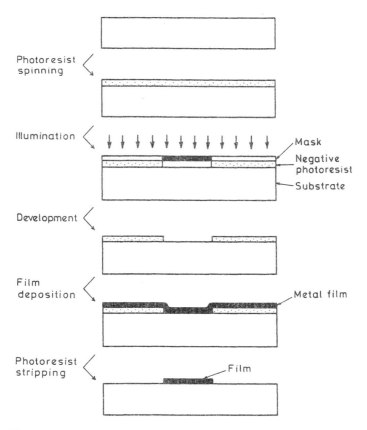

FIGURE 1.20 (continued). Processing steps of contact masking of thin films using (b) lift-off technique.

Raw materials for thick films are available in so-called "paste" form. Thick-film pastes consist of three main ingredients:

(1) Finely powdered functional phase (metal and/or metal-oxide powders) of carefully controlled size and composition. Conductive pastes contain precious metals (Pd, Pt, Au, Ag) or base metals (Cu), resistive ones — RuO_2, high permittivity dielectrics — $BaTiO_3$, and low permittivity insulating layers — different metal-oxides and glasses.

(2) Finely powdered oxide and glass phases, also of controlled size, shape, and composition — as inorganic binder materials

(3) An organic vehicle that suspends the inorganic constituents until the paste is fired, which must also give the proper rheological characteristics for the screen printing

The shaping technique of thick films is basically a screen-printing technique. The circuit is fabricated by successively printing each layer until the desired circuitry is achieved. The basic tools used in printing are the squeegee and the screen. The system that incorporates these basic tools and provides adequate means for controlling them precisely to make them useful in fabrication is the screen printer. The structure of a screen is shown in Figure 1.21, and Figure 1.22 shows the screen-printing process.

The holes of the stainless steel, polyester, or nylon screen are filled by a photolithographically shaped emulsion, which has openings according to the layout pattern. The substrate is placed under the screen and the squeegee presses the paste through the screen mask openings while travelling along.

After the film has been printed on a substrate, it should be allowed to "settle" since printing through the screen tends to produce mesh lines in the pattern. The settling time depends on the viscosity of the paste; however, times vary from about 5 min to approximately 20 min.

The thickness of the layer is influenced by many parameters, such as the emulsion thickness, screen mesh number, and paste viscosity, and by the most important parameters of the printing: travel speed and pressure of the squeegee and the so-called snap off parameter (distance between screen and substrate).

After settling, the films should be dried before firing. This removes only the volatiles, leaving behind the binders. Desirable drying temperatures depend on the solvents and vary from 100 to 150°C for 5 to 15 min. The resulting films should be tough enough to permit handling and subsequent printings. Several methods have been used for drying the films, viz., hot plates, ovens, heat guns, and infrared lamps.

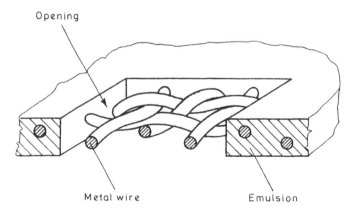

FIGURE 1.21. Cross section of a typical screen mask.

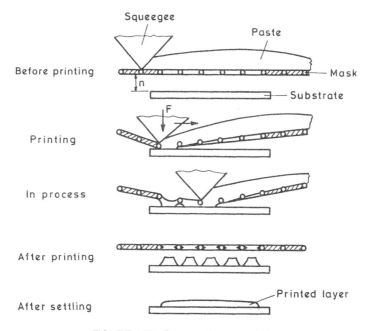

FIGURE 1.22. Process of screen printing.

Firing is the most critical step of thick-film technology. The most important parameters are the temperature profile and the atmosphere. Thick-film pastes mostly need the peak temperature in the range of 700–1000°C for 10 min (typical value is 850°C). Temperature should be controlled to better than ±3°C at the peak. For firing, belt furnaces are used. The schematic structure of a typical belt furnace is shown in Figure 1.23, and Figure 1.24 shows the firing profile. Thick films commonly need dry air atmosphere, but base metals (for instance copper) need inert (N_2) atmosphere for firing.

The firing of thick-film circuits is a rather complex process in which the organic binders and solvents burn out in the first phase, the metallic elements are sintered, and glassy components are melted to anchor the film to the substrate. Bonding through chemical reactions is also possible. Controlled cooling to room ambient temperature follows. The typical layer structure is shown in Figure 1.25.

Screen printing is not a unique layer fabrication method in thick-film technology. A newly developed tape transfer technique uses special glass ceramic materials cast on polyester support films. Thick-film dielectric layers are formed after punching and laminating the tapes onto the alumina ceramic substrates. Conductive layers and firing parameters are the same as in the classical thick-film technology. The process flowchart is given in Figure 1.26.

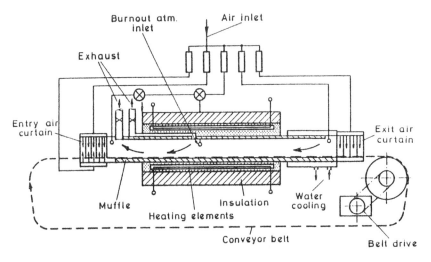

FIGURE 1.23. Belt furnace for thick-film firing.

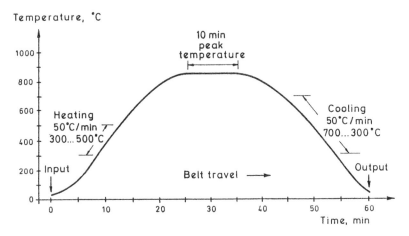

FIGURE 1.24. Typical firing profile for thick films.

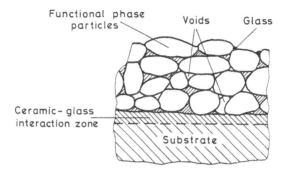

FIGURE 1.25. Typical microstructure of thick films.

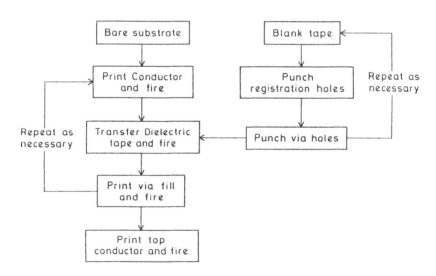

FIGURE 1.26. The tape transfer process.

33

1.3.3 Polymer Thick-Film Technology [10–12]

Polymer thick-film (PTF) technology is a technique for making conductors, resistors, and insulation in several layers on regular printed circuit board laminates, ceramics, anodized aluminium, flexible substrates, and moulded engineering plastics. These may form the structure of the product as well as serve as the printed circuit board (PCB) itself. For rigid boards, the same substrates are generally used as in regular PCBs (epoxy-glass, phenolic, etc.). For flexible boards and membrane switches, polyester, polycarbonate, polyethersulphone, or polyimide are used. For moulded circuits high-performance engineering plastics are used, such as polyethersulphone, polyetherimide, and polyacrylsulphone. PTF is fabricated by a sequence of screen-printing and curing steps similar to that in high-temperature thick-film technology.

PTF pastes contain active ingredients, polymer matrix, and solvents. The active ingredients for conductor pastes are particles of conducting elements. The polymer is the "glue" for the active ingredients binding the particles to the substrate and to each other. There are three main types: thermoplastic, thermosetting, and UV curable polymers. The former two types are cured by some form of heating. The role of the solvent in the paste is to dissolve the polymer, control the viscosity, and wet the substrate surface.

In the thermoplastics, the desired viscosity is obtained by the use of solvents. After printing, the paste hardens when the solvents dry off. If heated, the paste will soften reversibly. These pastes have poorer environmental properties than the thermosets. They are less resistant to solvents; adhesion to the substrate is poorer; they do not withstand high temperatures, but they are flexible, easy to use, and find extensive application in membrane panel switches and tactile switch sensors.

Polyesters, poly(vinyl acetate) (PVAc), and acrylics are thermoplastic polymers used in polymer thick films. Essentially, they have long molecular chains with weak bonds between the chains. These bonds can be broken by solvent interaction.

PTF layers using polyester binders have high flexibility, good adhesion to polyester substrates because of their closely fitting structures, high temperature stability, and inertness to most solvents. The latter, however, limits the choice of solvents for use in the paste.

The PVAc binder has good flexibility and also good adhesion to polyester sheets, which is probably a result of hydrogen bonding. It is also soluble in a wide range of solvents. On the other hand, the film has poorer high-temperature behaviour than the polyester structure and is more easily attacked by solvents.

The acrylic (methyl methacrylate) polymer can be dissolved in a wide range of solvents but is much less flexible than the former two systems. Thus, its application is limited in the flexible circuits.

Thermoplastic polyimides have more complex and rigid molecular chains with cyclic rings. Because deformation and rotation of the molecular chains is inhibited, the layer becomes stiffer and more resistant to deformation with a high melting point and glass transition temperature.

The thermosetting pastes contain polymers that are not completely cured. During the curing process, the polymer is irreversibly cross-linked, making a very strong and stable matrix resistant to solvent and rehating. The most common thermosetting polymers used in thick films are epoxies, phenolics, and thermosetting polyimides. Cross-linking can be achieved by homopolymerization using a catalyst or by copolymerization using a hardener. In both cases, the process is accelerated by heating. The degree of cross-linking will affect the physical properties of the film. If the polymer is not fully cured, then there will be chances for further cross-linking during the life of the layer, resulting in instabilities and drift of the electrical parameters. The function of the solvents in the paste is to provide the right viscosity and good printing properties. Other additives may be stabilizers, ceramic particles, catalysts, silicone, etc.

UV-curable pastes include UV-sensitive photoinitiators. They are used primarily for dielectrics and may be solvent-free.

The conducting elements used in PTF conductor and resistor pastes are silver, copper, nickel, or carbon (graphite). Processing temperatures for polymer thick films are too low to result in a sintering effect, and conduction relies on point to point contact between adjacent discrete particles. Further discussion will be given about the conduction mechanism in Section 3.2.1.

Silver-based conducting pastes are widely used. They have reasonably high conductivity, giving a sheet resistivity typically 20–70 mohm/square. It provides good contact to materials such as copper foil, even if they are slightly oxidized on the surface. Silver in PTF conductors, however, has a tendency to migrate in an electric field and at high relative humidity conditions, just as occurs in conventional "CERMET" thick-film hybrid circuits [13]. This means that silver ions move and grow dendrites under a DC electric field between two conductors close to each other when humidity is present and, in time, may give rise to a leakage current or a partial short circuit. In most cases a good, dense protective overcoat will prevent this problem. Normally, silver-based pastes are not directly solderable.

Copper-based pastes have a somewhat lower conductivity than the silver-based ones. Copper has little tendency to migrate but tends to oxidize under high-temperature processing, and precautions must be taken to avoid high-contact resistance layers. Some copper pastes are directly solderable.

Nickel paste has significantly lower conductivity than silver and copper but has good solderability and stability.

Carbon or a mixture of carbon and silver is used in pastes for keyboard contacts or conductor straps, where a relatively high resistance can be tolerated. Carbon is also used for resistor pastes possessing low conductivity and good stability. The resistivity depends on the concentration of carbon in the insulating matrix.

Insulating, or dielectric, pastes are composed of the same basic ingredients, except that they lack the conducting particles.

Thermal curing is achieved either in an ordinary heated cabinet ("box oven"), an infrared radiation (IR) furnace, or vapour-phase furnace at temperatures ranging from 80 to 120°C for thermoplastics and from 150 to 250°C for thermosetting pastes. Microwave heating is also possible. UV curing is done by exposing the board to intense ultraviolet radiation for a short time.

The resistance and stability of PTF resistors depend on curing type and conditions. Even for conductors and dielectrics, the curing conditions are most important. The different ingredients in a paste, the different pastes used in one process, and the substrate material must be compatible thermally (equal thermal expansion coefficients) and chemically.

Polyimide-based PTF systems are generally very stable and reliable. They are cured at 300–400°C and require ceramic or polyimide substrates.

1.4 PROCESSING OF MICROSENSOR POLYMER FILMS

In this book, it is simply impossible to give a detailed description about all polymer materials and their processing. Materials and structures are described mainly in Appendix 1 and also in Chapter 3 in connection with the operation of sensors.

In this section a general survey will be given about those polymer processes that are compatible or easily applicable with the different inorganic sensor technologies described in the previous sections.

There are two main possibilities to apply polymer films in sensors (see Figure 1.27):

(1) Preprocessed polymer films are synthesized and shaped into sheet forms separately from the sensor structures [see Figure 1.27(a)]. Thin polymer sheets are generally produced by extrusion and stretching. The films can also be metallized using vacuum evaporation with continuous movement of the sheet. The metal films can be shaped by photolithography and etching—as in the case of thin-film processing (see

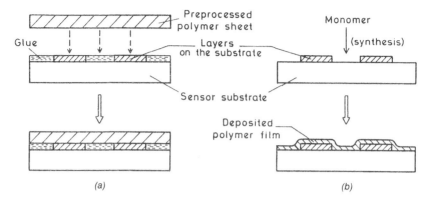

FIGURE 1.27. Polymer film application possibilities: (a) gluing preprocessed polymer film, (b) direct deposition of polymer film.

Section 1.3). Polymer films can be mounted into inorganic sensor parts: a typical method is gluing them to the surface of a previously fabricated microsensor part. An example for this technology is the application of preprocessed electropoled poly(vinylidene fluoride) (PVDF) foils in piezoelectric and pyroelectric devices [14].

(2) Polymerization directly on sensor surfaces [see Figure 1.27(b)] is the commonly used technique for fabrication of thin polymer layers. The synthesis and shaping occur in the sensor structure itself; thus, such methods are needed, which are compatible with inorganic sensor parts.

Photosensitive polymers for photolithography are widely used in integrated circuit technologies as photoresists. The method can be adopted for shaping sensor polymer films. Photosensitive polymers give a special advantageous possibility for technology: UV illumination means polymerization and shaping at the same time. Nonphotosensitive polymers can also be patterned photolithographically, but it is a more complicated process. The well-known addition and condensation polymer synthesis processes can be combined with a lot of different coating, patterning, and curing methods. Polymer mono- and bilayers can be obtained by the Langmuir-Blodgett method followed by a polymer synthesis. Electrochemical and plasma polymerization processes are also used in the fabrication of microsensors.

It must be mentioned here that polymers can also be used in sensors as

(1) Sensor substrates (polyimides, epoxies) made by conventional sheet technologies

(2) Conductive composite, isolation, and protective layers made by thin- and thick-film technologies, which are also used for sensing polymer films

(3) Light guiding and/or cladding materials in optical fibres: polymer-cladded silica core and fluorinated acrylic-cladded poly(methyl methacrylate) (PMMA) core optical fibres are often used. Their production technology is similar to the pulling method of glass fibres described in Section 1.2.4.

1.4.1 Processing of Photosensitive Polymers

A significant advantage of photosensitive polymers is the relative ease of processing using an existing integrated circuit or microsensor fabrication line. They can be applied and patterned with the same technology as photoresists without using the latter ones.

A photosensitive polymer is called positive when it becomes soluble because of the degradation (bond cleavage) during the photo exposure. The photosensitivity is called negative when the polymer is cross-linked during the exposure and it becomes unsoluble. Microsensor polymers are generally of the latter type because of their better stability during the device operation. The technology flowchart is given in Figure 1.28. Its first step is the application of adhesion promoter to the surface of the inorganic substrate. Adhesion generally must be improved by the use of a coupling agent. (For example, in the case of polyimide organosilanes and aluminium-chelate can be used as the coupling agent.)

The raw material for polymer layers is generally a liquid form solution of the monomers and additives, depending on the application (ionophores, conductive particles, etc.). This liquid can easily be applied by spin coating. The thickness of the layer depends on the solution viscosity, spin rate, and spin time. A high-viscosity solution gives a thicker film at the

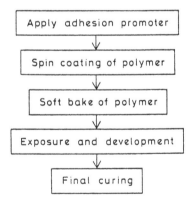

FIGURE 1.28. Processing flowchart of photosensitive polymer layers.

same spin rate. To get a smooth surface, the spin rate should be in the range of 3000–5000 r.p.m. for 10–100 sec. Other coating methods, such as dipping, casting, and immersion, are also applicable, but spinning is the most widely used technique for photoresist application in the IC technology. The soft baking process makes it possible to get a dry and partially cross-linked polymer layer.

The degree of polymerization can be determined by the curing temperature and time and has been investigated by several methods. Infrared absorption spectra analysis is one well known method.

The curing can also be done by IR annealing. After exposure, the development may be done by the classical methods: bath immersion, drop-spin, spraying, or shower developing. The patterned polymer layer generally needs a final curing to continue the cross-linking process and to get a stable dense layer.

An example for this technology is the photo-polymerization technique used for patterning photosensitive polyimide and poly(styrene sulphonate) in humidity sensors [15,16,17].

1.4.2 Photolithography of Nonphotosensitive Polymers

Nonphotosensitive polymers can also be patterned by photolithography, but the process is more complex. It is summarized in Figure 1.29. The greatest problem in this case is the prevention of the interaction between the soft baked polymer layer and the photoresist. The reaction can take place if the soft bake temperature is not high enough or the photoresist is applied onto a hot polymer film. It is important to select a suitable temperature to make the polymer hard enough to prevent the reaction with the photoresist, but not too hard for etching. Through experiments, the best soft baking temperature and time have to be determined. Also, a hard baking process is necessary for the photoresist after development; otherwise, it cannot withstand the polymer etcher. Figure 1.30 gives a comparison between the patterning technologies of photosensitive and nonphotosensitive polymer layers.

Photolithographic patterning is often used for shaping nonphotosensitive polyimide layers in sensors [15].

1.4.3 Printing of Polymers

Screen printing and curing are widely used technologies for polymer composite materials that are available in paste form. A detailed description is given about the so-called polymer thick-film (PTF) technology in Section 1.3.3.

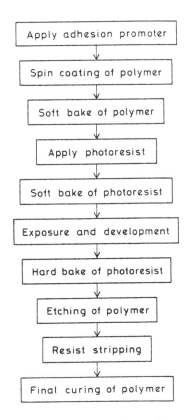

FIGURE 1.29. Process sequence for patterning nonphotosensitive polymer films.

Special printing processes have also been developed, and the main idea is shown in Figure 1.31 [18]. Metal mould masks are used instead of screen masks. The masks contain openings where the deposition of polymers onto the substrate is desired. A special squeegee with a rectangular pipe structure cross section is used. The pipe is filled with the solution of monomer. When the squeegee passes over the mask openings, the material is printed onto the substrate. The thickness of the printed layer is well controlled by the thickness of the metal mask. The method is often used for polyimide, epoxy, and silicone-resin coatings.

1.4.4 Electrochemical Polymerization

Conducting and semiconducting polymers can be synthesized and deposited onto a conductive surface part of a given substrate from monomer solutions by electrochemical polymerization. The advantage of this method is the precise flow control and rate of film deposition by varying the potential

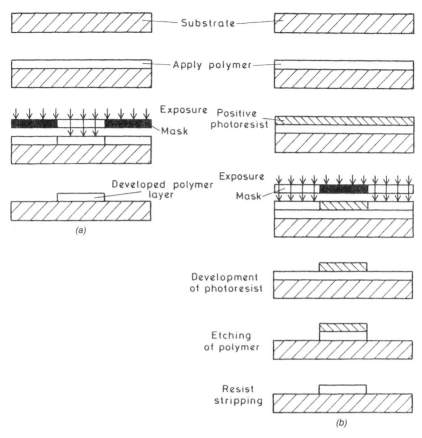

FIGURE 1.30. Comparison between the patterning of (a) photosensitive and (b) nonphoto-sensitive polymer layers.

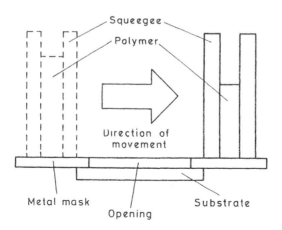

FIGURE 1.31. Special printing technique developed for polymers. (Redrawn from Takasago et al. [18], © 1987 IEEE).

41

of the working electrode in the system. The method gives a high-quality thin film and has great perspectives in biosensors, where enzymes may be entrapped into conducting polymer films [19].

Electropolymerization, like the well-known radical polymerization, is a chain process. While the growing end groups in the latter case are electrically neutral, they are charged in electropolymerization. During the synthesis, charge transfer also takes place. Polymerizing systems are electrically neutral. Hence, some negatively charged ions are present in cationically polymerized systems, while vice versa at anionically growing species. The participation of these counter-ions makes the polymerization intrinsically more complex than radical ones. Since the oppositely charged ions attract each other and strongly interact with the molecules surrounding them, a variety of species coexist in the polymerized medium. This is the basis of the conductivity modifying mechanism of electropolymerized layers.

There are two different ways in which an electrochemical system can be employed to perform the polymerization of a monomer:

(1) Polymerizations in which the monomer undergoes a direct activation at the electrodes

(2) Polymerizations promoted by initiators (free radicals, ions, organometallic species) produced by an electrochemical reaction on depolarizers different from the monomer

In many cases, paired electrochemical syntheses occur in which both anodic and cathodic processes contribute to the overall reaction. Therefore, the application of potential cycling is often necessary to obtain homogeneous synthesized layers.

Various aromatic compounds can be polymerized by electrochemical oxidation in a solution containing a supporting electrolyte. The reaction involves the subtraction of hydrogen from the monomer, forming fully conjugated polymers. The simultaneous electrochemical doping of electrolyte anions may render these polymers highly electrically conducting.

As an example, the electropolymerization mechanism for conducting polypyrrole (PPy) is shown in Figure 1.32 [20]. The electropolymerization process [see Figure 1.32(a)] is a simultaneous growth and doping, which enables negatively charged A^- species (acting as dopants) to be irreversibly captured.

The reaction is, in fact, a condensing oxidation, which follows an electronic stoichiometry, allowing a coulometric control of the film thickness to be deposited. For reasons of electroneutrality, the dopant counter-ions are inserted into the polymer matrix, which can be considered as a polymeric organic salt.

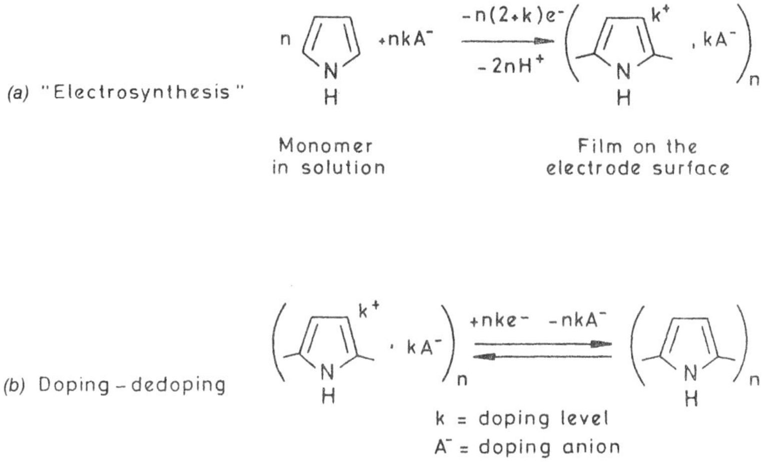

(a) "Electrosynthesis"

Monomer
in solution

Film on the
electrode surface

(b) Doping – dedoping

k = doping level
A⁻ = doping anion

FIGURE 1.32. Schematic electropolymerization mechanism for polypyrrole [20]: (a) synthesis and (b) doping-dedoping.

In the scheme, k represents the doping level, which can also be altered by a further electrochemical process. In fact, this latter doping-dedoping reaction is a completely reversible redox process [see Figure 1.32(b)] resulting in conductivity changes over several orders of magnitude.

In the practice, electrosynthesis is carried out by subsequent growing-doping and doping-dedoping processes using cyclic anodic-cathodic reactions. A typical cyclic voltammogram for poly(N-methylpyrrole) film growing is shown in Figure 1.33 [21]. The behaviour of the deposited films can be investigated using cyclic voltammetry (see Section 2.4.3).

1.4.5 Vacuum-Deposited Polymer Films

Vacuum deposition processes are also possible methods to obtain thin polymer films that have high density, thermal stability, and insolubility in organic solvents, acids, and alkalis. The layers can be deposited on any substrates that cannot be attacked by the vacuum processes. Polymer films can be growth by [22–26]:

- vacuum pyrolysis, which consists of a sublimation, a pyrolysis, and a deposition-polymerization process
- vacuum polymerization stimulated by electron bombardment
- vacuum polymerization initiated by ultraviolet irradiation
- vacuum evaporation using a resistance-heated solid polymer source and more effectively using an electron beam

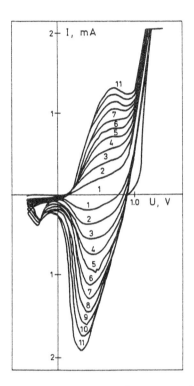

FIGURE 1.33. Electrochemical deposition of poly(*N*-methylpyrrole) by cycling at 50 mV/s. The first eleven cycles are shown. (Redrawn from the data of Bartlett et al. [21], with the kind permission of Elsevier Sequoia S. A., Lausanne, Switzerland, publisher of *Sensors and Actuators A* and *Sensors and Actuators B*).

- RF sputtering of polymer targets in a plasma composed of polymer fragments and with argon added to the plasma
- plasma or glow discharge polymerization of monomer gases or vapours

Figure 1.34 shows a schematic comparison of the different vacuum deposition processes. Although many papers have been published describing the vacuum deposition processes of polymer films, there is still considerable uncertainty about their growing mechanism and precise composition.

Vacuum pyrolysis [22] is a common technology for preparing parylene films. These thin films are deposited using a vacuum process in which the crystalline solid dimer (di-*p*-xylylene) is sublimed under vacuum at 100–250°C and then pyrolyzed at about 450–750°C to give a nearly quantitative yield of the reactive intermediate, *p*-xylylene. Although this is stable in the gas phase, it condenses and simultaneously polymerizes on any solid

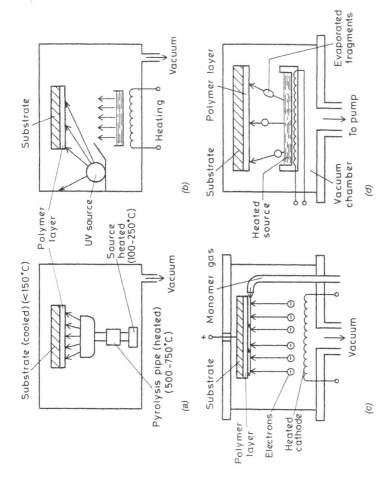

FIGURE 1.34. Schematic illustration of the vacuum deposition processes used for polymer films: (a) vacuum pyrolysis, (b) vacuum UV-polymerization, (c) electron beam polymerization, (d) vacuum evaporation, (e) RF sputtering, (f) glow discharge (plasma) polymerization.

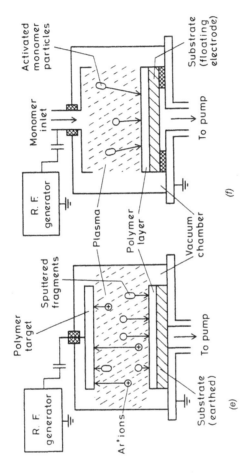

FIGURE 1.34 (continued). Schematic illustration of the vacuum deposition processes used for polymer films: (a) vacuum pyrolysis, (b) vacuum UV-polymerization, (c) electron beam polymerization, (d) vacuum evaporation, (e) RF sputtering, (f) glow discharge (plasma) polymerization.

surface at a temperature less than 120°C to give a continuous film of linear, high molecular weight poly(p-xylylene) called parylene.

Polymerization of simple organic monomer gases on surfaces under electron bombardment was observed many years ago, and it can be used for the deposition of thin polymer films [23]. It was proved experimentally that active sites are produced on the surface by the electron beam, and the polymer film grows by interaction of adsorbed monomer with these active sites rather than by direct collision of gas phase molecules. The process is often used to prepare polymers from butadiene or silicon oils, for example.

The mechanism of the UV-initiated polymer film deposition process is similar to the former one.

Vacuum evaporation and sputtering technologies have already been described in Section 1.3.1 in connection with inorganic thin-film deposition processes. These methods are also used to deposit polymer films. Obviously, if one thermally evaporates or ion impact sputters solid bulk polymers, the emission will be of fragmented material whose molecular weight is less than that of the parent polymer. Typical spectra for vacuum evaporated or sputtered PTFE show the existence of CF^+, CF_2^+, CF_3^+, $C_2F_3^+$, $C_2F_4^+$, $C_3F_5^+$, and $C_3F_6^+$ ions, for example [25,27].

During evaporation a thermal decomposition occurs in the heated polymer [24]. When the polymer is sputtered, molecular fragments are ejected from the target, and further molecular dissociation results from excitation of a plasma in the emitted material. Target emission in an RF sputtering process may not be solely from sputtering phenomena. If the polymer target is locally heated, then thermal evaporation could also occur. At both processes, the low-weight fragments arriving at the substrate build up a polymer film by recombination.

The glow discharge, or so-called plasma polymerization process [26,28,29], is similar to the plasma-enhanced chemical vapour deposition, and it can be performed in PECVD and RF sputtering equipment, which is widely used in the fabrication of electronic components and integrated circuits (see Sections 1.1 and 1.3). The reactor chamber has to be evacuated to 10^{-2} Pa, after which the monomer gas will flow in up to a pressure of 1–20 Pa. A high voltage RF source will then be switched on with a capacitive coupling of the glow discharge.

The duration of the process depends on the desired polymer film thickness. The pressure, voltage, and current of the plasma should be controlled during the reaction. If the monomer is a solid, the reactor should be heated up to the melting point. The process is very similar to the RF sputtering, with the difference that there is no target and that the low-weight molecular fragments are produced from the monomer gas molecules by decomposition in the plasma. In the case of PTFE plasma polymerization, CF_4 or C_3F_8 are generally used as monomer gases.

The properties of the plasma-deposited and RF-sputtered polymer layers are very similar; however, a few significant differences have also been discovered [26]. It is still not clear what similarities and differences the layer growing mechanisms have.

Evaporation is used for the fabrication of polyethylene layers [24], and all three latter methods are used for preparing PTFE layers, for example [25,26,27].

It is very important that polymer alloys and metal polymer composite thin films can also be deposited [30,31]:

- by evaporation from two sources
- by sputtering from composite targets that consist of metal foils superimposed by polymer foils or by cosputtering from different targets
- by simultaneous plasma processes, in which not only plasma polymerization, but plasma etching and sputtering from a built-in target, also occur at the same time

The combined processes are often used for the deposition of Au/Cu/Pd-PTFE composite thin films. As examples for the vacuum deposited layers from the field of sensors, the plasma polymerization of polytetrafluoroethylene (PTFE) and poly(hexamethyldisiloxane) (PHMDS) layers in chemical and humidity sensors can be mentioned [28,29]. Metal-polymer composite thin films can be used in gas sensors [32].

1.4.6 Other Polymer Film Technologies

There are a lot of conventional coating methods such as spin coating, dipping, casting, bath immersion, and various curing technologies using ovens, hot plate, vacuum oven, conveyor belt furnace, and infrared (IR) irradiation, which are widely used in polymer film application. A newly developed process is the *microwave assisted curing,* which gives an ease of and precise control of the temperature profile.

Cross-linking of highly oriented monomer films made by the Langmuir-Blodgett method (see Section 1.3.1) can simply be achieved by UV irradiation or by chemical reaction in many cases. The process is illustrated in Figure 1.35. Nowadays, the latter method is often used in chemical and biosensor applications [33]. Recently, an LB-polymerization process has also been developed for polyimide films [34].

CVD and gas state polymerization are principally the same processes as the CVD for growing inorganic layers (see Section 1.3.1): the polymer film deposition onto the substrate is due to the chemical reaction of gas components introduced into the reaction chamber [35].

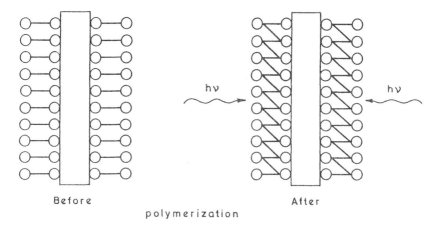

Before   After

polymerization

FIGURE 1.35. Polymerization of LB films.

In connection with plasma processing technologies (see Section 1.4.5), *plasma etching* of polymers must be mentioned, which seems to be a promising patterning technology for high-resolution polymer films [36]. Plasma etching has been utilized in silicon processing for many years. The technology generally applies an O_2/CF_4 plasma; the ionized particles bombard the polymer surface and remove its material fragments through chemical reactions.

Ion-beam treatment and ion implantation (see Section 1.1.6) techniques can be used to modify the properties of polymer films. Recently, polyimide films have been irradiated by well-controlled ion implantation methods [37]. Polyimide is a kind of thermally stable aromatic polymer with excellent insulation characteristics; however, after the implantation, its conductivity can be enhanced by several orders of magnitude. The electrical behaviour, in correlation with the Raman spectroscopy, showed that graphite precipitates are present in the implanted films. The appearance of the graphite phase in the modified films may be attributed to the transformation of the benzene rings into graphite cells, while the most noncarbon atoms (i.e., H, O, N) and atomic groups would form gas molecules and escape from the layer. Thus, implantation can be used as a synthesis of polymer composites from pure polymer films [37].

1.5 REFERENCES AND SUPPLEMENTARY READING

1. Sze, S. M., *VLSI Technology,* McGraw-Hill Book Co., New York (1985).
2. Elliott, D. J., *Microlithography, Process Technology for IC Fabrication,* McGraw-Hill Book Co., New York (1986).

3. Moulson, A. I. and Herbert, I. M., *Electroceramics. Materials, Properties, Applications*, Chapman and Hall (1990).

4. Baker, D., Koehler, D. C., Fleckenstein, W. O., Roden, C. E. and Sabia, R., *Physical Desing of Electronic Systems, Vol. II.*, Prentice-Hall, Inc., New York (1970).

5. Buchanan, R. C., *Ceramic Materials for Electronics*, Marcel Dekker, Inc., New York (1986).

6. Maissel, L. I. and Glang, R., *Handbook of Thin Film Technology*, McGraw-Hill Book Co., New York (1970).

7. Illyefalvi Vitéz, Zs. and Harsányi, G., *Electronics Technology*, Textbook for Undergraduate Students, TU Budapest (1992).

8. Harper, C. A., *Handbook of Thick Film Hybrid Microelectronics*, McGraw-Hill Book Co., New York (1974).

9. Agnew, J., *Thick Film Technology*, Hayden Book Co., NJ (1973).

10. Martin F. W., "The Use of Polymer Thick Film for Making Printed Circuit Boards," *Hybrid Circuits*, (1982), No. 1, p. 22.

11. Kirby, P. L., "Origins and Advances in Polymer Thick Film Technology," *Proc. of the 7th European Hybrid Microel. Conf.*, Hamburg (1989), p. 4.1.

12. Harsányi, G., "Screen Printed Polymer Thick Film Resistors on Epoxy-Glass Substrates," *9th Int. Spring Seminar on Electronics Technology*, Balatonfüred, Hungary (1986), pp. 126–129.

13. Ripka, G. and Harsányi, G., "Electrochemical Migration in Thick Film ICs," *Electrocomponent Sci. and Techn.* (1985), 11, pp. 281–290.

14. Münch, W. V. and Thiemann, U., "Pyroelectric Detector Array with PVDF on Silicon Integrated Circuit," *Sensors and Actuators A*, 25–27 (1991), pp. 167–172.

15. Endo, A., Takada, M., Adachi, K., Takasago, H., Yada, T. and Onishi, Y., "Material and Processing Technologies of Polyimide for Advanced Electronic Devices," *J. Electrochem. Soc.*, (1987), Vol. 134, No. 10, pp. 2522–2527.

16. Boltshauser, T. and Baltes, H., "Capacitive Humidity Sensors in SACMOS Technology with Moisture Absorbing Photosensitive Polyimide," *Sensors and Actuators A*, 25–27 (1991), pp. 509–512.

17. Hijikigawa, M., Miyoshi, S., Sugihara, T. and Jinda, A., "A Thin-Film Resistance Humidity Sensor," *Sensors and Actuators*, 4 (1983), pp. 307–315.

18. Takasago, H., Takada, M., Adachi, K., Endo, A., Yamada, K., Makita, T., Gofuku, E. and Onishi, Y., "Advanced Copper/Polyimide Hybrid Technology," *IEEE Transactions on CHMT*, 10, No. 3 (1987), pp. 425–432.

19. Yon Hin, B. F. Y. and Lowe, C. R., "Amperometric Response of Polypyrrole Entrapped Bienzyme Films," *Sensors and Actuators B*, 7 (1992) pp. 339–342.

20. Bidan, G., "Electroconducting Conjugated Polymers: New Sensitive Matrices to Build up Chemical or Electrochemical Sensors. A Review," *Sensors and Actuators B*, 6 (1992), pp. 45–56.

21. Bartlett, P. N., Gardner, J. W. and Whitaker, R. G., "Electrochemical Deposition of Conducting Polymers onto Electronic Substrates for Sensor Applications," *Sensors and Actuators A*, 21–23 (1990), pp. 911–914.

22. Taylor, R. C. and Welber, B., "Laser-Monitored Deposition of Parylene Thin Films," *Thin Solid Films*, 26 (1975), pp. 221–226.

23. Hill, G. W., "Electron Beam Polymerization of Insulating Films," *Microelectronics*, Vol. 4, No. 3 (1965) pp. 109–116.

24. Luff, P. P. and White, M., "The Structure and Properties of Evaporated Polyethylene Thin Films," *Thin Solid Films,* 6 (1970), pp. 175–195.

25. Morrison, D. T. and Robertson, T., "R. F. Sputtering of Plastics," *Thin Solid Films,* 15 (1973), pp. 87–101.

26. Holland, L., Biederman, H. and Ojha, S. M., "Sputtered and Plasma Polymerized Fluorocarbon Films," *Thin Solid Films,* 35 (1976), L19–L21.

27. Murakami, Y. and Shintani, T., "Vacuum Deposition of Teflon-FEP," *Thin Solid Films,* 9 (1972), pp. 301–304.

28. Leimbrock, W., Landgraf, V. and Kampfrath, G., "An Extended Site-Binding Model and Experimental Results of Organic Membranes for Reference ISFETS," *Sensors and Actuators B,* 2 (1990), pp. 1–6.

29. Sugimoto, I., Nakamura, M. and Kuwano, H., "Molecular Sensing Using Plasma Polymer Films," *Sensors and Actuators B,* 10 (1993), pp. 117–122.

30. Bruschi, P., Nannini, A. and Massara, A., "Low Temperature Behaviour of Ion-Beam-Grown Polymer-Metal Composite Thin Films," *Thin Solid Films,* 196 (1991), pp. 201–213.

31. Kay, E. and Hecq, M., "Metal Clusters in Plasma Polymerized Matrices: Gold," *J. Appl. Phys.,* 55 (2) (1984), pp. 370–374.

32. Bruschi, P., Cacialli, F. and Nannini, A., "Sensing Properties of Polypyrrole-Polytetrafluoroethylene Composite Thin Films from Granular Metal-Polymer Precursors," *Sensors and Actuators A,* 32 (1992), pp. 313–317.

33. Karymov, M. A., Kruchinin, A. A., Tarantov, Y. A., Balova, A., Remisova, A., Sukhodolov, N. G. and Yanklovich, A. I., "Langmuir-Blodgett Film Based Membrane for DNA-probe Biosensor," *Sensors and Actuators B,* 6 (1992), pp. 208–210.

34. Imai, Y., Nishikata, Y. and Kakimoto, M., "Preparation and Microelectronic Applications of Langmuir-Blodgett Films of Polyimides and Related Polymers," *Proceedings of the 32nd Midwest Symp. on Circ. and Syst.,* Champaign, Illinois (1989), pp. 735–748.

35. Ojio, T. and Miyata, S., "Highly Transparent and Conducting Polypyrrole-Poly(vinyl alkohol) Composite Films Prepared by Gas State Polymerization," *Polymer Journal,* Vol. 18, No. 1 (1986), pp. 95–98.

36. Schubert, P. J. and Nevin, J. H., "A Polyimide-Based Capacitive Humidity Sensor," *IEEE Transactions on Electron Devices,* ED-32, No. 7 (1985), pp. 1220–1223.

37. Xu, D., Xu, X. and Zou, Sh., "Ion-Beam-Modified Polyimide as a Novel Temperature Sensor: Fundamental Aspects and Applications," *Rev. Sci. Instrum.,* 63 (1) (1992), pp. 202–206.

CHAPTER 2

Sensor Structures with Sensitive Polymers

As described in the previous chapter, sensors operating with sensitive polymers generally consist of an inorganic part and a polymer film that interacts with the environment. The properties of the polymer film vary during this interaction, and this change is detected by the inorganic part which provides or modifies an electrical (or optical) signal. The structure is shown in Figure 2.1.

A survey will be given in this chapter about the typical device structures that can be used to measure the changes in the properties of the polymer films. The most important structures can be categorized into the following groups:

- Impedance-type sensors follow the changes of the measurand with capacitance and/or with resistance changes.
- Semiconductor-based types are more complicated; the sensing effect can change the most important characteristics and/or parameter of the devices, for example, shift the diode characteristics, alter the threshold voltage of the FET, etc.
- At resonant sensors, the shift of the resonance frequency can be measured as a function of the quantity in question.
- Electrochemical cells are widely used in chemical sensors: the electrode potential, the cell current, and/or the cell resistance can be measured as a function of the analyte.
- Calorimetric sensors are based on the measurement of a temperature difference caused by physical, or rather chemical, reactions caused by the analyte.
- Fibre-optic sensors represent a new generation of sensors: they are based on the changes of the light propagation, absorption, and/or emission properties and give an optical output signal.

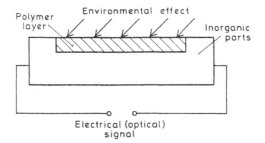

FIGURE 2.1. General structure and operation scheme of polymer-based sensors.

2.1 IMPEDANCE-TYPE SENSORS

The structure and operation principles of impedance-type sensors are the most simple. Measurand changes alter the permittivity and/or resistivity of the polymer layer, which, in turn, produces the variation of capacitance and/or resistance values of the sensor.

Figures 2.2 and 2.3 show the simplest device structures: the sheet capacitor and the film resistor, respectively. The basic equations are also given in the figures. These expressions are also used in the design of discrete and integrated passive components.

Figure 2.4 shows the interdigital device structure, which is rarely used in integrated devices because only small capacitance and resistance values can be realized. It is, however, a very popular structure in the sensorics because of the large free surface, which can interact with the environment.

Neglecting the edge effects, the capacitance of an interdigital capacitor can be calculated from the capacitance of a two-dimensional unit cell, C_u, using the formula:

$$C = C_u \cdot (n - 1) \cdot l \tag{2.1}$$

where n is the number and l the length of the electrodes. The capacitance of a unit cell, C_u, is determined not only by the geometry and the permittivity of the polymer layer, but also by the permittivity of the substrate and the gas environment.

There are two special cases of particular interest: when the polymer layer is a thin film ($t \ll a,b$) or when it is a thick film with relative high permittivity value (10–20). In the former case, the unit capacitance can be expressed using a complete elliptic integral of the first kind, $K[x]$ [1]:

$$C_u = \epsilon_0 \cdot \frac{(\epsilon_s + \epsilon_g)}{2} \cdot \frac{K\left\{\left[1 - \left(\frac{a}{a+b}\right)^2\right]^{1/2}\right\}}{K\left(\frac{a}{a+b}\right)} + \epsilon_0 \cdot \epsilon_p \cdot \frac{t}{a} \tag{2.2}$$

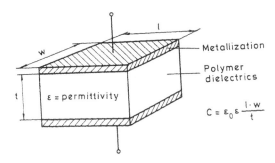

FIGURE 2.2. Structure and parameters of sheet capacitors.

$$C = \varepsilon_0 \varepsilon \frac{l \cdot w}{t}$$

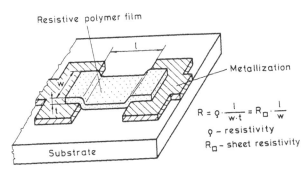

FIGURE 2.3. Structure and parameters of film resistors.

$$R = \varrho \cdot \frac{l}{w \cdot t} = R_\square \cdot \frac{l}{w}$$

ϱ – resistivity

$R_\square$ – sheet resistivity

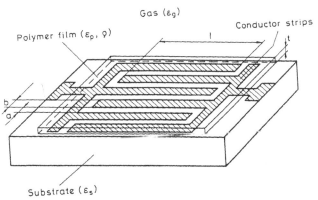

FIGURE 2.4. Interdigital structure used in both capacitor- and resistor-type sensors.

(The symbols are explained in the figures.) In the second case, the electric field is concentrated into the substrate and the polymer film. Thus, the capacitance of the unit cell is

$$C_u = \epsilon_0 \cdot \frac{(\epsilon_s + \epsilon_p)}{2} \cdot \frac{K\left\{\left[1 - \left(\frac{a}{a+b}\right)^2\right]^{1/2}\right\}}{K\left(\frac{a}{a+b}\right)} \qquad (2.3)$$

When the interdigital structure is used as resistor, the resistance value can be estimated as

$$R = R_\square \cdot \frac{a}{(n-1)\cdot(l + 2\cdot a + b)} \qquad (2.4)$$

At AC investigations of the interdigital structure, both the capacitive and the resistive behaviour must be taken into account [2]. Figure 2.5 shows the equivalent circuit model, including interface impedance. In the figure, R_b is the bulk resistance of the polymer; R_e and C_e are the polymer/electrode interface resistance and capacitance, respectively; and C_g is the geometric capacitance, which consists of the bulk polymer capacitance and the stray capacitance. The parameters can be measured experimentally by the complex impedance spectra method using the so-called Cole-Cole plots. Figure 2.6 shows a typical complex impedance diagram for interdigital structures using polymer layers with relatively low permittivity and large resistivity. The horizontal axis is the resistance, i.e., the real part of the impedance (ReZ), and the vertical axis is the reactance, i.e., the imaginary part of the impedance (ImZ). In the cases when the polymer is an electrolyte, for example, C_g–R_b and C_e–R_e pairs are dominating different frequency ranges.

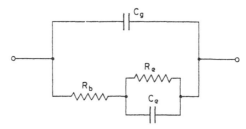

FIGURE 2.5. Equivalent circuit to interpret complex impedance spectra: C_g, geometrical capacity; R_b, bulk resistance; R_e, polymer/electrode interface resistance; C_e, polymer/electrode interface capacity.

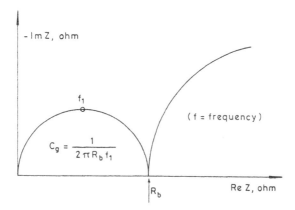

FIGURE 2.6. Complex impedance spectra of the equivalent circuit.

Therefore, R_b and C_g can be determined simply from the complex impedance spectra, as shown in Figure 2.6.

2.2 SEMICONDUCTOR-BASED SENSORS

Semiconductor-based sensors use silicon or compound semiconductors for fabricating inorganic parts. Sensitive polymer layers may be insulating, semiconducting, or conductive materials in these structures.

2.2.1 Field Effect Principle

In sensitive electronic devices based on the field effect principle, the semiconductor surface potential is modulated by potential or electric change variations elsewhere in the structure, usually called a gate, which is actually an insulator-(metal)-sensitive polymer-environment interface. In Section 1.1.7 and Figure 1.15, the structure and principle of operation of the conventional metal oxide semiconductor field effect transistor (MOSFET) has already been discussed qualitatively.

Figure 2.7 shows the cross section of the conventional PMOS transistor and the typical structure used in FET-based sensors. In both structures, the two p-type regions are known as the source and drain and can be interchanged without affecting the electrical behaviour of the device.

The thin insulating layer (SiO_2 and/or Si_3N_4) and the covering (metallization and/or sensing polymer) layer is called the gate, and it is this electrode that mainly controls the action of the transistor. In the conventional MOS transistor, a bias voltage is applied to the gate, which is called the control-

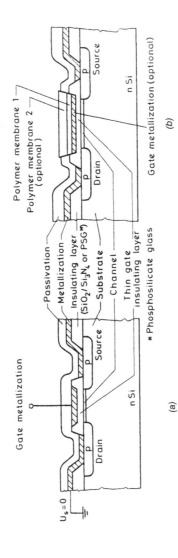

FIGURE 2.7. Field effect device structures used in (a) transistors (MOSFETs) and (b) sensors.

58

ling gate voltage. In the sensor FET devices, the gate metallization and/or polymer coating environment interaction produce the controlling potential, which is a function of the quantity to be measured and can be detected only as a potential difference according to a reference.

In order to understand the sensor FET's operation, it is necessary to complete the description of MOSFET with a few mathematical equations. If an increasing negative bias is applied to the gate, a positive charge is first induced at the silicon surface, and electrons are repelled away from this region. A depletion region results, which will deepen as the bias increases. A point called the turn-on (or threshold) voltage, U_t, is reached when free holes are induced at the surface.

Inversion is said to have taken place and a thin *p*-type channel is formed between the source and drain, and holes can flow between them via the channel. Removal of the bias restores the original condition. This is characteristic for the "enhancement" PMOS transistor. The most important equations describing the behaviour of MOSFETs are as follows. The drain-source and transfer characteristics are given in Figure 2.8. The drain-source current can be expressed as follows [3]:

$$I_{ds} = -\mu \cdot c_i \cdot \frac{w}{l} \cdot \left[(U_{gs} - U_t) \cdot U_{ds} - \frac{U_{ds}^2}{2} \right] \tag{2.5}$$

where

μ = the charge carrier mobility in the channel
c_i = the insulator capacitance per unit area at the gate
w = the channel width
l = the channel length

At the drain current saturation, when

$$|U_{ds}| \geq |U_{gs} - U_t| \tag{2.6}$$

the equation for the current becomes

$$I_{ds} = -\frac{\mu \cdot c_i}{2} \cdot \frac{w}{l} \cdot (U_{gs} - U_t)^2 \tag{2.7}$$

The turn-on or threshold voltage of the channel can be expressed as [3]:

$$U_t = -\frac{Q_{ss} + Q_b}{c_i} + 2 \cdot \Phi_f + \Phi_{ms} \tag{2.8}$$

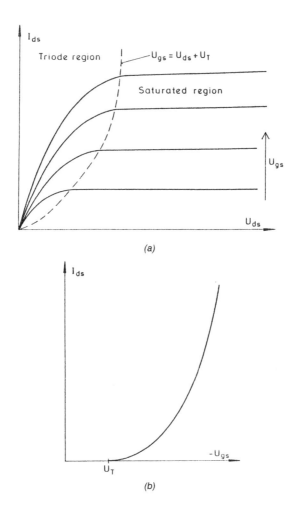

FIGURE 2.8. Typical characteristics of the MOS transistors: (a) drain characteristics and (b) transfer characteristic in saturation.

where

Φ_f = the Fermi-potential of the bulk semiconductor
Φ_{ms} = the work function difference between the gate and the semiconductor
Q_{ss} = the surface state and oxide charge per unit area
Q_b = the bulk charge per unit area in the depletion region

Probably, the most sensitive parameter to manufacture and to environment effects of the MOS transistors is the turn-on voltage. This is a big difficulty for designers and technologists; however, it can be utilized successfully in sensors. In sensor FETs, the drain-source current is controlled not by the external gate voltage, but by charge and work function conditions of the free gate metallization or simply by the insulator polymer interface, respectively. There are a lot of special sensitive FET structures, but their names are not always used consequentially. In the following section, a brief survey will be given about the typical abbreviations and the typical structures, including their special application and electrical connection possibilities.

IR photosensitive FET can be built up using a pyroelectric polymer film over the gate metallization. A transparent metallization must be deposited as a counter electrode on the top of the polymer [4]. In fact, the device is an integrated version of the sensitive capacitor and the FET [5] [see Figure 2.9(a)] so it is similar to the douplex-gate IGFET (insulated-gate FET). The latter is used in humidity sensors with humidity-sensitive polymer dielectrics.

The chemically sensitive field effect transistors are called *CHEMFETs*. The name is often used for the modified versions of ISFET (discussed later).

TMOSFET [5], *thin-film metal-oxide-semiconductor,* structures can be used as gas-sensitive FETs [see Figure 2.9(b)]. For thin porous metal films, charged species or strong dipoles located on the surface of the metal or on the insulator between the metal grains are detected through a capacitive coupling to the semiconductor surface. A shift of the threshold voltage can be measured on the transistor, which is due to the change of the surface charge and work function difference between the metal and semiconductor caused by the dipole layer. Hydrogen molecules that dissociate on the metal surface and deliver free hydrogen atoms to the metal contribute an additional sensing mechanism. The hydrogen atoms diffuse through the metal grains and build up a dipole layer of hydrogen atoms at the metal-silicon dioxide interface, which causes a threshold voltage shift in the characteristics. The sensitivity and selectivity of gas sensors with catalytic metal gates can be influenced by polymers on the top of a thin metal gate, which can act as membranes, i.e., only certain gases can penetrate the membrane while others are hindered by it.

The polymer can also react with a certain gas and may deliver reaction species that can be detected. Dipole molecules of gases might change the work function properties of the polymer as well. Another group of gas sensors uses [6] a polymer layer between the oxide and metallization film [see Figure 2.9(c)]. The applied polymers can change their relative permittivity when adsorbing or absorbing large dipole moment gas molecules. The per-

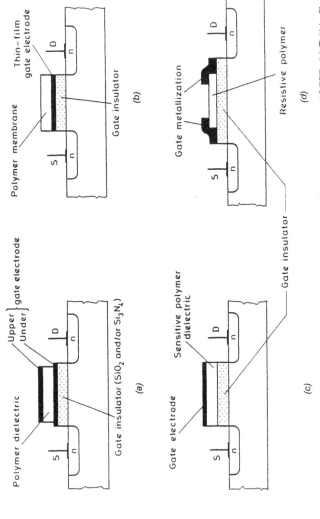

FIGURE 2.9. Typical sensitive FET structures using polymer films: (a) duplex-gate IGFET; (b) T (thin-film) MOSFET; (c) capacitively controlled FET; (d) CFT (charge flow transistor).

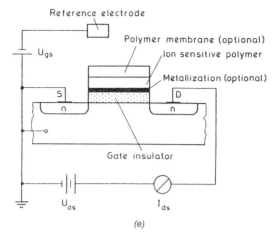

(e)

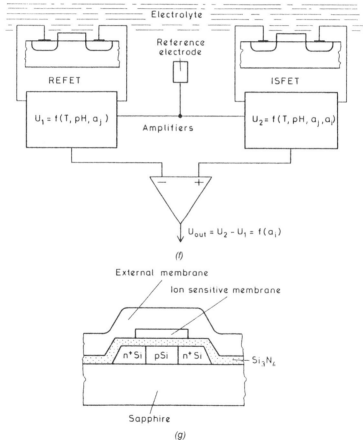

(f)

(g)

FIGURE 2.9 (continued). Typical sensitive FET structures using polymer films: (e) structure and connection of ISFET; (f) differential ISFET/REFET measurement set-up; (g) sensing SOS (silicon on sapphire) FET.

mittivity change can be detected by a shift in the FET characteristics [see Equations (2.7) and (2.8)].

CFT, the *charge-flow transistor* [7], is an FET with a resistive polymer gate [see Figure 2.9(d)] connected to two metal electrodes. There are a lot of conductive polymer types that show a resistivity change under different environmental conditions (see Chapter 3). The polymer-oxide-substrate region of the CFT can be simulated as an RC transmission line. There is a time delay between the application of the gate-to-source voltage and the appearance of drain current in the completed channel. Thus, the CFT acts as a transducer, converting a sheet-resistance measurement into the measurement of a delay time.

The structure of the *ion sensitive field effect transistor* (ISFET) is similar to the conventional Insulated-Gate FET (IGFET). The metal gate electrode is substituted by a multilayer structure containing ion-selective and/or permselective insulating and sometimes metallization layers. The top layer is directly exposed to the solution to be analysed. Figure 2.9(e) shows the structure and operation of the ISFET. Since the first reports about ISFET, various devices of this type have been developed and described [8–10].

The potential difference, E_c, between the solution and the gate insulator surface depends upon the ion activity of the solution. Under electrochemical equilibrium, E_c is given by the Nernst-Nicolsky-Eisenman equation:

$$E_c = E_0 + \frac{R \cdot T}{z_i \cdot F} \cdot \ln\left[a_i + \sum_j K_{ij} \cdot (a_j)^{z_i/z_j}\right] \qquad (2.9)$$

where E_c is the e.m.f. measured on the ion-selective electrode, E_0 is the standard potential of the ion-selective electrode, a_i is the activity of the primary ion, a_j is the activity of the interfering ions, K_{ij}'s are the selectivity coefficients of the primary ion over the interfering ions, R is the universal gas constant, T is the absolute temperature, z_i is the valency of the primary ion, z_j is the same of the secondary ions, and F is the Faraday constant.

The practical application of ISFET similar to other potentiometric sensors requires a reference electrode [see Figure 2.9(e)]. The static behaviour of the ISFET can be derived by taking into account the potential differences among the interfaces in Figure 2.9(e), and it results in the threshold voltage:

$$U_t(\text{ISFET}) = E_{\text{ref}} - E_c + U_t \qquad (2.10)$$

where E_{ref} is the electrode potential of the reference electrode and E_c and U_t are given by Equations (2.8) and (2.9).

In a practical measurement set-up, constant voltages are applied to the drain and reference electrode, respectively, while the substrate and the source are grounded [see Figure 2.9(e)]. Changes in the gate insulator–solution potential are monitored as changes in the drain-source current, which is often measured with a current-voltage converter amplifier unit. Applying an appropriate feedback circuit, the gate voltage can be varied to keep the current constant. Thus, the former potential changes can be directly followed.

Normally, conventional glass electrodes are used to obtain a constant voltage drop at the reference electrode-electrolyte interface. Such a reference system has the following disadvantages [9]:

- The conventional glass electrode and the ISFET cannot be integrated.
- The lifetime is limited by the regeneration period necessary for the glass electrode.

These problems can be overcome by contacting the liquid with a noble metal electrode, e.g., a gold electrode. As the potential at a metal-electrolyte interface is not stable, a differential measurement set-up consisting of an ISFET and a nonsensitive reference-FET called REFET has to be established [11]. As shown in Figure 2.9(f), additional common mode disturbances, such as temperature effects or the influence of interfering ions, can be suppressed by using a differential ISFET/REFET pair.

The lack of selectivity of an ISFET might be overcome by the application of specific ionophores. Synthetic macrocyclic polyether-type complexes with cationic guest species have tremendously stimulated the molecular design and synthesis of receptor molecules that selectively recognize ions and even neutral molecules (see Chapter 3). Attachment of these receptor molecules to the gate oxide surface of an ISFET results in a SURFET. An alternative approach is to deposit a membrane also, incorporating the receptor molecules. The structure is called membrane-modified FET (MEMFET) [12].

A lot of work has been done on sensing effects in which the pH change or other chemical change detected by a field effect chemical sensor is induced by the presence of an enzyme [13]. In many cases, an enzyme is immobilized in a layer on top of the sensing gate surface of an ISFET to enable measurement of the presence of the enzyme substrate in the solution into which the device is placed. These field effect devices are called *ENFETs.*

ImmunoFET (immunosensitive field effect transistor) is based on the phenomenon that immunoreactions can cause a change in charge density in an antibody loaded membrane deposited on an ISFET [14].

SOS (silicon on sapphire) technology can be used for SOS/ISFET, shown in Figure 2.9(g), which is a special solution [13].

2.2.2 Other Semiconductor Sensors

Metal semiconductor diodes, the so-called Schottky-barrier devices, build up from n-GaAs, and discontinuous platinum films can be used as detectors of different gases over a wide range of temperatures. The sensitivity of the device can be measured by the change of capacitance when it is exposed to gases. The sensitivity and selectivity can be modified by a polymer layer on top of the device, which acts as a semipermeable membrane for gas separation [15] (see Figure 2.10).

Interesting device structures can be fabricated using semiconducting or conductive polymers, which can change their resistivity value under different environmental conditions (exposure to gases, solutions, etc.). Such devices also use silicon-based structures, but the polymer here is an active part of the devices, not only an operation modifying layer or membrane. Figure 2.11 shows a silicon/(conductive polymer)/gold metallization device structure, which can be used as a diode that changes its characteristics as a function of the quantity to be measured [16].

In Figure 2.12 a semiconducting, polymer-based FET has been illustrated [17]. Its structure is the counterpart of CHEMFETs. The silicon basis gives the gate electrode; the drain and source are metal films covered by the semiconductor polymer material. The conducting channel is induced in the latter part. The characteristic of this device is a function of the resistivity of the polymer, which depends on different environment effects.

2.3 RESONANT SENSORS

Resonant sensors generally have an electro-mechanical resonance frequency, which is a function of the measurand. The output is a quasidigital frequency signal, which is much less prone to noise and interference and can be measured directly in digital systems. This is a great advantage over conventional analog sensors. A common approach for obtaining a fre-

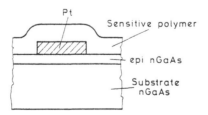

FIGURE 2.10. Schematic cross section of a Pt/GaAs-polymer Schottky barrier sensor.

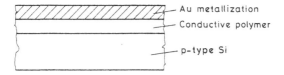

FIGURE 2.11. Schematic cross section of an Au/(conductive polymer) diode structure.

quency output is the use of electronic oscillators in which the sensor element itself is the frequency-determining element. Piezoelectric resonant sensors use the elastic properties of different piezoelectric materials (see Section 3.1.1) to sense and measure a physical or chemical phenomenon. They are based on the propagation of elastic bulk or surface waves, and therefore, they can be divided into three main categories:

- bulk acoustic wave (BAW) sensors
- surface acoustic wave (SAW) sensors
- Lamb wave (LW), acoustic plate mode (APM), and Love wave sensors

Resonant sensors that apply sensitive polymer layers consist of an inorganic resonator covered by the polymer film. The polymer acts as a selective sorbent that adsorbs or absorbs particles from the surroundings. The sensor's operation is based on the gravimetric principle: mass and density change due to the adsorption and absorption results varies in acoustic wave propagation properties; therefore, a shift in the oscillation frequency can be detected. The selectivity is determined by the polymer material (see Section 3.4).

2.3.1 BAW Sensors

BAW (otherwise called thickness-shear mode, or TSM) sensors have been applied to a wide variety of mass, chemical, and biochemical meas-

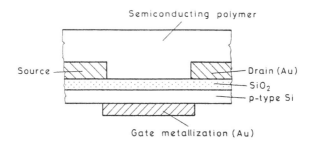

FIGURE 2.12. Structure of the semiconducting polymer-based FET.

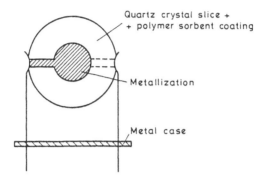

FIGURE 2.13. Piezoelectric crystal plate used as quartz microbalance sensor.

urement applications. Gas sensors operate by binding molecules to the device surface via one of several mechanisms, depending on the properties of the covering polymer film [18]. For BAW sensors, small quartz crystal disks of 10–15 mm diameter and 0.1–0.2 mm thickness are used (see Figure 2.13). The resonance frequencies are between 6 and 20 MHz. The historic development of quartz resonator sensors originates from the use of quartz resonators as time bases for frequency control. The high reliability and stability of quartz oscillators rely on the stability of a mechanical resonance in a structure composed of single-crystal quartz.

It has been well known that the deposition of a small mass of material on the surface of a quartz microbalance (QMB) lowers its resonance frequency. QMB sensors are commonly used for thin-film thickness monitoring in vacuum evaporation equipment (see Section 1.3). The gravimetric sensor principle is based on changes in the fundamental oscillation frequency, f_0, upon adsorption or absorption of particles from the surrounding gas phase. In the simplest case it can be described through the Sauerbrey equation [19]:

$$\Delta f = -f_0^2 \cdot \frac{c_f}{A} \cdot \Delta m \qquad (2.11)$$

where

Δm = the mass change
Δf = the frequency shift
A = the coated crystal area
f_0 = the operating frequency
c_f = the mass sensitivity

A sensitivity of about 2.3×10^{-10} m²/gs can be reached with an AT-cut quartz crystal operating with vibration in the thickness shear mode. For a 10-MHz crystal, the detection limit is at the nanogram level. From Equation (2.11), it can be easily shown that the relative frequency shift is equal to the relative mass change:

$$\frac{\Delta f}{f_0} = -\frac{\Delta m}{m} \qquad (2.12)$$

Thus, the detection limit is determined by the mass of the whole crystal. The operating frequency of the QMB sensors is determined partly by the thickness of the bulk crystal slice, so it has an upper practical limit of 20 MHz, which also determines the detection limit at a value of 0.1 ng/mm². This is still the greatest disadvantage of the BAW sensors.

2.3.2 SAW Sensors

The phenomenon of waves that occur on the surface of solids was described by Lord Rayleigh long ago [21]. White and Voltmer developed the interdigital transducer (IDT), which allows the convenient generation of surface acoustic waves (SAW) in piezoelectric solids (see Section 3.1.1) [22]. A large number of different devices for signal processing has been developed since then. Examples of these are filters, delay lines, resonators, and, recently, sensors. Both quartz- and LiNbO₃-based SAW devices were evaluated for their performance as gas detectors when coated with sensitizing sorbent organic polymer films.

In SAW devices the acoustic wave energy is constrained to the surface. The wave propagates along the surface of a solid medium. SAWs normally include different waves, the most important of which are the Rayleigh waves. The displacement of the particles near the surface of the solid that is propagating the Rayleigh wave has two components: a longitudinal and a shear vertical component [23]. The superposition of these two components results in surface trajectories that follow a retrograde elliptical path around their quiescent positions.

Waves can be generated quite easily in piezoelectric substrates using an IDT electrode, which can be fabricated microlithographically from thin-film metallization. The time-varying voltage will result in a synchronously varying deformation of the piezoelectric substrate and the subsequent generation of a propagating Rayleigh surface wave. The counterpart of the phenomenon can also occur: SAWs can generate alternating voltage in another IDT called a receiver. Delay line is a configuration consisting of two (or more) IDTs and a propagation path between them (see Figure 2.14). The

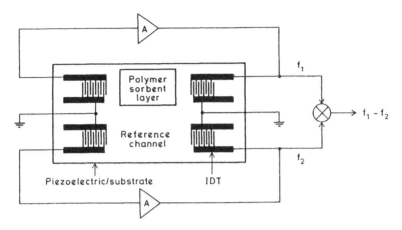

FIGURE 2.14. Structure and connection of the dual delay line SAW sensors.

first interdigital transducer excites an SAW whose frequency is mainly determined by the elastic constants of the piezoelectric material and the geometrical sizes of the generator IDT. An SAW will be received by the receiver IDT after travelling along the surface of the substrate. The propagation path is the sensitive area. All changes in the boundary conditions for SAW propagation lead to a variation of the SAW received by the second IDT. This well-known fact can be used for sensor applications.

The SAW resonator is, in fact, a delay line with an amplified feedback (see Figure 2.14) [24]. It can be used as a sensor in a way similar to the delay line; standing waves are formed by reflection from gratings. Any change in the environment leads to a change in the resonance frequency and can be registered [24].

The SAW sensors typically operate at high frequencies up to the GHz range. Because the sensitivity of the device increases as the square of the fundamental frequency, SAW sensors have greater potential sensitivity than BAW types. Many SAW sensor applications use a dual delay line configuration; one delay line can be coated with the sorptive or reactive polymer film, while the other remains inert or protected from the environmental effects (see Figure 2.14). Typically, the frequency difference will be measured, which is in the order of kHz and can easily be sampled. The uncoated or protected resonator acts as a reference to compensate undesired effects, for instance, frequency fluctuations caused by temperature or pressure changes. The response behaviour for an SAW device coated with a thin isotropic nonconducting polymer film is [23]:

$$\Delta f = (k_1 + k_2) \cdot (m/A) \cdot f_0^2 - k_2 \cdot t \cdot f_0^2 \cdot c_1 \qquad (2.13)$$

where

m/A = the mass per unit area of the overlayer
t = the thickness of the film
k_1, k_2 = material constants for the substrate
c_1 = material constant
f_0 = the unperturbed oscillation frequency
Δf = the frequency shift

The first half of the equation yields the shift resulting from mass loading, whereas the second half describes the effect of changes in the elastic properties of the film on the resonant frequency. For soft, rubbery polymers the second term becomes negligible; thus, frequency shift is predicted to be directly proportional to the mass changes due to the sorbate. This may be expressed by a simplified equation [23].

$$\frac{\Delta f}{f_0} = (k_1 + k_2) \cdot f_0 \cdot (\Delta m/A) \qquad (2.14)$$

where Δm is the mass change. When a good coating is available, it is usually possible to detect vapours at the 10–100 ppb concentration level, with a selectivity of 1000:1 or more over commonly encountered interfaces [25]. The mass detection limit is in the range of 0.05 pg/mm^2.

2.3.3 Other Resonator-Type Sensors

A new resonator sensor concept similar to SAW sensors, but employing Lamb waves (LW), was first published in 1988 [26]. Lamb waves propagate in the bulk of plates whose thickness is small compared to the ultrasonic wavelength: the thickness/wavelength ratio is less than one. Particle motions of Lamb waves are similar to those of Rayleigh waves; however, in a thin plate, the waves give rise to a series of symmetric and antisymmetric plate modes. The lowest order antisymmetric mode has a unique flexural character, hence, the name "flexural plate wave," or FPW.

Figure 2.15 shows a schematic cross section of an FPW sensor device [27]. The core of the device is an ultrasonic delay line consisting of a composite plate of a low stress insulation, metallization, and piezoelectric film. IDTs on the piezoelectric layer launch and receive the waves and, together with the amplifier, form a feedback oscillator whose output frequency depends on the mass per unit area of the membrane, including the chemically sensitive film. The advantages are in the low-MHz operation frequency range and in the possibility of operation while immersed in a liquid. Theo-

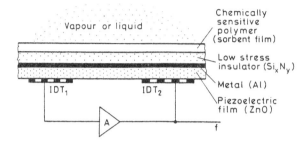

FIGURE 2.15. Schematic cross section of an FPW sensor.

retical and experimental results also show that, compared to BAW and SAW gravimetric sensors, LW sensors and especially FPW devices, are more sensitive to added mass. They are very attractive for chemical and biological sensing [27].

Recent advances in the design and construction of acoustic plate mode (APM) devices have allowed their use as sensors [28,29]. Figure 2.16 shows the cross section of a typical APM sensor. The sensor consists of a piezoelectric plate, a receptor polymer film, a fluid containment cell, and appropriate signal processing electronics. The piezoelectric plate supports APMs that are primarily horizontally polarized shear waves. Therefore, the device is called a shear horizontal acoustic plate mode (SH-APM) sensor. Typical plates are a few acoustic wavelengths thick. The waves are generated and detected by IDTs on the lower surface and reflect between the upper and lower surface of the substrate. This beam interacts with the receptor polymer film and viscously coupled fluid sample when it reflects from the upper surface. Changes in the film due to mass loading or viscoelastic stiffening will alter the phase of the reflected wave and result in a phase shift in the electrical signal. Similar to SAW sensors, a dual delay line configuration can be used to compensate nonspecific responses and undesired en-

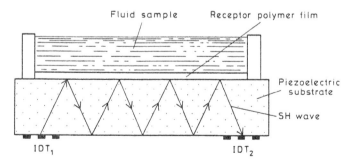

FIGURE 2.16. Cross section of an SH-APM sensor.

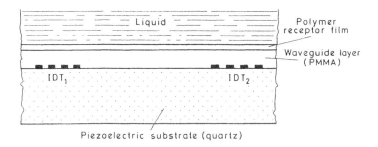

FIGURE 2.17. Typical geometry used in Love wave sensors.

vironmental effects. The advantage of the APM sensor is that electrical connections can be made on the surface of the device, which is not immersed in solution [29]. Typical APM device frequencies are 25–200 MHz.

Since a wave reflection mechanism occurs, acoustic energy cannot be concentrated at the surface unless very thin plates are used. An alternative approach reported by Gizeli et al. is to apply another geometry to excite another type of guided SH wave, the Love wave [28,30]. Mass loading of the surface skimming bulk wave (SSBW) device results in a composite acoustic device, known as Love plate, with an SH wave guided at the interface of the deposited polymer overlayer and the SSBW substrate. SSBWs are acoustic waves with only a shear displacement component propagating just below the surface of a piezoelectric substrate. SSBW suffers from a considerable acoustic loss. A thin overlayer of a dielectric material on the surface may convert the SSBW into a guided SH or Love wave, increasing the coupling coefficient of the wave and reducing the losses.

Generally, devices using thin films to help guide the waves are called Love wave devices. Figure 2.17 demonstrates the geometry used in Love plates. The substrate is generally piezoelectric quartz crystal; the overlayer is silica or poly(methyl methacrylate) (PMMA), which exhibit much lower shear acoustic velocities than quartz, and so fulfill the necessary requirement for the guidance of a Love wave. PMMA shows much lower viscoelasticity than a lot of other polymers; thus, it shows low acoustic losses. The application in the field of sensors is based on a similar effect to the sensing mechanisms of SII-AMP sensors. Changes in the receptor film (generally LB-films or LB-polymers) due to mass loading or viscoelastic stiffening alters the phase of the reflected wave. Phase shifts ($\Delta\Phi$) can be measured as a function of the mass-load ($\Delta m/A$) [30]:

$$\Delta\Phi = C \cdot (\Delta m/A) \tag{2.15}$$

where C is a constant.

Typical sensitivities that can be reached by this type of sensor are in the range of 5–30 deg/(10^{-6} g/cm^2), depending on the applied materials and layer thicknesses.

2.4 ELECTROCHEMICAL CELLS

Electrochemical cells are widely used as sensors for the measurement of chemical quantities, typically ion- or gas-molecule concentrations in different media. In the simplest cases, an electrochemical cell consists of a minimum of two electrodes and an ionic conductive material, called an electrolyte, between them (see Figure 2.18).

Electrochemical sensors are based upon potentiometric, amperometric (more generally, voltammetric), or conductivity measurements. The different principles always require a specific design of the electrochemical cell. The operation of the sensors is based on the reactions and their equilibria at the interfaces between electronic and ionic conductors on the electrode surfaces.

Electrochemical cell-type sensors can be classified into three main categories according to the characteristics of the electrode reaction [31]:

- Type A is characterized by a direct participation of the mobile ions, which are originally present in the electrolyte.
- Type B cells use the electrolytes as "solvents" for the charged products formed by redox reactions from neutral molecules that are present in the electrolyte or in its surroundings.
- Type C cells are characterized by competing or several step electrode reactions.

The measurement principles will be summarized separately in the next sections.

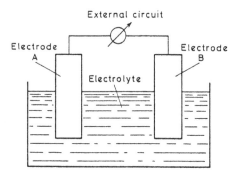

FIGURE 2.18. General structure of an electrochemical cell.

2.4.1 Potentiometric Sensors

In potentiometric sensors, the potential difference between the reference electrode and the working electrode is measured without polarizing the electrochemical cell, that is, very small current is allowed. In that case the equilibrium electrode potential difference can be monitored, which is given by the equation mentioned in Section 2.2 [see Equation (2.9)] [32]:

$$E = E_0 + \frac{R \cdot T}{z_i \cdot F} \cdot \ln\left[a_i + \sum_j K_{ij} \cdot (a_j)^{z_i/z_j}\right] \qquad (2.16)$$

where the symbols are the same as in Equation (2.9). In the case of ideal ion-selective electrodes (ISEs), the cross-sensitivity constants, the K_{ij}'s, could be neglected. In practice, they must be taken into consideration. The ion activity can be handled as a linear function of the ion concentration:

$$a_i = k_i \cdot c_i \quad (k_i < 1) \qquad (2.17)$$

where k_i is the activity coefficient of the ion in question.

The conventional electrodes use generally metal, metal-salt, or metal-electrolyte-glass structures. Recently, for ISE purposes another generation of electrodes—the membrane electrodes—are also used. These contain an internal reference electrode and an internal reference electrolyte in an isolation tube, which is closed by an ion-selective membrane that is generally a special polymer material. One side of the membrane is in contact with the analyte solution to be probed. The structure is shown in Figure 2.19(a). The polymer-sample solution interactions can be divided into the following groups:

- The polymer may contain active grains or ionic sites that are the basis of ion exchange and/or complexation within the membrane (and are called ionophores). By reversible exchange of the ions between the analyte solution and the membrane phase or by a reversible penetration of the ions accompanied by the complexation process, a membrane potential will be developed, which can also be described by the Nernst equation. Recently, sensor electrodes with nonionic grains, so-called neutral charge carriers, are preferably used for complexation processes with ions of the analyte.
- A surface charge and potential shift may also be developed by the sorption of the ions onto and/or into the polymer membrane. The process is similar to the adsorption of H^+ ions to the surface of the glass electrode.

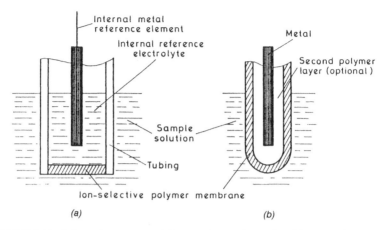

Internal metal
reference element

Internal reference
electrolyte

Metal

Second polymer
layer (optional)

Sample
solution

Tubing

Ion-selective polymer membrane

(a)   *(b)*

FIGURE 2.19. Schematic structure of the membrane electrodes: (a) conventional liquid electrolyte-based type; (b) solid-contact electrode.

Figure 2.19(a) shows the cell configuration for a conventional liquid junction ion-selective electrode. A reversible ion or electron transport mechanism is present not only at the membrane-analyte solution interface, but also at every phase boundary, resulting in well-defined interfacial potentials. Electrodes of this type usually exhibit little potentiometric drift [33].

A variety of other types of ion sensors were also developed in which permselective or ion-selective polymeric membranes are deposited directly onto solid electrode surfaces with no internal electrolyte solution. Figure 2.19(b) illustrates the cell configuration for the latter type called solid-contact ISE, which is similar to a coated-wire electrode. Often, in this type of ISE, the membrane/solid interface is ill-defined, leading to significant potentiometric drift. A stable and reversible transition from electronic conductivity in the metal electrode to ionic conductivity in the polymeric membrane can be maintained by using double-membrane structures. Polymer electrolytes are generally combined with permselective membranes. The mechanisms responsible for the improvement are still the subject of debate. Both adhesion and reaction mechanisms may influence the behaviour of the electrodes.

The application of electroconducting conjugated polymers (ECPs) results in many new possibilities to build up electrochemical sensors (see Chapter 3). In these materials, a semiconducting polymer contains ionic dopants and/or specific molecules covalently grafted to the ECP [34]. The great perspective for ECPs is the phenomenon that the ion exchange or ion sorption occurs directly in the electroconductive material. The ion exchange process directly changes the doping level and also the conductivity and work function of the semiconducting polymer. Thus, ECPs can be used directly as sensitive electrodes.

The detection of electro-inactive ions by ECPs is also possible, which is based on the "ion-sieving" or memory effect. This means that it is possible to vary the permeability of electropolymerized films via their controlled anodic growth. The cut-off size of ions can be determined by the size of the anion that was used for the electropolymerization, removing them by an electrochemical dedoping process (see Section 1.4.4).

By suitable choice of electropolymerization conditions, enzymes acting as negative dopants can be entrapped through coulombic interactions. This is a technique for immobilization of enzymes at the surface of electrodes, which is the basis of the fabrication of enzyme electrodes.

2.4.2 Amperometric Sensors

Amperometry is a method of electrochemical analysis in which the signal of interest is a current that is linearly dependent upon the concentration of the analyte. As the chemical species approach the working (or sensing) electrode, electrons are catalytically transferred from the analyte to the working electrode or to the analyte from the electrode. The direction of flow of electrons depends upon the properties of the analyte and can be controlled by the electric potential applied to the working electrode.

To maintain charge neutrality within the sample, a counter reaction occurs at a second electrode, the counter electrode. If a constant potential difference is maintained by the external circuit, a continuous current flow can be measured. Linear current ion-concentration characteristics can be obtained by amperometry at diffusion-controlled processes in the so-called "limiting current operating mode." Diffusion control can occur at several interfaces in the electrochemical cell:

- Diffusion of ions from the electrolyte to the surface of the electrode is used in the analytical method called polarography [35]. Applying the Fick law of diffusion and the Faraday's law for the charge transfer, it can be shown easily that the current density, i, is proportional to the concentration, c, of the electroactive component in the electrolyte:

$$i = \text{constant} \times c \tag{2.18}$$

- Diffusion of molecules is possible from the surrounding gas or liquid through a capillary into the electrolyte and/or to the surface of the electrolyte counter-electrode interface, where they will be ionized. In that case, of course, the electrical current density is proportional to the partial pressure, *p,* of the analyte:

$$i = \text{constant} \times p \qquad (2.19)$$

- Diffusion will also control the charge transfer process when particle flow limiting membranes are used on the surface of the electrolyte. This phenomenon is the basis of concentration measurements in the Clark-type O_2 sensors [36]. The process can be described similarly to the former one.
- Particle flow limiting membranes can also be used on the surface of electrodes or within the electrolyte [37].

Figure 2.20(a) shows typical voltage-current characteristics of diffusion-controlled processes. Figure 2.20(b) gives the typical limiting current value and analyte concentration relationship. Setting the operation mode of the sensor in the limiting current region and keeping the voltage on a constant value, a current proportional to the concentration can be measured.

To guarantee the constant potential difference between the working and counter electrodes independently from the current that polarizes the electrodes, the application of a potentiostat and a reference electrode is often necessary. Figure 2.20(c) shows a simple circuit diagram [38]. The function of the circuit is to set the electrode potential of the working electrode to a given value (U_{in}) with respect to the reference electrode. The electrode potential of the reference electrode is independent of the current flowing through the cell. It is also inert to the analyte to be examined. At the same time, the cell current is converted into an output voltage signal using a simple current-voltage converter circuit.

Polymer materials can be used for different purposes in the amperometric sensors:

- polymer solid electrolytes between the electrodes [39]
- diffusion-limiting polymer membranes [38]

2.4.3 Other Types of Electrochemical Measurements

The amperometric method is a special case of the *voltammetric measurements* where the whole potential-current diagrams are used for the analysis. Generally, the potential of the working electrode controlled by a potentiostat is changed continuously with a constant scan rate. Any reaction at the electrode surface can usually be detected as a current superimposed to

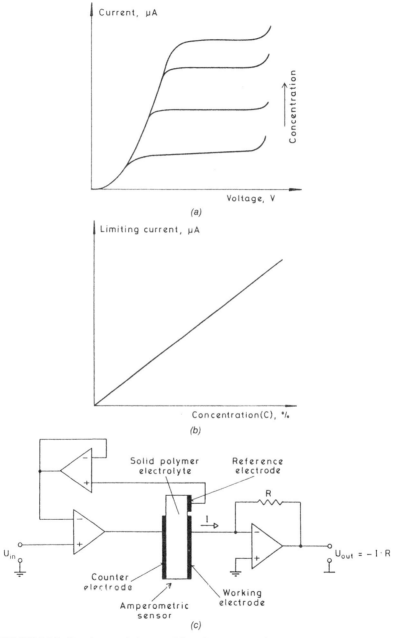

FIGURE 2.20. Structure and characteristics of amperometric sensors: (a) voltage-current characteristics; (b) limiting current–concentration characteristic; (c) schematic structure and circuit connection.

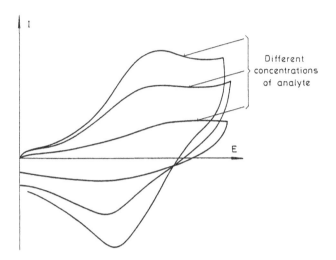

FIGURE 2.21. Typical cyclic voltammograms showing oxygen reduction using graphite rods coated with polymer [40].

the base current due to double-layer charging. Thus, in these voltammograms, current peaks can be observed. The peak potential values can be used for qualitative analysis, and the height of the peaks is a function of analyte concentration.

The method can be combined with electrodeposition, alternating the direction of current flow and following the response by cyclic voltammetry (see Section 1.4.4). Figure 2.21 shows typical cyclic voltammograms at different parameters [40].

The structure of voltammetric sensors is the same as that of the amperometric sensors. They are generally fabricated with reference, auxiliary, and working electrodes. Most polymer-modified electrode systems require the presence of a solution electrolyte, which is generally the sample itself. However, electrochemical cells in voltammetric sensors can operate with polymer electrolytes to measure, for instance, electroactive gases, as usual with amperometric sensors.

Conductimetric sensors are based on the measurement of electrolyte conductivity, which is changed when the material is exposed to a different environment. The sensing effect is based on the change of the number of mobile charge carriers in the electrolyte. If the electrodes are prevented from polarization, the electrolyte shows ohmic behaviour. As a first approximation the conductivity can be expressed as [41]

$$\sigma = \sum_i n_i \cdot z_i \cdot e \cdot \mu_i \qquad (2.20)$$

where

μ_i = the mobility of the different ions
z_i = the valency
n_i = the volume number of ionic carriers
e = the elementary electric charge

According to Equation (2.20), the conductivity is a linear function of the ion concentration; therefore, it can be used for sensor principles. Conductimetric sensors often apply polymer electrolytes, the behaviour of which is described in Section 3.3. The structure of the sensors and their measurement techniques have been summarized in Section 2.1 in connection with the impedance-type sensors.

2.5 CALORIMETRIC SENSORS

Calorimetric sensors are widely used for the detection of reactive gas components. Their operation principle is quite different from other sensor principles and results in a unique solution that cannot be discussed in the other sections.

Calorimetric sensors (often called pellistors) generally utilize the heat generation or consumption during catalytic chemical reactions of reactive gas molecules for their detection [41]. Their operation principle and general structure can be seen in Figure 2.22. The sensor consists of a substrate that is heated continuously by the heating resistor. Its temperature is different from the ambient temperature, of course.

This temperature difference results in a heat flow from the substrate to the environment. In a stationary state, the temperature difference is constant, and the power dissipation is equal to the heat flow. The temperature of the substrate can be monitored using a temperature sensor (thermistor, thermoelement, or Pt-resistor). The polymer catalyst coating can adsorb and absorb gas components from the environment, which can take part in catalytic chemical reactions. The reaction heat generated or consumed by the

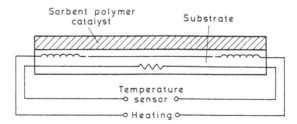

FIGURE 2.22. Schematic structure of the calorimetric sensors.

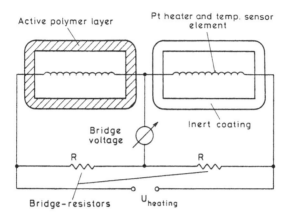

FIGURE 2.23. Application of calorimetric sensors in dual structure.

chemical reaction can disturb the stationary state heat flow from the sensor to the environment, and a temperature change can be detected, which is a function of the partial pressure of the gas component in question.

Calorimetric sensor applications generally use dual sensor configuration; one substrate is coated with the catalyst polymer film, while the other remains inert or protected from the environment effects (see Figure 2.23). The uncoated or protected substrate acts as a reference to compensate undesired effects, for instance, temperature fluctuations caused by the ambient temperature, heat conductivity, or ambient gas velocity changes.

In the case of adiabatic sensor type, the power dissipation is the same at both substrates, and the temperature difference will be measured. In the isothermic-type sensors, the temperatures are held at the same value, and the heating power difference necessary to keep the same temperatures will be monitored.

Conventional calorimetric sensors utilize thermodynamically controlled catalytic reactions. In this case, steady-state signals are obtained under continuous flow conditions at constant partial pressure of chemically active components. These signals are proportional to the generated catalytic reaction heat obtained for constant partial pressure of the given component [41].

Some new type sensors, in contrast, monitor kinetically controlled reactions between the sensor material and the detected molecules under chopped flow conditions. Only changes in the equilibrium concentration of adsorbed or absorbed molecules lead to time-dependent nonequilibrium heat generation, which only occurs during pressure variations. Fast and rectangular-shaped partial pressure variations should be produced, and the maximum temperature change that occurs during heat generation upon adsorption or desorption of molecules has to be determined [42]. Great efforts have been made to develop calorimetric sensors that can operate even at room temperature and therefore do not need special heating.

Calorimetric sensors generally suffer from lack of selectivity and from interference caused by water vapour. Various methods have been developed to overcome these serious problems, one of them being in the application of selective polymer catalysts [43].

2.6 FIBRE-OPTIC SENSORS

In optical communication links, it is the optical fibre that provides the transmission channel. The fibre consists of a solid cylinder of transparent material, the core, surrounded by a cladding of similar material (see Figure 2.24). Light waves propagate through the core in a series of plane wave-fronts, or modes; the simple light ray path used in elementary optics is an example of a mode. For this propagation to occur, the refractive index of the core must be larger than that of the cladding. There are two basic structures that have this property: step-index fibres with two different refractive index values for core and cladding and graded-index fibres providing a continuous transition of refractive index from core to cladding. There are multimode fibres that allow a great number of modes to propagate and single-mode fibres that only allow one mode to propagate. The propagation in the fibres can be described with the effective refractive index, which depends on both core and cladding refractive indices and type and thickness of the fibre.

The attenuation and bandwidth of an optical fibre will determine the maximum distance that signals can be sent. Attenuation is usually expressed in dB/km, while bandwidth is usually quoted in terms of the bandwidth length product, which has units of GHzkm, or MHzkm. Attenuation is dependent on impurities in the core, and so the fibre must be made from very pure materials. However, this phenomenon can be used even in fibre-optic sensors when the impurities originate from the environment.

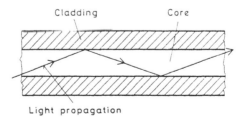

FIGURE 2.24. Typical structure of optical fibres.

Optical fibres can be made basically from two different materials: silica glass (SiO_2) and plastic polymer, generally poly(methyl methacrylate) (PMMA). The addition of certain dopants to the glass will vary the refractive index. All-plastic fibres use different plastics for the core and cladding. All-glass fibres exhibit very low losses and high bandwidths, which make them ideal for use in long-haul telecommunication routes. Large core fibres for use in medical and industrial applications are generally made of plastic, making them more robust than all-glass fibres, and are cheaper to produce. However, the very high attenuation and low bandwidth of these fibres tend to limit their uses, for instance, to the field of sensors.

Optical fibres offer the same advantages to transducer systems that they have in telecommunication: low signal attenuation, high information-transfer capacity, elimination of electromagnetic interference problems, and flexibility. Another advantage of optical fibre sensing is the extraordinary growth of optical telecommunication. Most sensors are still electrical or electronic. Having an electronic sensing and an optical data handling system, the optical signals must be converted to electronic ones and then back again. With the application of fibre-optic sensors, this disadvantage can be avoided. Moreover, they are also noncorrosive and biocompatible, which is important in the application field of medicine and environment protection [44,45].

Optical fibres can be made sensitive to a wide variety of phenomena, and there are many combinations where polymers can be used as sensitive layers or membranes.

Fibre-optic sensors can be categorized in many ways. A basic division is into intrinsic and extrinsic. In the first, the measured parameter causes a change in the transmission properties of the fibre itself; in the second the fibre acts just as a light conductor to and from the sensor.

There are a lot of interactions that can modulate the properties of the light transmitted through or reflected by the sensor causing phase, intensity, polarization, or spectral changes; even the generation of secondary light waves is possible. There are also various different sensor structures available in that field. A few typical examples using sensitive polymers are given in the following sections. The measurement techniques and detection modes are closely related to the sensor structures; therefore, they are discussed together.

2.6.1 Optrode-Style Sensors

Figure 2.25 shows a typical structure of chemical or biochemical fibre-optic sensors called an optrode. It consists of two (may be one or more than two) fibres for the input and output light. The operation is based on spectral

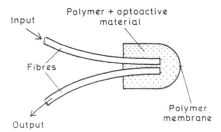

FIGURE 2.25. Schematic structure of the optrode-style fibre-optic sensors.

change caused by different physical and chemical effects at the tip of the sensor:

- Colourimetric detection is based on the colour changes of indicators entrapped in the polymer material at the sensor tip [46]. Practically the reflected light spectra are to be measured. A special case of that method is the simple reflectance measurement using a monochromatic light source or a special wavelength region [48]. Sometimes the spectral change is caused directly by the colour change of the analyte of interest.
- Fluorimetric detection is based on the phenomenon of fluorescence quenching caused by the component to be detected. A secondary light wave is generated, the spectrum of which differs from the incident light spectrum [45].
- Catalysed light emission caused by chemiluminescence or bioluminescence reaction can also be used. There is no need for incident input light in this case [47].

Optrodes can be used for the detection of gases, ions, and biochemical compounds. The optoactive polymer material and the membrane at the tip can contain ionophores, indicators, fluorescent dyes, immobilized chemi- or bioluminescence enzymes, organic adsorbents, etc., according to the application and operation principles. Chemical sensitivity and selectivity can be increased by immobilizing reagents and/or applying permselective membranes.

2.6.2 Core-Based Sensors

In core-based sensors, the core-analyte interaction is directly utilized as the basis of operation.

Most core-based sensors use porous fibres with chemically sensitive reagents immobilized physically or chemically on the surfaces of the pores to

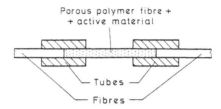

FIGURE 2.26. Typical core-based sensor using porous polymer core.

sensitize them to the analyte of interest. Figure 2.26 shows the mentioned structure. Porous polymer fibre can be made by heterogeneous cross-linking polymerization of methyl methacrylate (MMA) and triethylene glycol dimethacrylate (TGDM) [45,49]. After polymerization, the inert solvent can be removed, leaving an interconnected porous structure. It exhibits very high gas permeability and liquid impermeability so it can be used for the detection of gases in liquids. Vapours permeating into the porous zone can produce a spectral change in transmission. For better sensitivity and selectivity, colourimetric reagents can be trapped within the polymer matrix. The observed light intensity is a function of the concentration of the analyte to be measured, c, and can be expressed:

$$I = I_0 \cdot \exp[-h \cdot l \cdot f(c)] \qquad (2.21)$$

where I_0 is the intensity of the incident light; h is the extinction coefficient, both of light absorption and of light scattering; l is the light passing length of the sensing segment; and $f(c)$ is the concentration of the absorption and scattering centres.

Evanescent-wave refractometric core-based sensors are those in which the optical fibre transmission properties depend on the refractive index of the medium they are in [50]. Such sensors generally use short segments of fibre from which the cladding has been removed. For such fibres the loss depends on the refractive index of the surrounding medium, which may be a function of its composition. In fact, the optical transmission efficiency of the core is changed instead of absorbance changes.

2.6.3 Coating-Based Sensors

Coating-based chemical fibre-optic sensors can also be made using microporous or other type sensitive claddings [50] (see Figure 2.27).

Evanescent-wave refractometric cladding-based sensors detect the absorption of the species in the polymeric cladding, which leads to a variation of its refractive index, thus, also to a variation of the transmission efficiency. Porous polymer cladding-type sensors can be used, for instance, for humidity measurements [51]. The quantity of moisture absorbed by the microporous cladding varies with humidity. The optical power level of the transmitted light in the core varies according to the moisture absorption quantity because of the refractive index change of the cladding. Another example is the application of organopolysiloxanes as cladding materials since they can change their refractive index when they adsorb and/or absorb molecules from the surrounding medium.

Evanescent-wave spectroscopic sensors are based on the absorbance changes or fluorescence effect in the sensitized cladding caused by the species to be detected. If the refractive indexes of the waveguide and the cladding material are nearly equal, then the lightwave is reflected not at the fibre/cladding interface, but at the cladding/air interface. If the light is absorbed as it passes through the cladding, then the device can be used for spectroscopic analysis, often called internal reflection spectrometry. When the fluorescence effect occurs in the cladding, the energy of a generated secondary lightwave is pumped back into the core [50].

2.6.4 Interferometric Sensors

Conventional fibre-optic and integrated optical interferometers can also be used as sensors. Their operation is based on the optical interference between reference and sensitive lightwaves. According to the well-known rules of interference, the resultant intensity is a function of the difference in optical path length of the branches, which can be modulated by a change of either the refractive index or the geometrical length difference between the paths. Figures 2.28(a) and 2.28(b) show a transmission type and a reflection type interferometer structure, respectively.

The Mach-Zehnder interferometer [52] consists of two Y-branches [see Figure 2.28(a)]. The incoming light is split into two parts, which are guided in the two branches of the interferometer. For sensor applications, one branch is usually affected by the quantity to be measured, while the other provides the reference phase. If one of the branches of the integrated

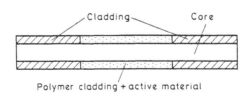

FIGURE 2.27. Fibre-optic sensor with sensitive cladding.

structure is covered with a sensitive polymer film, the refractive index of which is a function of its surrounding medium, the effective refractive index of this waveguide branch is a function of the quantity to be measured. Thus, the interferometer acts as a sensor based on the refractive index modulation caused by the evanescent field effect in the branch covered with the sensitive polymer.

Figure 2.28(b) illustrates the operation of reflection-type interferometer sensors [53]. The superposition of the two partial beams reflected from the waveguide/polymer and polymer/air interfaces, respectively, can be influenced either by swelling of the film caused by permeation of gases and liquids or by adsorption of particles on the top of the film, which will introduce an additional reflection. Moreover, an introduced analyte can interact with the polymer film, thus influencing the value of the refractive index. Spectral interferometry allows these effects to be discriminated to a certain extent.

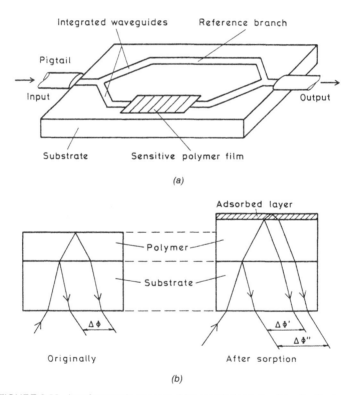

FIGURE 2.28. Interferometric sensors: (a) transmission type; (b) reflection type.

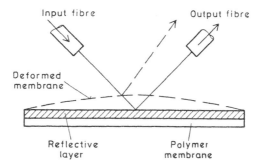

FIGURE 2.29. Fibre-optic sensor for mechanical quantities using intensity modulation.

2.6.5 Other Fibre-Optic Sensor Types

Extrinsic fibre-optic sensors can be fabricated for the detection of physical (mostly mechanical) quantities (displacement, force, pressure, acceleration, acoustic, etc.) from polymer fibres and polymer membranes (see Figure 2.29). These sensors are based on the modulation of reflected light intensity measurements. The most important difference from the former polymer-based fibre-optic sensor types is that the behaviour of this present type sensor is determined by the mechanical properties of the polymer membrane and not by the the the optical parameter changes in the light guiding or reflecting media.

2.7 REFERENCES AND SUPPLEMENTARY READING

1. Mansour, J. and Brown, P. M., *Field Analysis and Electromagnetics,* McGraw-Hill Book Company, Inc., New York (1963).
2. Tsuchitani, S., Sugawara, T., Kinjo, N. and Ohara, S., "A Humidity Sensor Using Ionic Copolymer and Its Application to a Humidity-Temperature Module," *Sensors and Actuators,* 15 (1988), pp. 375–386.
3. Wolfendale, E., editor, *MOS Integrated Circuit Design,* Butterworth & Co. Publishers Ltd., London (1973).
4. Ruppel, W., "Pyroelectric Sensor Arrays on Silicon," *Sensors and Actuators A,* 31 (1992), pp. 225–228.
5. Hedborg, E., Spetz, A., Ninquist, F. and Lundström, I., "Polymer Membranes for Modification of the Sensitivity of Field-Effect Gas Sensors," *Sensors and Actuators B,* 7 (1992), pp. 661–664.
6. Endres, H. E., Drost, S., Obermeier, E., Hutter, F., Haas, K. H. and Schmidt, H., "Gassensitive MOS-Feldeffekttransistoren mit Gatebeschichtung aus Ormosil," *Sensoren—Technologie und Anwendung, Fachtagung,* Bad Neuheim (1986), pp. 60–65.

7. Senturia, S. D., Sechen, G. M. and Wishneusky, J. A., "The Charge-Flow Transistor. A New MOS Device," *Applied Physics Letters*, Vol. 30, No. 2 (1977), pp. 106–108.

8. Bergveld, P., "Development, Operation and Application of the Ion Sensitive Field Effect Transistor as a Tool for Electrophysiology," *IEEE Trans. Biomed. Eng.*, BME-19 (1972), pp. 342–351.

9. Matsuo, T. and Esashi, M., "Methods of ISFET Fabrication," *Sensors and Actuators*, 1 (1981), pp. 77–96.

10. McBride, P. T., Janata, J., Comte, P. A., Moss, S. D. and Johnson, C. C., "Ion-Selective Field Effect Transistors with Polymeric Membranes," *Analitica Chimica Acta*, 101 (1978), pp. 239–245.

11. Müller, E., Woias, P., Hein, P. and Koch, S., "Differential ISFET/REFET Pairs as a Reference System for Integrated ISFET-Sensor-Arrays," *Int. Conf. on Solid State Sensors and Actuators (Transducers '91)*, San Francisco, CA (1991), pp. 267–470.

12. Reinhoudt, N., "Application of Supramolecular Chemistry in the Development of Ion-Selective CHEMFETs," *Sensors and Actuators B*, 6 (1992), pp. 179–185.

13. Saito, A., Miyamoto, Sh., Kimura, J. and Kuriyama, T., "ISFET Glucose Sensor for Undiluted Serum Sample Measurement," *Sensors and Actuators B*, 5 (1991), pp. 237–239.

14. Schasfoort, R. B. M., Keldermans, C. E. J., Kooyman, R. P. H., Bergveld, P. and Greve, J., "Competitive Immunological Detection of Progesterone by Means of the Ionstep Induced Response of an ImmunoFET," *Sensors and Actuators B*, 1 (1990), pp. 368–372.

15. Lechuga, L. M., Calle, A., Golmayo, D., Briones, F., De Abajo, J. and De La Campa, J. G., "Ammonia Sensitivity of Pt/GaAs Schottky Barrier Diodes. Improvement of the Sensor with an Organic Layer," *Sensors and Actuators B*, 8 (1992), pp. 249–252.

16. Remaki, B., Jullien, D. and Jouve, C., "Barrier Properties of Conducting Poly(ethylene) with Gold and Silicon," *Sensors and Actuators A*, 33 (1992), pp. 85–89.

17. Tsumura, A., Koezuka, H., Tsunoda, S. and Ando, T., "Chemically Prepared Poly(N-methylpyrrole) Thin Film. Its Application to the Field-Effect Transistor," *Chem. Lett.* (1986), p. 863.

18. Lucklum, R., Henning, B., Hauptmann, P., Schierbaum, K. D., Vaihinger, S. and Göpel, W., "Quartz Microbalance Sensors for Gas Detection," *Sensors and Actuators A*, 25–27 (1991), pp. 705–710.

19. Sauerbrey, G., "Verwendung von Schwingquarzen zur Microwagung," *Z. Phys.*, 155 (1959), pp. 206–222.

20. Endres, H. E., Mickle, L. D., Kösslinger, C., Drost, S. and Hutter, F., "A Gas Sensor System with Dielectric and Mass Sensors," *Sensors and Actuators B*, 6 (1992), pp. 285–288.

21. Lord Rayleigh, "On Waves Propagated along the Plane Surface of an Elastic Solid," *Proc. London Math. Soc.*, 17 (1885), pp. 4–11.

22. White, R. M. and Voltmer, F. W., "Direct Piezoelectric Coupling to Surface Elastic Waves," *Appl. Phys. Lett.*, 7 (1965), pp. 314–316.

23. Wohltjen, H., "Mechanism of Operation and Design Considerations for Surface Acoustic Wave Device Vapour Sensors," *Sensors and Actuators*, 5 (1984), pp. 307–325.

24. Grate, J. W., Martin, S. J., White, R. M., "Acoustic Wave Microsensors Part II, to be published in *Analytical Chemistry* (1993).

25. Rapp, M., Binz, P., Kabbe, I., Van Schickfus, M., Hunklinger, S., Fuchs, A., Schrepp,

W. and Fleischmann, B., "A New High-Frequency High-Sensitivity SAW Device for NO_2 Gas Detection in the Sub-ppm Range," *Sensors and Actuators B,* 4 (1991), pp. 103–108.

26. Wenzel, S. W. and White, R. M., "A Multisensor Employing an Ultrasonic Lamb-Wave Oscillator," *IEEE Trans. Electron Devices,* 35 (1988), pp. 735–743.

27. Wenzel, S. W. and White, R. M., "Flexural Plate-Wave Gravimetric Chemical Sensor," *Sensors and Actuators A,* 21–23 (1990), pp. 700–703.

28. Grate, J. W., Martin, S. J., White, R. M., "Acoustic Wave Microsensors Part I," *Analytical Chemistry,* Vol. 65, No. 21 (1993), pp. 940A–948A.

29. Andle, J. C., Vetelino, J. F., Lade, M. W. and McAllister, D. J., "An Acoustic Plate Mode Device for Biosensor Applications," *1991 Int. Conf. on Solid State Sensors and Actuators (Transducers '91),* San Francisco, CA (1991) pp. 483–485.

30. Gizeli, E., Goddard, N. J., Lowe, C. R. and Stevenson, A. C., "A Love Plate Biosensor Utilising a Polymer Layer," *Sensors and Actuators B,* 6 (1992), pp. 131–137.

31. Wiemhöfer, H. D. and Göpel, W., "Fundamentals and Principles of Potentiometric Gas Sensors Based upon Solid Electrolytes," *Sensors and Actuators B,* 4 (1991), pp. 365–372.

32. Sudhölter, E. J. R., Van der Wal, P. D., Skawronska-Ptasinska, M., Van den Berg, A. and Reinhoudt, D. N., "Ion-Sensing Using Chemically-Modified ISFETs," *Sensors and Actuators,* 17 (1989), pp. 189–194.

33. Goldberg, H. D., Cha, G. S., Meyerhoff, M. E. and Brown, R. B., "Improved Stability at the Polymeric Membrane/Solid Contact Interface of Solid-State Potentiometric Ion Sensors," *Proc. of the 1991 Int. Conf. on Solid State Sensors and Actuators (Transducers '91),* San Francisco, CA (1991), pp. 781–784.

34. Bidan, G., "Electroconducting Conjugated Polymers: New-Sensitive Matrices to Build up Chemical or Electrochemical Sensors. A Review," *Sensors and Actuators B,* 6 (1992), pp. 45–56.

35. Buttner, W. J., Maclay, G. J. and Stetter, J. R., "An Integrated Amperometric Microsensor," *Sensors and Actuators B,* 1 (1990), pp. 303–307.

36. Clark, L. C. Jr., "Monitor and Control of Blood and Tissue Oxygen Tensions," *Trans. Am. Soc. Art. Int. Organs,* 2 (1956), pp. 41–45.

37. Van den Berg, A., Grisel, A., Koudelka, M. and Van der Schoot, B. H., "A Universal on Wafer Fabrication Technique for Diffusion-Limiting Membranes for Use in Microelectrohcemical Amperometric Sensors," *Sensors and Actuators B,* 5 (1991), pp. 71–74.

38. Schelter, W., Gumbrecht, W., Montag, B., Sykora, V. and Erhardt, W., "Combination of Amperometric and Potentiometric Sensor Principles for On-Line Blood Monitoring," *Sensors and Actuators B,* 6 (1992), pp. 91–95.

39. Yan, H. and Lu, J., "Solid Polymer Electrolyte based Electrochemical Oxygen Sensor," *Sensors and Actuators,* 19 (1989), pp. 33–40.

40. Tieman, R. S., Heineman, W. R., Johnson, J. and Seguin, R., "Oxygen Sensors Based on the Ionically Conductive Polymer Poly(dimethyldiallylammonium chloride)," *Sensors and Actuators B,* 8 (1992), pp. 199–204.

41. Walsh, P. T. and Jones, T. A., "Calorimetric Chemical Sensors," in *Sensors: A Comprehensive Survey, Vol. II.* Eds., Göpel, W. et al. VCH Verlag, Weinheim (1991).

42. Schierbaum, K. D., Gerlach, A., Haug, M. and Göpel, W., "Selective Detection of Organic Molecules with Polymers and Supramolecular Compounds: Application of

Capacitance Quartz Microbalance and Calorimetric Transducers," *Sensors and Actuators A,* 31 (1992), pp. 130–137.

43. Marcinkowska, K., McGauley, P. M. and Symons, E. A., "A New Carbon Monoxide Sensor Based on a Hydrophobic CO Oxidation Catalyst," *Sensors and Actuators B,* 5 (1991), pp. 91–96.

44. Wolfbeis, O. S., "Biomedical Applications of Fibre Optic Chemical Sensors," *Int. J. Optoelectronics,* Vol. 6. No. 5 (1991), pp. 425–441.

45. Zhou, Q., Kritz, D., Bonell, L. and Siegel, Jr., G. H., "Porous Plastic Optical Fiber Sensor for Ammonia Measurement," *Applied Optics,* Vol. 28. No. 11 (1989).

46. Sadaoka, Y., Matsuguchi, M. and Sakai, Y., "Optical-Fibre and Quartz Oscillator Type Gas Sensors: Humidity Detection by Nafion Film with Crystal Violet and Related Compounds," *Sensors and Actuators A,* 25–27 (1991), pp. 489–492.

47. Gautier, S. M., Blum, L. J. and Coulet, P. R., "Alternate Determination of ATP and NADH with a Single Bioluminescence-Based Fibre-Optic Sensor," *Sensors and Actuators B,* 1 (1990), pp. 580–584.

48. Czolk, R., Reichert, J. and Ache, H. J., "An Optical Sensor for the Detection of Heavy Metal Ions," *Sensors and Actuators B,* 7 (1992), pp. 540–543.

49. Zhou, Q. and Siegel Jr., G. H., "Detection of Carbon Monoxide with a Porous Polymer Optical Fibre," *Int. J. Optoelectronics,* Vol. 4. No. 5 (1989), pp. 415–423.

50. Lieberman, R. A., "Recent Progress in Intrinsic Fiber-Optic Chemical Sensing II," *Sensors and Actuators B,* 11 (1993), pp. 43–55.

51. Ogawa, K., Tsuchiya, S., Kawakami, H. and Tsutsui, T., "Humidity-Sensing Effects of Optical Fibres with Miro-Porous SiO_2 Cladding," *Electronic Letters* Vol. 24. No. 1 (1988), pp. 42–43.

52. Brandenburg, A., Edelhauser, R. and Hutter, F., "Integrated Optical Gas Sensors Using Organically Modified Silicates as Sensitive Films," *Sensors and Actuators B,* 11 (1993), pp. 361–374.

53. Gauglitz, G., Brecht, G., Kraus, G. and Nahm, W., "Chemical and Biochemical Sensors Based on Interferometry at Thin (Multi)layers," *Sensors and Actuators B,* 11 (1993), pp. 21–27.

CHAPTER 3

Sensing Effects and Sensitive Polymers

Sensing effects are those physical and chemical phenomena that are the basis of the sensors' operation; i.e., they provide an electrical or optical signal as a function of the quantity to be measured. The structure and behaviour of the applied materials and their most important parameters are described in this chapter.

It is difficult to give a proper categorization of the sensing effects due to the large variety of material structures and physical/chemical interaction phenomena that can be used to fabricate sensors based on sensing polymer films.

Considering the electric sensing devices that are based on the change of their electrical parameter, it is well known that there are insulating and (semi)conductive polymers, but the possible conduction mechanisms within the latter type are less known. Depending on the type of charge carriers, there are ionically conductive polymers, often called polyelectrolytes, and electronically conductive ones.

Electronically conducting and semiconducting polymers have attracted a great deal of interest as sensing layers applied in sensors. There are two types: composites that contain an electrically insulating polymer matrix loaded with a conductive filler and polymers that are intrinsically conducting or can be made so by doping. The latter type materials have a very special structure containing one-dimensional organic backbones based on the alteration of single and double bonds and therefore are called electroconducting conjugated polymers.

Polymers can also be used as selecting materials: as selective sorbents for both molecules and ions or as permselective membranes. The optical behaviour of various polymers can also be used in fibre-optic sensors.

According to the mentioned properties, the sensing effects in polymers can be described using the following grouping of materials:

- dielectrics
- conductive composites
- electrolytes
- sorbents
- membranes with receptor molecules
- permselective membranes
- ion-exchange membranes
- optically sensitive polymers
- electroconducting conjugated polymers

3.1 SENSING EFFECTS BASED ON DIELECTRIC BEHAVIOUR

Dielectrics are insulating materials; that means they theoretically do not contain movable electrical charges. However, the gravities of electrically charged particles can be shifted under certain conditions. The dielectric behaviour can be described by two important physical parameters: one is the permittivity that describes the "polarizability" in electric fields, and the other is the vector of spontaneous polarization, which exists without an electric field as well.

Polarization can be changed by mechanical stress or by temperature variation; the former is called piezoelectric and the latter pyroelectric effect. Permittivity changes may also be detected due to the absorption of molecules with high dipole moments or due to the swelling, even when sorbing nonpolar molecules. Swelling is a pure geometrical effect that can be detected by capacitance changes. Dielectrics with high spontaneous polarization, called electrets, can also be used in capacitive sensors.

The following sections describe the mentioned phenomena in detail.

3.1.1 Piezoelectric Effect and Piezoelectric Polymers

Piezoelectric effect is the production of electricity by pressure [1,2]. It occurs only in insulating materials and is manifested by the appearance of charges on the surfaces of a single crystal that is being mechanically deformed. It is easy to see the nature of the basic molecular mechanism involved. The application of stress has the effect of separating the centre of gravity of the positive charges from the centre of gravity of the negative charges, producing a dipole moment. Clearly, whether or not the effect occurs depends upon the symmetry of the distributions of the positive and negative ions. This restricts the effect so that it can occur only in those material structures that are not having a centre of symmetry. For a centrosymmetric crystal, no combination of uniform stresses will produce the necessary separation of the centres of gravity of the charges.

It is clear that the converse effect also exists. When an electric field is applied to a piezoelectric crystal, it will strain mechanically. There is a one-to-one correspondence between the piezoelectric effect and its converse. Those materials in which strain produces an electric field will strain when an electric field is applied. In piezoelectric materials the electric polarization, P, is related to the mechanical stress, T, or, conversely, the development of a mechanical strain, S, is related to an applied field, E. One can define a piezoelectric coefficient, d, relating polarization to stress and strain to field, respectively, by

$$d = \left(\frac{\partial P}{\partial T}\right)_E \qquad d^* = \left(\frac{\partial S}{\partial E}\right)_T \qquad (3.1)$$

where the subscript E indicates that the field is held constant and the subscript T that the stress is held constant. In other words, the piezoelectric coefficient is given by the rate of change of polarization with stress at a constant field or the rate of change of strain with field at constant stress. The units of d and d^* will be coulombs per newton or meters per volt. It can be shown by the laws of thermodynamics that $d = d^*$.

Because the polarization and field are vector quantities and mechanical stress and strain are second rank tensors, d must be a third rank tensor. To a first approximation, the linearity can be supposed:

$$P_l = \sum_{i,k} d_{ikl} \cdot T_{ik} \qquad (i,k,l = 1,2,3) \qquad (3.2)$$

where 1, 2, 3 represent the directions of the different geometrical axes: x,y,z, and d_{ikl} are the piezoelectric coefficients. There will be twenty-seven of them, but $T_{ik} = T_{ki}$ and $d_{lik} = d_{ikl}$; thus, there are fifteen independent ones. How many of these are nonzero depends on the symmetry of microstructure of the material in question. An alternative piezoelectric coefficient, g, may be defined as

$$g = -\left(\frac{\partial E}{\partial T}\right)_P = \left(\frac{\partial S}{\partial T}\right)_T \qquad (3.3)$$

where the subscript P indicates constant polarization and T constant stress. The dimension of g will be in $m^2 C^{-1}$.

The relationship between g and d can be seen, by inspection of Equations (3.1) and (3.3) to be

$$d = \epsilon \cdot g \tag{3.4}$$

where ϵ is the dielectric constant.

In practical applications, the important property of a piezoelectric material is its effectiveness in converting electrical to mechanical energy or vice versa. This is given by its coupling coefficient k, which is defined by

$$k^2 = \frac{\text{Electrical energy converted to mechanical energy}}{\text{Input electrical energy}}$$

$$\tag{3.5}$$

$$k^2 = \frac{\text{Mechanical energy converted to electrical energy}}{\text{Input mechanical energy}}$$

The magnitude of k is proportional to the geometric mean of the piezoelectric coefficients d and g and is a measure of the ability of the material both to detect and to generate mechanical vibrations. There are three main types of piezoelectric materials: single crystals, ceramics, and polymers.

Among the thirty-two classes of single crystal materials, eleven possess a centre of symmetry and are nonpolar. For these, an applied stress results in symmetric ionic displacements so that there is no net change in dipole moment. The other twenty-one crystal classes are noncentrosymmetric, and twenty of these exhibit piezoelectric effect. The single exception in the cubic system possesses symmetry characteristics, which combine to give no piezoelectric effect. Polycrystalline and amorphous materials in which the axes of the dipole moments are randomly oriented show no piezoelectricity. If the axes can be suitable aligned, piezoelectric polycrystalline ceramics or polymers can be produced.

A polar direction can be developed in a material by applying a static field: this process is known as *poling*. There is, of course, no question of rotating the grains themselves, but the dipole axes can be oriented by reversal or by changes through other angles that depend on the microstructure involved, so that the spontaneous polarization has a component in the direction of the poling field. Electrodes have to be applied to the material for the poling process and these also serve for most subsequent piezoelectric applications. The exception is when the deformation is to be a shear. In this case the poling electrodes have to be removed, and electrodes are applied in planes perpendicular to the poling electrodes.

Electrical voltage generated by mechanical stress in piezoelectric materials decays due to the charge dissipation. Thus, it increases with applied force but drops to zero when the force remains constant. Voltage drops to a negative peak as pressure is removed and subseqauently decays to zero, as shown in Figure 3.1. That is why piezoelectric effect can be used for dynamic processes but not for static measurements.

Piezoelectric crystals are widely used to control the frequency of electronic oscillators. The crystal is cut in the form of a thin plate, which has a sharp mechanical resonance frequency determined by the dimension of the plate. In a suitable circuit, this resonance can be exited by an applied alternating voltage, the frequency of which it then controls, giving a very stable electronic oscillator working at a fixed frequency. Such circuits are universally used to provide the fixed frequency "clock" pulses in computers and watches and to control the frequencies of radio transmitters. Piezoelectric-based oscillators are also used in sensors where the resonance frequency is a function of the quantity to be measured (see Section 2.3).

The piezoelectric behaviour of poly(vinylidene fluoride) (PVDF or PVF_2) has been well known for more than two decades [3]. This is a polymer having the basic monomeric unit $-CH_2-CF_2-$ and is similar to polytetrafluoroethylene (PTFE), which has a monomeric unit $-CF_2-CF_2-$ and is chemically very inert. Piezoelectricity in PVDF polymers arises because they have regions where the polymer chains are ordered and form localized crystalline phases surrounded by amorphous regions.

The common form of PVDF contains numerous randomly arranged chains of carbon atoms. A closely coupled pair of hydrogen atoms is attached to every other carbon atom in each chain, and a closely coupled pair

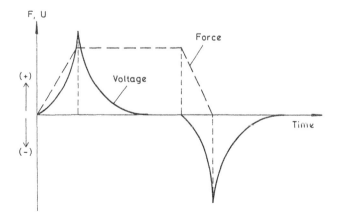

FIGURE 3.1. Typical output signal of piezoelectric devices.

of fluorine atoms joins the other carbon atoms in the chains (see Figure 3.2). This structure, the alpha phase, forms when PVDF polymers solidify from a molten state. This phase does not exhibit piezoelectric behaviour. To develop this quality, alpha phase material has to be converted to the beta phase in which the carbon chains are aligned in parallel strips and planes. This phase is typically produced by mechanically stretching a film of the material at temperatures between 50 and 100°C, producing extensions in lengths of 400–500%. The stretching can be performed biaxially or uniaxially and produces regions of the beta phase, where the pairs of hydrogen and fluorine atoms are still distributed randomly around the axes of each chain.

In order to exhibit piezoelectricity, the material must be still poled. This can be done by depositing aluminium electrodes on the top and bottom of the sheet, heating to about 100°C, and applying a large field of 8×10^7 V/cm.

The electrical field causes pairs of hydrogen atoms to rotate around the chain and assume a permanent position of the high potential side of the carbon atoms, while the fluorine atoms take up a permanent position of the low potential side (see Figure 3.2).

The method of poling mentioned above is just one of the possible solutions. The technologies that are often applied are as follows (see Figure 3.3) [4,5].

(1) *Heat and field method:* Both surfaces of the sheet must be metallized and a large field is applied while the material is heated.

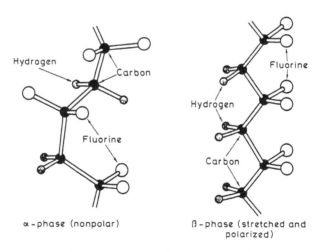

α-phase (nonpolar)

β-phase (stretched and polarized)

FIGURE 3.2. Crystalline forms of PVDF.

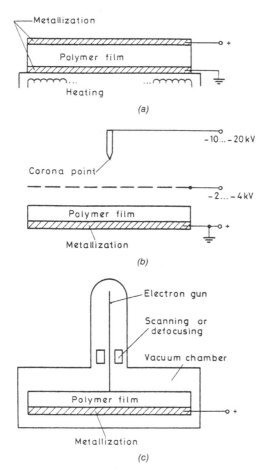

FIGURE 3.3. Poling methods of piezoelectric polymers: (a) heat and field, (b) corona poling, and (c) electron-beam poling.

(2) *Corona discharge poling* applies one side metallized polymer sheet and high voltages exceeding breakdown in the sample. The counter electrode is a corona point at about -10 to -20 kV potential. The discharge process and the electron injection into the insulating sheet is controlled by a wire mesh grid connected to voltages of -2 to -4 kV.

(3) *Electron beam poling* is an effective method for piezoelectric sheet formation, where the charge implantation is done directly by electron beam radiation.

In cases (2) and (3) not only dipole moment alignment, but also electric charge injection, occurs; thus, the materials may have permanent polarization after the process. These samples are called *electrets,* and if they also show piezoelectricity, they are called *piezoelectric electrets.*

Samples polarized over part of their thickness and unpolarized over the remainder of the volume are called monomorphs, while samples having a polarization that changes its sign in the midplane of the film are called bimorphs. The poling of bimorphs can be achieved by irradiating first one, then the other, sample surface with electron beams penetrating approximately to the midplane of the samples.

The major axes of the piezoelectric film are numbered conventionally according to the fabrication technology: the length of "1" axis is that in which the film was stretched. The width is the "2" axis, and the thickness—the direction of poling—is the "3" axis (see Figure 3.4). The most important parameters of PVDF are summarized in Table 3.1, compared with other piezoelectric materials [6]. Typical parameters of a piezopolymer film are given in Table 3.2 [6]. The piezoelectric parameters depend on the different parameters of the technology. Figure 3.5 shows an example of this relationship [4].

Until the 1980s, the almost entirely used piezoelectric polymer was PVDF. Recently, copolymers of PVDF with trifluoroethylene P(VDF-TrFE) have been investigated, and they possess better piezoelectric parameters: $d_{31} = 30$ pC/N and $d_{31} = 20\%$. Optimized corona poling scheme and spin casting technology have been developed for the fabrication of P(VDF-TrFE) thin films, which are compatible with silicon microcircuit processing [7]. The characteristic parameters depend on the composition of the opolymer, of course (see, for example, Figure 3.6).

Polymer ceramic composites can offer an attractive solution when looking for materials that combine flexibility and ease of manufacturing characteristics of polymers, with the high piezoelectric coefficients of the elec-

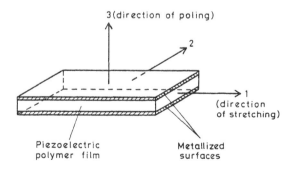

FIGURE 3.4. Marking axes in a piezoelectric polymer film.

TABLE 3.1. The Comparison of Piezoelectric Materials [6].

Property	Unit	PVDF	PZT*	BaTiO$_3$
Density, ϱ	10^3 kg/m^3	1.78	7.5	5.7
Relative permittivity, ϵ_r	–	12	1200	1700
d_{31} constant	10^{-12} C/N	23	110	78
g_{31} constant	10^{-3} Vm/N	216	10	5
k_{31} constant	% (at 1 kHz)	12	30	21

* PZT = Pb(Zr,Ti)O$_3$.

TABLE 3.2. Typical Properties of a PVDF Film [6].

t	Thickness	9, 28, 52, 110, 800 $\times$ 10^{-3} mm
d_{31}	Piezoelectric strain constant	23 pC/N
d_{33}	Piezoelectric strain constant	-33 pC/N
g_{31}	Piezoelectric stress constant	216 $\times$ 10^{-3} Vm/N
g_{33}	Piezoelectric stress constant	$-339 \times$ 10^{-3} Vm/N
k_{31}	Electromechanical coupling factor	19% (at 1 kHz)
Y	Elasticity constant	2 $\times$ 10^9 N/m^2
r	Volume resistivity	10^{13} Ω m
T	Operating temperature range	$-40°$C to 100°C
c	Water absorption	0.02% H$_2$O
E_{max}	Breakdown electric field	3 $\times$ 10^{-5} V/m

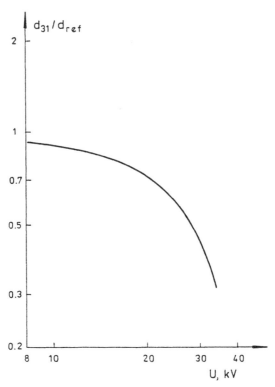

FIGURE 3.5. Typical dependency of d_{31} of PVDF films on electron beam acceleration voltage. (Redrawn from the data of Sessler and Berraissoul [4], ©1989, IEEE, with permission.)

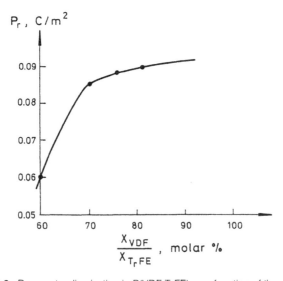

FIGURE 3.6. Remnant polizarization in P(VDF-TrFE) as a function of the composition.

tronic ceramics. For example, a ceramic powder of lead-titanate can be hot rolled and mixed with a copolymer of P(VDF-TrFE) [9]. As another interesting example, a PZT/polymer composite material can be mentioned, which is laminated from sintered PZT ceramic and polyethylene (PE) sheets, resulting in a sandwich-layer sample structure [10]. The piezoelectric constants and other physical parameters indicate these composites to be good candidates for ultrasonic applications.

3.1.2 Polymer Electrets

In the previous section, piezoelectric electrets were briefly discussed; however, it must be emphasized that not only piezoelectric electrets exist. An electret is a piece of dielectric material that has permanent polarization. Polymer electret sensors find a widespread application as microphones and ultrasonic transducers [7]. Poling methods of electrets are the same as those of piezoelectric materials and are summarized in the previous section.

It has been shown that the conductivity of the penetrated areas arises at charge injection poling methods, but after the process, this conductivity decreases by several orders of magnitude and the materials would require years for charge redistribution. The charged particles are closed into conduction traps. Levels of breakdown strength of the polymer, rather than the density of trap sites, seem to limit capabilities for charge storage. The relative charge density persisting behaviour is a function of the material, as shown in Figure 3.7 [11]. The PTFE system traps retain charge indefinitely,

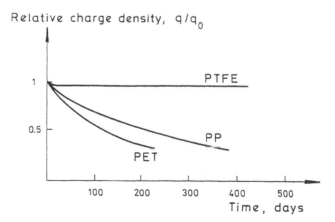

FIGURE 3.7. Relative charge density persisting in various polymers. (Redrawn with permission from the data of Baker [11], from *Proc. of the Symp. at the 2. Chemical Congress of the North American Continent,* Las Vegas, 1980, ©1981 American Chemical Society.)

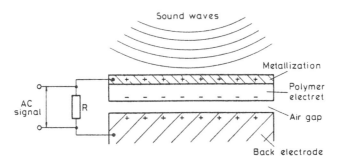

FIGURE 3.8. Operation principle of electret-based acoustic sensors.

whereas those in the poly(ethylene terephthalate) (PET) and polypropylene (PP) decay relatively quickly.

The basis of the sound detection using electrets is illustrated in Figure 3.8. (A detailed description will be given in Section 4.3.1.)

3.1.3 Pyroelectricity in Polymers

When piezoelectric materials are under stress, the centres of gravity of the positive and negative charges are separated, forming an electrostatic dipole and hence a polarization of the film. In electrets, the centres of gravity of the positive and negative charges are separated even without a stress being applied. These will exhibit spontaneous polarization, which means that there must be permanent electrostatic charge on the surfaces of the film, with one face positive and another negative, depending on the direction of the polarization vector. However, the observer would not, in general, be aware of these charges because the atmosphere normally contains sufficient free positive and negative ions to neutralize the free surface charge by being attracted to and absorbed on the surface.

The spontaneous polarization will be a strong function of temperature, since the atomic dipole moments vary as the crystal expands or contracts. Heating the crystal will tend to desorb the surface neutralizing ions, as well as changing the polarization, so that a surface charge may then be detected. Thus, the crystal appears to have been charged by heating. This is called the *pyroelectric effect* [1,12].

The electric field developed across a pyroelectric crystal can be remarkably large when it is subjected to a small change in temperature. We define a *pyroelectric coefficient, p_e*, as the change in flux density, D, in the crystal due to a change in temperature, T, i.e.,

$$p_e = \frac{\partial D}{\partial T} \tag{3.6}$$

the units of which are coulombs per square meter per degree. The pyroelectric voltage signal is the film's developed charge divided by the capacitance. Thus,

$$U = p_e \cdot t \cdot \Delta T / \epsilon \qquad (3.7)$$

where ϵ is the permittivity and t the thickness of the pyroelectric film.

When the temperature changes, an excess of charge appears on one of the polar faces and a current flows in the external circuit. The sense of the current flow depends on the direction of the polarization change. After the initial surge, the current dies away exponentially with time and eventually falls to zero until another temperature change comes along.

Pyroelectric films can be used to detect any radiation that results in a change in temperature of the film but are generally used for infrared detection. Because of its extreme high sensitivity, a temperature rise of less than one-thousandth of a degree can be detected. The detector must be designed so that the heat generated by the radiation does not flow away too quickly. Such detectors are widely used in burglar alarms, which detect the thermal radiation from a human body. By using a pyroelectric array similar to the sensitive screen in a television camera tube, infrared images can be detected from the different heat radiation sources of the object being viewed, so that the operator can "see" in the dark. These are used in a wide variety of satellite and military applications. Infrared imaging has an important role in medical diagnostics as well.

The performance characteristics of the sensors depend on the sensing material, on the preparation and geometry of the electrodes, the use of absorbing coatings, the thermal design of the structure, and the nature of the electronic interface as well. Since the output is rate-of-change-dependent, the radiant flux incident upon the sensor must be chopped, pulsed, or otherwise modulated. The sensors are normally operated at room temperature or other ambient temperatures without artificial cooling. The spectral response of pyroelectric sensors extends from below the far UV to beyond the far IR but is often limited to a band of wavelengths in the IR region by means of a window.

The voltage responsivity of a pyroelectric sensor is characterized by a figure of merit, F, which for a freely suspended sensor element is given by [13]:

$$F_{free} = p_e T / (c_{th} \cdot \epsilon_r) \qquad (3.8)$$

where

ϵ_r = the relative dielectric permittivity
p_e = the pyroelectric coefficient
c_{th} = the heat capacity per volume

When the sensor is connected to a heat sink, its figure of merit must be replaced by expressions that take into account the thermal conductivity, g_{th}, of the layer. If the electric capacity of the pyroelectric element is much higher than the total load capacity of the electrical circuit, the so-called "voltage figure of merit" can be applied [14]:

$$F_{sink} = p_e/(g_{th} \cdot \epsilon) \qquad\qquad (\epsilon = \epsilon_r \cdot \epsilon_0) \qquad\qquad (3.9)$$

Another characteristic parameter is the penetration depth, l, of the temperature wave into the pyroelectric at a given modulation frequency (typically 25 Hz). Table 3.3 lists the most important parameters for some pyroelectric materials [13,14].

PVDF, like other piezoelectric materials, also exhibits a pyroelectric effect. The pyroelectric behaviour of PVDF was discovered in 1971. In the following fifteen years, however, the interest in this material gradually decreased as it turned out that its sensitivity was relatively small (see Table 3.3). Due to the increased demands for IR "people sensors" in alarm appliances and in automatic door operation systems, for instance, cost has become more important, and PVDF returned to the focus of interests as a possibility for cheap solution. To overcome the problem of low sensitivity, integrating the sensor with the readout electronics seems to be the best option. These integrated constructions are discussed in Section 4.4.2.

The other possibility is to develop polymers with better parameters. One example is the P(VDF-TrFE) with a pyroelectric coefficient of $p_e = 3.6 \cdot 10^{-5}$ cm^{-2}K^{-1}, which is a little higher than that of PVDF and with a low dielectric constant value of about 6 [15].

Pyroelectric thin polymer films made by the Langmuir-Blodgett technique were also developed. Using appropriate deposition methods, the layers do not need poling processes to get spontaneous polarization. The

TABLE 3.3. Material Data for the Common Pyroelectric Materials [13,14].

Property	PVDF	NaNO$_2$	LiTaO$_3$	TGS*
p_e (10^{-4}Cm^{-2}K^{-1})	0.25	0.4	1.8	2.8
Relative permittivity, ϵ_r	9	4	47	38
Dissipation factor, tgδ	0.03	0.02	0.005	0.01
c_{th} (10^6 Jm^{-3} K^{-1})	2.3	2.2	3.2	2.3
g_{th} (W/m K)	0.14	2.2	3.9	0.65
l(10^{-6} m, at 25 Hz)	28	110	90	55
F_{free} (10^{-12} mV^{-1})	0.87	4.3	1.0	4.6
F_{sink} (10^6 V/W)	2.2	0.53	0.11	1.3

* TGS = Triglycine sulphate.

alternating layer technique realizes multilayers from two different materials, one material possessing a much larger dipole than the other. The materials are deposited alternately with their polar head groups adjacent to one another. Even though the dipoles may oppose one another, a net dipole remains, since one is much larger than the other. If the correct materials are chosen, the dipoles may even face in the same direction in the multilayer. Generally, the deposition process is followed by a UV-polymerization.

It has been found that acid/amine and acid/oxyaniline alternating multi-layers can form films that exhibit pyroelectric behaviour. However, the sensitivity of the pyroelectric thin polymer films ($p_e = 3.4 \cdot 10^{-7}$ cm^{-2}K^{-1}) is still one to two orders of magnitude less than coefficients for PVDF [16].

3.1.4 Sensing Effects Based on Permittivity Changes

Dielectric sensors for detecting molecules by capacitance changes are of general interest since many interaction mechanisms change the relative permittivity. Adsorption or absorption of molecules with a high dipole moment such as H_2O and SO_2 leads to a change in the dielectric constant. Hydrocarbon molecules, even nonpolar ones, can also be detected due to the strong swelling effect. Although these effects are principally based on the same phenomena, humidity sensing films will be discussed separately because of their specialties.

3.1.4.1 HUMIDITY SENSITIVE BEHAVIOUR

There are several special definitions in connection with humidity measurement techniques; therefore, it is necessary to clarify them.

Humidity (H) is defined as the mass of vapour carried by a unit mass of vapour-free gas. The partial pressure of each species in the mixture is in direct proportion to its molar fraction. Thus

$$H = (M_A \cdot p_A)/[M_B \cdot (1 - p_A)] \tag{3.10}$$

where M is the molecular weight and p is the partial pressure of the constituents, component A is the water vapour and component B is the air. Relative humidity (RH) is defined as the ratio of partial pressure of water vapour to the saturated vapour pressure (p_S) of the water at the gas temperature:

$$RH = p/p_s \tag{3.11}$$

The dew point can also be used as a characteristic parameter for the water vapour content in the air. It is defined as the temperature (T_d) at which the gas becomes saturated during cooling:

$$p(T_d) = p_S(T) \tag{3.12}$$

The connections between the mentioned parameters are given by the laws of psychrometry [17].

A lot of polymer types are known to absorb moisture. It is well known that water molecules undergo chemisorption and physisorption on the solid surfaces [17]. The chemisorption, which is the stronger process, causes dissociation of water molecules to form surface hydroxyls. Physisorption then takes place on top of this. The water molecules in the first physisorbed layer are double hydrogen bonded to two surface hydroxyls, while they are singly hydrogen bonded in the second and succeeding layers.

If there are pores, capillary condensation of water takes place in addition to the adsorption. Capillary condensation causes the presence of water liquid in all pores with radii up to the critical value, given by the Kelvin's law [18]. A large number of models exist to describe the absorption of water vapour into a solid; however, there is a lack of accurate theoretical models.

The absorbed amount of water can be estimated by the Dubinin's equation, which describes the absorption of a vapour into micropores. This formula is a semi-empirical in nature, requiring three empirical fitting parameters, and includes standard thermodynamic functions [19]. According to this law, the fractional volume of water, V, can be expressed as

$$V = V_{om} \cdot \exp\left[-(R \cdot T \cdot \ln RH/W)^n - A \cdot (T - T_0) \right] \tag{3.13}$$

where

V_{om} = the maximum fractional volume of absorption at the temperature T_0
T = the absolute temperature
A = the thermal coefficient of limiting absorption
W = the free energy of absorption
R = the universal gas constant
n = an empirical factor determined by trial and error

Generally, a porous humidity-sensitive polymer is considered to be composed of two phases, i.e., solid polymer material and water absorbed and condensed in it (called quasi-liquid water). The relative permittivities of polymers used in humidity sensors are known to range from three to ten, whereas pure water has a far larger value of about seventy-eight at 25°C. Thus, the capacitance of polymer layers changes sensitively with the absorption of water. The humidity sensitive characteristics can be derived according to the phase law of two mixed phases, using one of the versions of the effective medium theory [20]. There are several mixture formulas for the mentioned problem. Because the volume fraction of absorbed water in polymers is quite small, all equations are close to linear over the range of interest. Therefore, the experimental control is quite critical. A semi-empirical formula by Looyenga for the dielectric constant of the mixture is given by [19]:

$$\epsilon = [v_w \cdot (\epsilon_w^{1/3} - \epsilon_p^{1/3}) + \epsilon_p^{1/3}]^3 \qquad (3.14)$$

where indices w and p represent the water and polymer phases, respectively, and v_w is the volume fraction of water absorbed in the film.

The most common types of polymers used in humidity sensors are cellulose acetate and polyimides. The first commercially available polymer humidity sensors applied cellulose acetate as sensitive layers. Its permittivity value is roughly proportional to the ambient humidity; it shows a small nonlinearlity in a range over 80 % RH. Thin layers (with a max. 1-μm thickness) have hysteresis less than 2 % and response times less than 1 sec [21]. The disadvantage is the relatively strong frequency and temperature dependence.

Recent improvement in polymide processing for integrated circuit and interlay dielectric applications (see Section 1.4) enables humidity sensors to be fabricated on integrated circuit structures [22]. Using this technique, small, low-cost humidity sensors are available, and, most importantly, it becomes possible to integrate the humidity sensor with other sensors and signal-processing circuitry on the same substrate. The application of photo-sensitive polyimides (see Section 1.4) reduces the number of patterning process steps. Polyimides offer the following further advantages: an almost linear response to humidity changes, good absorption of water, resistance to chemical attack, mechanical strength, and high temperature capability. The sensitivity and response depend upon the sensor geometry, including the layer thickness and the geometry of the electrodes as well.

The typical normalized capacitance-humidity characteristics of PI layers are shown in Figure 3.9(a) [23,24]. Figure 3.9(b) shows a typical structural unit of a preimidized, planarizing, and negatively photoimageable polyimide type used in humidity microsensors [20]. The layers from this polyimide type have linear uptake by weight of more than 2.8 % at 50 % RH, which is also nearly independent of the temperature. The dielectric constant varies from approximately 3.0 in the dry air to 4.2 at 100 % RH for frequencies up to 10 MHz. The diffusion constant is relatively large, which implies response times in the order of seconds.

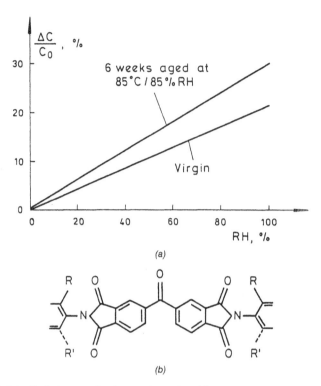

(a)

(b)

FIGURE 3.9. Typical normalized capacitance-humidity characteristics of (a) PI layers [23,24] and (b) the structural unit of a typical photosensitive polyimide.

The offered advantages stated give an explanation for the recent developments on polymer humidity sensitive layers dealing with the improvement possibilities of PI films. Very little work has been done to investigate the chemical degradation of PI and its effects on dielectric behaviour, although it has been well known that the presence of water in the layers can lead to long-term reliability problems and chemical instability. Microanalytical examinations indicated oxidation and deimidization of the PI surface with aging in high-humidity atmospheres. The chemical changes may lead to a change in morphology, resulting in a decreased surface to volume ratio, increased water absorption, and a change in sensitivity of the layer [24].

The application of stabilizing burn-in cycle, a cross-linking agent, or a suitable overcoat may help to stabilize the morphology and surface chemistry of the films. Refreshing heat-up cycles after exposures to high humidity may also rebuild the original morphology through a reimidization process [24]. The typical aging-shift in the characteristics can also be seen in Figure 3.9(a). The shift can result in a failure of 15–20% RH.

Recently, a cured acetylene-terminated polyisoimide oligomer was used to prepare a capacitive-type humidity sensor [25]. Tha amount of sorbed water is small, and the sorbed water did not form clusters. This is due to the hydrophobic nature and rigid structure of the film. The sorption behaviour reflects its excellent sensor properties, i.e., low hysteresis, low-temperature coefficient, and long-term stability in a high temperature and humid atmosphere.

Like other polymers, plasma polymeric films grown from malonic acid dinitrile, acetonitrile, and chlorobenzene also show defined absorption of water from humid surroundings. Such thin films also can be used as an active part of humidity sensors. They have high-linearity capacitance-humidity characteristics. Composite polymers containing hygroscopic salts (for example, BaF_2) show an increased sensitivity at the humidity value of the saturated salt solution. Thus, a sharp increase in the capacitance value can be observed in given humidity ranges [26].

3.1.4.2 GAS-SENSITIVE BEHAVIOUR

As mentioned previously, polymers may also change their permittivity when they absorb different inorganic and/or organic gas components with high dipole moment molecules, which can change the molecular polarizability of the polymer due to physical and chemical interactions. Molecules with lower or even with zero permanent dipole moment may also be detected if a significant swelling occurs. In that case, the polymer should contain strong polar groups. During the swelling, the gravity of the different electrical charges will be shifted oppositely.

Polysiloxanes have been studied in details for a few years as gas-sensing sorbent films. They show relatively good selectivity and change several physical properties, including permittivity, fractional volume, refractive index, etc., with the gas concentration. Thus, they can be used to realize a lot of gas sensor types, such as capacitive, resonant, calorimetric, and fibre-optic ones.

Polysiloxanes have a structure that is similar to the one of organically modified silicates containing sidegroups, which act as a adsorption or chemisorption centres.

Organically modified silicates consist of a three-dimensional non-crystalline inorganic network of siloxane bonds as shown in Figure 3.10(a). Functional organic groups (R) can be connected to this network by silicon-carbon bonds, or functional compounds may be incorporated into the network by gel entrapment [27].

Polysiloxanes may contain one, two, and/or three-dimensional structural units, $-SiR_2O-$, $-SiR(O-)_2$, and $-Si(O-)_3$, with different numbers of

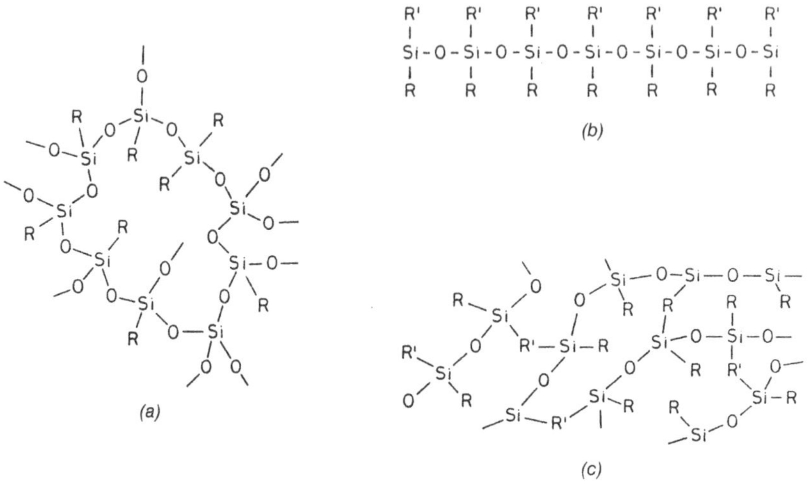

FIGURE 3.10. Typical structure of (a) organically modified silicates [27], (b) polysiloxane oligomers, and (c) cross-linked polysiloxanes [29].

oxygen bonded to one silicon atom. Oligomeric chains can be built up from $-SiR_2O-$ units, as shown in Figure 3.10(b). A three-dimensional network can be achieved by cross-linking the oligomers with organic substituents, as shown in Figure 3.10(c). Additionally, polysiloxanes may be utilized as matrices for inorganic compounds such as inorganic salts, which modify their specific molecular detection properties. Moreover, macrocyclic cage compounds as supramolecular structures can be bonded into the polysiloxane framework.

As shown in Figure 3.10, organically modified silicates and polysiloxanes have very similar structures; thus, similar physical and chemical behaviour can be expected. Both material types are extensively investigated as sensing layers, and many similarities have been found. However, organically modified silicates are rather inorganic materials, while polysiloxanes mean an intermediate area and are often called "inorganic polymers." Accordingly, organically modified silicates are already out of the scope of this book, though polysiloxanes are discussed as a new area of sensing polymer films.

Various polysiloxanes with different organic groups have been studied over the last several years for the application in capacitive sensing devices. Only a few examples with great perspectives are mentioned here.

High sensitivity to inorganic molecules with a high dipole moment such as SO_2 was detected by Endres and Drost using heteropolysiloxanes containing tertiary amino groups as adsorption centres [28]. In this polymer, the organic residue consists of the molecular group $-CH_2-CH_2-CH_2-N(CH_3)_3$, which acts as an adsorption centre for SO_2. Interdigital capacitors applying this type of dielectric layers show relatively big capacitance changes when exposed to a 0–200 vpm SO_2 concentration in the air.

Reversible changes of thickness due to swelling of polymers have also been measured by capacitance variations. The largest effect was found at poly(cyanopropylmethylsiloxane) in an atmosphere containing *n*-hexane or ethanol [29]. It is interesting that different sensing mechanisms were found at the two materials. While *n*-hexane causes a clean swelling effect with a sensitivity of $1.3 \cdot 10^{-4}$ μm/Pa, which is resulted from the interaction with the polar $Si-O$ bonds in the polymer, ethanol, in contrast, alters the permittivity of the layer due to the additional contribution from the permanent dipole moment of the OH group.

The latter type behaviour was also found at different conventional (carbon-based) polymers interacting with strong polar organic molecules. For example, a capacitance change of 20% can be observed at poly(ethylene glycol) layers when the partial pressure of dimethylformamide is 152 Pa [29].

3.2 SENSING EFFECTS IN CONDUCTIVE POLYMER COMPOSITES

Electrically conducting or semiconducting polymers have attracted a great deal of interest applied as sensing layers in sensors. There are two types of conducting polymers:

(1) Polymers that are intrinsically conducting or can be made so by doping
(2) Composites that contain an electrically insulating polymer matrix loaded with a conductive filler

In this section the behaviour of the latter type conductive polymers will be discussed.

The bulk materials are known in practice as conductive rubbers. Composite films can be fabricated by spin coating, by thick-film, and by thin-film technologies. A number of precursors are known as polymer thick-film (PTF) pastes. Their physical and chemical properties are similar to those of CERMET thick films (see Sections 1.3.2 and 1.3.3). Thin-film composites are deposited by simultaneous evaporation or RF sputtering of metals and polymers. The often used filler materials are metals (Cu, Pd, Au, Pt), carbon black, and semiconducting metal-oxides (V_2O_3, TiO_2, etc.). The most important polymers that can be used as matrices are: polyethylene, polyimides, polyesters, poly(vinyl acetate) (PVAc), PTFE, polyurethane, poly(vinyl alcohol) (PVA), epoxies and acrylics, e.g., poly(methyl methacrylate).

3.2.1 Theory of the Conduction Mechanism

In order to understand the electrical behaviour of polymer composites, it is necessary to give a short summary about the conduction models. The concept of percolation [30] can be used to understand the change in sensitivity as a function of filler concentration in metal-insulator composites. The percolation threshold is defined as the filler volume fraction at which the resistivity begins to decrease. For low filler concentrations, it is basically equal to the resistivity of the polymer matrix (see Figure 3.11). The critical threshold, v_c, is associated with the concentration at which the filler particles begin to form conductive paths. As it is increased further, more conductive paths are created through the composite, resulting in a resistivity drop by several orders of magnitude. Above v_c the function is saturated, and the resistivity approaches monotonously the resistivity of the conducting phase. Typical percolation curves are shown in Figure 3.12 [30].

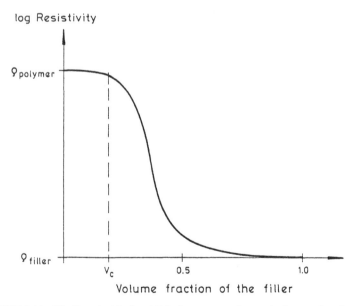

FIGURE 3.11. Effective electrical resistivity for a composite conductor as a function of the filler volume fraction.

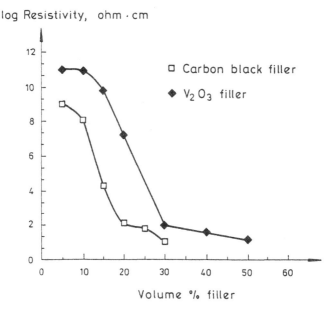

FIGURE 3.12. Percolation curves for polyethylene composites. (Redrawn from the data of Ruschau et al. [30], with the kind permission of Elsevier Sequoia S. A., Lausanne, Switzerland, publisher of *Sensors and Actuators A* and *Sensors and Actuators B*.)

A detailed theoretical description of the conduction mechanism in composite conductors is based on several effects: percolation, quantum mechanical tunneling between the conducting particles, and thermally activated hopping of electrons through localized deep energy levels in the band gap of the polymer matrix, and in some cases, field emission may also be supposed [31]. Percolation always dominates the resistivity when the filler concentration is far above v_c. If it is comparable to or smaller than v_c, tunneling and hopping have important consequences. The extra current provided by this mechanism decreases the resistivity relative to the case of pure percolation, but the amount is very sensitive to the temperature, the composition, a number of geometrical factors, and environmental effects that influence the current density between conductive particles.

The transmission-probability, P, of electrons through an energy barrier of height, W_0, and width, a (which is also the mean distance between neighbouring conductive grains) is (see Figure 3.13)

$$P \propto \exp\left[- \frac{4 \cdot \pi \cdot a}{h} \cdot (2 \cdot m_e \cdot W_0)^{1/2} \right] \tag{3.15}$$

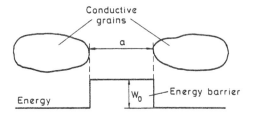

FIGURE 3.13. Model for electron tunneling between conductive grains.

where m_e is the electron mass and h the Plank constant. The volume resistivity is inversely proportional to the transmission probability:

$$\varrho \propto \exp(B \cdot W_0^{1/2} \cdot a) \qquad (3.16)$$

where

$$B = (2 \cdot m_e)^{1/2} \cdot 4 \cdot \pi/h \qquad (3.17)$$

3.2.2 Pressure and Deformation Sensitivity

Direct pressure dependence of composite resistors can be measured when the samples are immersed in a fluid high-pressure media and compression is given on the latter [32]. The pressure-coefficient of resistance can be defined as

$$PCR = (1/R) \cdot dR/dp \qquad (3.18)$$

where p is now the pressure and R the resistance.

Its value is varying in the range of -1 to $-10\,\text{GPa}^{-1}$ (-0.1 to -1%/bar), depending strongly on the volume fraction of the filler. A high *PCR* can be achieved when the volume fraction is near the critical value, v_c. The results can easily be explained by the percolation model. The copressability of the filler (graphite or metal powder) is much smaller than that of any polymer, and when the composite polymer is compressed, the actual volume fraction of the filler increases. The number of conducting filaments through the polymer body also increases, leading to a strong decrease in the resistivity. Since the latter is an extremely nonlinear function of the volume fraction of filler near v_c, we also obtain a nonlinear decrease with increasing pressure, as shown in Figure 3.14 [32].

The percolation model, however, cannot describe the almost linear behaviour of composite film resistors (see Figure 3.15). A tunneling effect must also be considered in that case.

It has been well known for several decades that metal films, semiconductor, and CERMET film resistors are characterized by a resistance variation proportional to the applied stress and the deformation, respectively. The phenomenon is called piezoresistive effect.

More recently, it has been pointed out that polymer thick-film resistors screened and cured on epoxy-glass or polyimide substrates also present a notable sensitivity to deformation (see Figure 3.15). If a mechanical strain, S, is induced on a resistor by an applied stress, the resistance change can be expressed as [33,34]

$$(R - R_0)/R_0 = \Gamma \cdot S \qquad (3.19)$$

where Γ is the piezorestivity tensor. In the case of film resistors, Equation (3.19) can be simplified, and the gauge factor, G, of a resistor can be defined as the ratio of the fractional change in resistance to the fractional change in geometrical sizes

$$G = (dR/R)/(dl/l) \qquad (3.20)$$

where l is the resistor length. If the current is parallel to the strain, the lon-

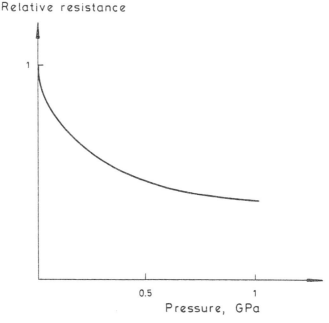

FIGURE 3.14. Relative resistance changes for a graphite-filled polymer resistor as a function of pressure. (Redrawn from the data of Lundberg and Sundqvist [32] with permission of the American Institute of Physics.)

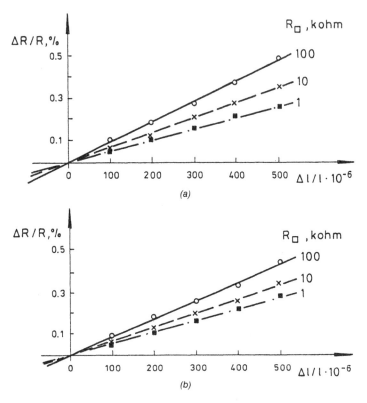

FIGURE 3.15. Relative resistance changes of PTF resistors as a function of strain: (a) longitudinal effect; (b) transverse effect.

gitudinal G_L, and if the current is perpendicular to the strain, the transversal G_T gauge factors can be defined. For isotropic materials their difference is determined by the Poisson ratio of the substrate and is independent from the physical properties of the resistor material:

$$G_L - G_T = 2 \cdot (1 - v) \tag{3.21}$$

where v is the Poisson's ratio.

In the most simple cases the resistance value is determined by the volume resistivity and by the ratio of the length to the cross section area, A:

$$R = (\varrho \cdot l)/A \tag{3.22}$$

Thus, the fractional change of resistance can be expressed as

$$dR/R = (d\varrho/\varrho) + (dl/l) - (dA/A) \tag{3.23}$$

In metal film resistors, the change in resistivity is negligible, but the high gauge factor (about ten) of polymer thick-film resistors cannot be explained by a pure geometrical effect.

Piezoresistive effect in PTF resistors can be explained with a tunneling process of electrons between the conductive grains embedded in the polymer matrix. The resistivity can be expressed by Equation (3.16). The applied strain will change the distance a between the nearest conductive grains, inducing a relative change of resistivity

$$G = (d\varrho/\varrho)/(da/a) = B \cdot W_0^{1/2} \cdot a \qquad (3.24)$$

where B is the constant defined by Equation (3.17). Applying the same raw materials (filler and polymer) in the composite resistors, an increase of the resistivity corresponds to a lower volumetric concentration of the conductive particles and consequently to a larger distance between the grains, which, in accordance with Equation (3.24), produces a higher gauge factor. The relationship is illustrated in Figure 3.16. Table 3.4 gives the most important parameters for PTF resistors compared to those of other types of piezoresistors [33,34].

The simplest way for measuring the gauge factors of film resistors is the cantilever technique (see Figure 3.17). A cantilever beam is clamped at one end, and the resistors to be investigated are located at the same distance from the clamped edge so that longitudinal and transversal gauge factors can be measured. The cantilever beam is made from a substrate plate, and the strain can be determined by the bending of the cantilever.

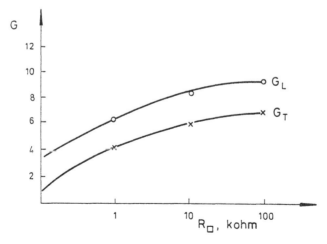

FIGURE 3.16. Characteristic guage factors for PTF resistors as a function of the sheet resistivity.

TABLE 3.4. Properties of PTF Resistors Compared
with Other Piezoresistor Types [33,34].

Resistor Type	G	TCR* (ppm/°C)	Long-term Stability	Relative Cost
Continuous metallic films	2	20	excellent	high
Thin films	50	0–20	very good	high
Semiconductors (diffused or implanted)	50	1500	good	medium
Thick films	10	50	very good	medium
Polymer thick films	10	500	poor	low

* TCR = Temperature coefficient of resistance.

Recently, a very interesting new phenomenon has been discovered and reported by Lachinov [35], i.e., an anomalously high sensitivity of the electrical conductivity to small uniaxial pressures in poly(3,3-phthalidilidene-4,4-biphenylilene). It has been established that in films less than 1 μm thick, a pressure of about 0.1–1 MPa leads to a change in the conductivity of 10^{10}–10^{12} Ω. The film shows metal-like conductivity. As the pressure is removed, the original dielectric state is restored. This phenomenon may be accounted for the electron instability occurring in certain structural regions of the polymer films.

3.2.3 Temperature Dependence

The temperature dependence of the resistivity of composite polymers shows different behaviour at low and high temperatures. Figure 3.18 shows

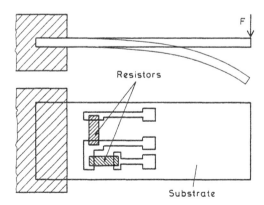

FIGURE 3.17. The cantilever method for measuring guage factors.

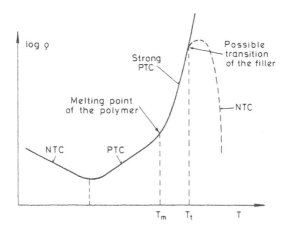

FIGURE 3.18. Typical resistivity-temperature behaviour of conducting polymer composites.

the character of the function; however, the real shape can vary in a wide range [32,36]. Generally, the polymer composites have an NTC effect at low and a strong PTC effect at high temperatures.

The PTC behaviour of the resistivity is easily given by a qualitative explanation in terms of percolation and thermal expansion [32]. The volume thermal expansion coefficient of the polymer matrices are generally higher by a few orders of magnitude than those of the conductive filler material. With T increases, the volume fraction of the filler will be decreased, and it slowly approaches the critical value, v_c. The increase in the resistivity due to the decrease in the filler concentration is large enough to dominate the effect. The combination of the increasing volume thermal expansion and the ever close approach to v_c leads to a strong PTC effect over a large temperature range.

The phenomenon is more complicated when the insulating phase is a crystalline polymer (such as polyethylene or a nylon). In this case, an especially sharp PTC effect can be measured near the transition temperature, T_m. Near T_m, there is a relatively large change in the specific volume, and the magnitude of the volume change is a function of the degree of crystallinity of the polymer. This effect has already been applied widely in composite polymer-based "switching" PTC resistors.

Combining the PTC effect with a transition in the filler grains, for example, from a semiconducting into a metallic phase, a sharp NTC effect can be observed at high temperature (see Figure 3.18). This behaviour may be obtained in the case of semiconducting metal-oxide filler materials, such as V_2O_3, TiO_2, etc.

At low temperatures, it can be supposed that the filler concentration is above the critical v_c value. In this range, the conduction is dominated by the filler grains. The NTC behaviour is understandable in the case of semiconductor grains, which have exponential type, $\exp(E/kT)$, resistivity dependence. The situation is different in the case of metal grains, which have positive TCR value. In that case, the NTC effect cannot be explained by the temperature dependence of the grains' resistivity. A thermally activated intergrain conduction mechanism should be supposed with an activation energy determined by the materials and a number of geometrical and physical parameters of the grain surfaces [37].

Recent studies suggest that the conduction mechanism is an electric field–dependent internal field emission of electrons between the conductive particles [39]. In the case of conductive carbon black, the particles tend to agglomerate in chains with a high degree of branching, and these form a large number of conduction paths. Along the chains a thermally activated electron hopping conduction mechanism may be supposed with a temperature characteristics of $\exp(T_0/T)^{1/4}$, which fits much better to the experimental results than the exponential type function [32].

3.2.4 Gas Sensitivity of Metal Polymer Composites [30,32]

It was discovered in 1985 that carbon-filled polyethylene composites have chemical sensing capabilities, which is in close connection with the percolation. A lot of experiments on various conductive polymer composites have shown that the resistance of the samples increases by several orders of magnitude when they are exposed to common hydrocarbon solvent vapours. The response of the resistors to liquid solvent is very similar to the response for solvent vapours. Table 3.5 summarizes typical resistance response values for a few composites and hydrocarbon types [30,32]. All data are in the form of resistance ratios R/R_i, where R_i is the initial resistance value and R is the saturated value during vapour exposure. After removing samples from the solvent liquid or vapour, the resistances return to the initial value.

All features of the experimental results can be qualitatively explained using the percolation theory. Clearly, the polymer matrix dissolves the hydrocarbons. During this swelling process, the matrix expands, decreasing the effective filler volume fraction, causing a dramatic increase in the resistivity near the percolation threshold. The result is eventually the breakdown in the percolation pathways through the polymer matrix. The response time depends on the thickness of the films. The maximum response is determined by the parameters of the matrix and the behaviour of

TABLE 3.5. Resistance Responses between Saturated and Initial States for Various Polymer Composites Exposed to Different Solvent Vapours [30,32,38,38].

Polymer Matrix	Filler Material	Solvent	Volume Change, %	R/R_i
Polyethylene	C black	Pentane	11.3	1700
		CCl_4	18.0	1.3×10^5
		Ethanol	0	1
	V_2O_3	Pentane	11.3	5.0×10^5
		CCl_4	18.0	5.0×10^6
		Ethanol	0	30
Polyurethane	C black	Pentane	0	1
		CCl_4	9.5	200
		Ethanol	1.1	17
	V_2O_3	Pentane	0	1
		CCl_4	9.5	1.5
		Ethanol	1.1	1
PTFE	C black	Pentane	–	1.21
		CCl_4	–	1
Poly(vinyl alcohol)	C black	Pentane	0	1
		CCl_4	0	1
		Ethanol	16.2	150
Polypyrrole (PPy)	Cu	Ethanol	–	1.05
		Methanol	–	1.1
	Pd	Methanol	–	1.12
PTFE-PPy	Pd	Ethanol	–	1.1
		Methanol	–	1.15
RTV silicone elastomer	C black	Hexane	–	10^9
		Ethanol	–	10^3

the various solvents. In some cases the resistivity of the solvent itself can play an important role in the resistance response. Polymer composites are suitable for use in monitoring organic atmospheres. A certain degree of selectivity can also be obtained by a selection of different polymer matrices.

More recently, a new type of film was developed which is sensitive to liquid phase chemical compounds, based on the swelling of an electrically conductive polymer composite prepared with a two-part RTV silicone elastomer and carbon black filler [39]. Best sensitivities were measured at 14% carbon black content. In the presence of a solvent with an intensive solvent-polymer interaction, a resistivity change of nine to ten orders of magnitude can be observed. The films were especially sensitive to low molecular weight hydrocarbons. The sharp increase in the resistivity values is useful in chemically actuated electronic switching devices.

3.2.5 Humidity-Sensitive Behaviour

Humidity sensing properties are generally described separately from the gas-sensitive characteristics; however, the sensing mechanism is the same at the metal- and/or semiconductor-polymer composites. The swelling of the polymer due to water absorption causes the breakdown of ohmic contacts between dispersed conductive particles, and thus the resistance of the film increases sharply as the relative humidity approaches 100%. These films belong to the group of the electronic conduction-type humidity-sensitive layers.

The water vapour dependency of resistivity has been demonstrated at RF sputtered Cu/Pd-PTFE-PPy metal-polymer composites by Bruschi et al. [38]. A humidity-sensitive layer was also developed from cross-linked hydrophilic acrylic polymer in which carbon particles were dispersed. Figure 3.19 shows the typical resistance-humidity characteristics of this

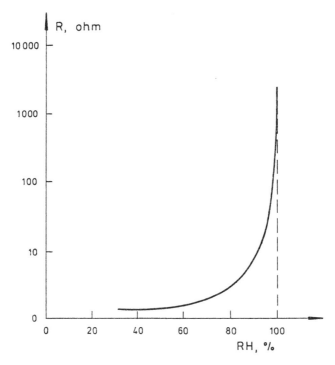

FIGURE 3.19. Resistance-humidity characteristics of carbon-polymer layers. (Redrawn from the data of Yamazone and Shimizu [40], with the kind permission of Elsevier Sequoia S. A., Lausanne, Switzerland, publisher of *Sensors and Actuators A* and *Sensors and Actuators B.*)

layer [40]. The sharp increase in resistivity over 80% RH is a property that is especially useful in dew point sensors.

3.3 SENSING EFFECTS OF POLYMER ELECTROLYTES

A group of organic polymers having constituent ionic monomer or inorganic salt exhibits ionic conductivity; thus, they are called polymer electrolytes. Their ionic conductivity can be modulated by several parameters of the environment, which is the basis of their application possibilities in sensors.

3.3.1 Model for Conduction and Sensing Mechanisms

If a free electrical charge is introduced into a polymer macromolecule, there are a large number of sites at which the charge can attach itself firmly to a molecular subunit. The strong polarization of the molecular units by the presence of the charge itself results in the formation of deep electron states in which the electrons will then be trapped. Stable "molecular ions" are thus readily formed. The trapped electrons are rather unlikely to hop between the molecular ions. Therefore, the conduction mechanism of these polymer salts may generally be controlled by the migration of the introduced movable counterions.

Ionic conductivity in polymer electrolytes can be expressed as [41]:

$$\sigma = \sum_i n_i \cdot z_i \cdot e \cdot \mu_i \qquad (3.25)$$

where

σ = the conductivity
μ_i = the ionic mobility
n_i = the number of ionic carriers per unit volume
z_i = the valency of the ionic carriers
e = the elementary electric charge

The conduction in polymer electrolytes is generally due to the movement of monovalent ions, thus $z_i = 1$. The ion transport is governed by the jump probability of ions into their trapping/hopping sites. In sensitive electrolyte films, an increase in the conductivity can be generated by increasing

(1) The number of ionic carriers by the addition of ions from the environment

(2) The degree of dissociation in the polymer electrolyte

(3) The mobility of ionic carriers based on a modification of the density and energy levels of trapping and/or hopping sites or on a change of the ionic carriers effective size and mass

There are a lot of recent studies dealing with theoretical and experimental descriptions of high ionic conductive polymer electrolytes; however, there is still no available general quantitative model for the explanation of experimental results. In practice, complex impedance spectra are used (see Section 2.1) for the estimation of bulk electrolyte resistance. From the temperature dependency of the resistance, the activation energy, E, can be calculated according to the Arrhenius equation [41,42]

$$R = R_0 \cdot \exp(E/k \cdot T) \qquad (3.26)$$

where

R_0 = the pre-exponential factor
k = the Boltzmann's constant
T = the absolute temperature

Because of the complexity of the physical processes in the electrolyte, E cannot be handled as a constant: various values can be measured in different temperature ranges under different environmental conditions. The activation energy functions give the possibility to set up qualitative models for the conduction mechanism [43].

According to the origin and type of the charged particles, polymer electrolyte types can be divided into the following groups:

(1) *Protonic conducting polymer electrolytes:* in these polymers, the protons, H^+ ions, are the majority charge carriers. Theoretically, they are similar to polymer salts, but because of their special properties due to the presence of H^+ ions, they are generally discussed separately. Polymers containing pendant acidic functional sidegroups such as sulphonic acid (SO_3H) or carboxylic (COOH) groups show protonic conduction behaviour [44]. Poly(styrene sulphonate) is an often used polyelectrolyte. Acid-form Nafion® (DuPont product), which is a tetrafluoroethylene polymer containing sulphonic acid groups is also a protonic electrolyte [45]. Acid/polymer complexes can also show protonic conductivity.

(2) *Polymer salts* are obtained when the H^+ ions of the acidic sidechains are replaced by metal, mainly alkali ions. The electric conduction is dominated by these cations. Sodium styrene sulphonate and alkali-form Nafion® are examples for this type.

(3) *Alkali salt–polymer complexes* [41,42]: it is well known that polyethers (also cyclic ethers, the so-called crown ethers) are effective complexing agents for alkali-metal cations. The dominant interaction is the coordination of ether oxygen atoms to alkali cations. The formation of solid polyether salts is possible. When the inorganic salt is purely dispersed in the polymer matrix, both cations and anions of the salt may take part in the conduction charge transfer. If the alkali salt has large anions, the probability of the polyether-cation complex formation will be high. In the latter case, the anion is completely surrounded by the ether oxygens; the unpaired electrons on the other oxygen act as a hopping site, and the cation is able to migrate through the space surrounded by the ether oxygens. In this case the conduction is dominated by the cation. Typical polyethers often used as electrolytes are poly(propylene oxide) (PPO) and poly(ethylene oxide) (PEO); typical alkali salts are $LiClO_4$, $LiCl$, $LiCF_3SO_3$, $LiSCN$, etc.

(4) *Copolymers of ionic and nonionic monomers:* the ionic copolymer is similar to a polymer salt, where one type of ion can migrate, and the counterparts are built into the polymer network. The conduction is possible by the migration of the movable ions. As an example, the copolymer of sodium styrene sulphonate (NaSS) and hydroxyethyl methacrylate (HEMA) can be mentioned [46].

(5) *Quaternized polymers* [47] where the quaternizing reagent is an alkyl halide, for instance: the alkyl groups are bound to the cyclic pyridyl sidegroups that become positively charged, forming a polymer salt with the halogen ions. Sometimes the cross-linking agent brings the halide anions into the structure, resulting in a simultaneous cross-linking and quaternization process. The latter is illustrated by Figure 3.20 in the case of PVPy [poly(4-vinyl pyridine)] cross-linked with 1,3-dichloropropane [47].

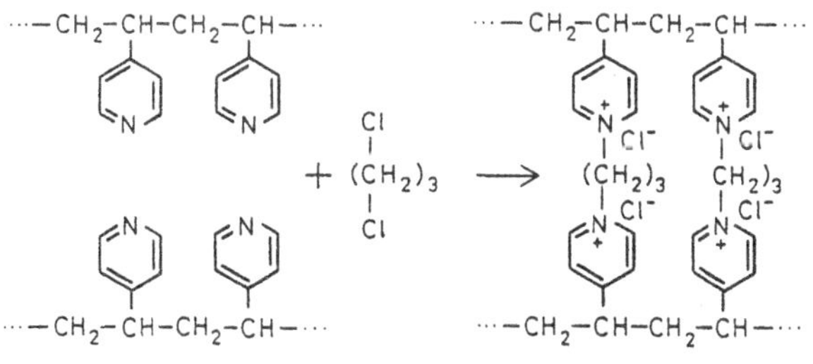

FIGURE 3.20. Reaction scheme of simultaneous cross-linking and quaternization of poly(4-vinyl pyridine) with 1,3-dichlorobutane. (Redrawn from the figure of Sakai et al. [47], with permission of the Electrochemical Society, Inc.)

(6) *Hydrogels,* or water-containing gels, [48] are cross-linked hydrophylic polymers, which swell in water to equilibrium water contents in the range of 30 to 99% by mass. A number of workers have studied electrical conduction in synthetic hydrogels and found that the conductivity reflects ionic mobilities in the water-swollen gel. However, at high water contents, protonic conduction may also play an important role according to the proton transmission mechanism, i.e.,

$$H_2O + H_3O^+ \rightleftharpoons H_3O^+ + H_2O \tag{3.27}$$

A hydrogel copolymer can be prepared from HEMA and *N,N*-dimethylamino-ethyl methacrylate (DMAEMA) by free radical solution polymerization [49].

(7) *Insulating sorbent polymers:* an insulating polymer film can be transformed to an ionic conductive electrolyte due to the sorption of gas molecules, which can decompose to ionized particles. As an example, polystyrene layers can be mentioned upon exposure to NO_2 gas [50] (see Section 3.3.2).

3.3.2 Chemically Sensitive Polymer Electrolytes

Ion-conductive polymers can be widely used in electrochemical cells with different electrodes and measuring methods such as solid electrolytes (see Section 2.4) for the detection of various gas and ionic components, respectively.

Nafion®, which is a polymer containing sulphonic acid groups, can be used to produce cation exchange membranes and seems to be a good candidate for standard electrolyte sensor membranes. Protonic Nafion® is suitable for carbon monoxide, nitric oxide, and oxygen detection. The operation principle of the oxygen sensor, similar to other electrolyte-type oxygen sensors, is the electrochemical reaction:

$$O_2 + 4H^+ + 4e^- \rightarrow 2H_2O \tag{3.28}$$

which takes place on a porous Pt electrode surface. Yan and Lu reported about a Nafion®-based limiting current type oxygen sensor [45].

Maseeh et al. [51] also used H^+-form Nafion® electrolyte with gold electrodes to build up an amperiometric NO sensor. The operation of the sensor is based on the reaction:

Sensing electrode:

$$NO + 2H_2O \rightarrow NO_3^- + 4H^+ + 3e^- \tag{3.29}$$

Counter electrode:

$$O_2 + 4H^+ + 4e^- \rightarrow 2H_2O \qquad (3.30)$$

Otagawa et al. [52] also used Nafion® films acidified in order to exchange Na^+ ions with H^+ ions so as to facilitate the following reaction at Pt sensing electrode:

$$CO + H_2O \rightarrow CO_2 + 2H^+ + 2e^- \qquad (3.31)$$

The counter-electrode reaction occurs again according to Equation (3.29). Based on these electrochemical reactions an amperometric CO sensor was developed.

A novel NO_2 sensing mechanism was discoverd by Cristensen et al. [50]. The sensor operates on an entirely new principle. It is based on a reversible change in conductivity in an otherwise insulating film of polystyrene upon exposure to NO_2 gas. An insulator-electrolyte transition occurs because the film is loaded with ionic charge carriers. Thus, a several orders of magnitude increase in conductivity can be measured when a NO_2/N_2 mixture comes into contact with the film. The behaviour is believed to be due to the self-ionization of N_2O_4, the form of NO_2 within the film, to $NO^+NO_3^-$, with the appropriate anodic and cathodic electrochemistry. Thus, the neutral N_2O_4 provides a nitrosonium nitrate salt bridge through the polystyrene, enabling conduction of charge by percolative ion exchange within the film. The sensor's operation principle is extremely specific to NO_2; therefore, extreme good selectivity can be expected.

Sheppard et al. [47] used a polymer hydrogel electrolyte for conductimetric pH detection. Hydrogel copolymers were prepared from HEMA/DMAEMA by free-radical polymerization using tetraethylene diacrylate cross-linker. The sensor can detect changes in conductivity resulting from pH-dependent swelling of the gel. The mechanism for the swelling is that tertiary amines in the gel become protonated as the pH is reduced, and counter-ions drawn into the gel to maintain electroneutrality provide an osmotic driving force for swelling. Due to the presence of ionized amines, a lower pH value is in the gel and the conductivity of the aqueous solution within the gel is increased relative to that of the solution external to the gel. The increase in the region of low pH can be explained by an increased role of protonic current transport.

Protonic conductor electrolytes can also detect small amounts of H_2 (CO, NH_3). Potentiometric, as well as amperometric, solid-state gas sensors using antimonic-acid ($Sb_2O_5\text{-}2H_2O$)/poly(vinyl alcohol) (PVA) complex as proton-conductor were developed by Miura et al. [53] for detecting small amounts of hydrogen in air at room temperature.

3.3.3 Humidity-Sensitive Behaviour of Polymer Electrolytes[1]

Various types of polymer electrolytes have been used for humidity sensors. When the polymer electrolytes sorb water vapour from the atmosphere, their electrical resistivity decreases. As a result, one can tell the humidity of the air surrounding the polymer by measuring the resistance impedance. The humidity sensors utilizing the variation of resistivity of the materials with humidity have been referred to as resistive-type humidity sensors.

The first resistive-type humidity sensor using polymeric materials was invented by Dunmore [54] in 1938. It was composed of polyvinyl acetate dispersed with LiCl. This sensor has sufficient sensitivity to humidity and has been used for many years. However, this sensor has the shortcoming effect that it cannot be used at high humidities because LiCl leaches out when the polymer matrix swells with sorbed water.

The ionic conductivity in a polymer electrolyte increases with an increase in water adsorption due to increases in the ionic mobility and/or charge carrier concentrations. An important problem encountered in adopting these materials to sensors is water resistivity. Polymer electrolytes are generally hydrophilic and soluble in water, so that they have a poor durability against water and dew condensation. Copolymerization with a hydrophobic polymer, cross-linking, graft polymerization on hydrophobic polymer surfaces, the use of interpenetrating polymer networks, the application of chlorinated and fluorinated polymers, and/or sidechains are suggested solutions to make a hydrophilic polymer resistive.

Key points of the conduction model for humidity-sensitive polymer electrolytes are as follows:

- At low humidity, electric conduction has an important role.
- At medium humidity, ionic conduction plays a major role in the humidity-sensitive resistance; with an increasing water content, the migration of hydrated ions is also possible.
- In the high humidity region, where the amount of sorbed water is large, protons become the majority carriers.

Since 1983, new types of humidity sensors have been developed using random copolymers of ionic and nonionic monomers. The ionic components have either a quaternary ammonium group or sulphonate group [55,56]. In order to elucidate the sensing mechanism, these copolymers are not adequate because of the random sequence of ionic and nonionic units in

[1]This section has been contributed by Prof. Yoshiro Sakai, *Dept. Applied Chemistry, Ehime University, Matsuyama 790, Japan. He became a professor in 1974. His current research topics include humidity sensors and gas sensors composed of polymeric materials.*

the polymer chains. Sakai et al. [57,58] have developed graft copolymers consisting of hydrophobic polymer films such as PTFE and hydrophilic branch polymers. When a thick film of PTFE was used, the hydrophilic polymers such as poly(styrene sulphonate) or quaternized poly(vinyl pyridine) (PVPy) were grafted on the surface of the PTFE film [58]. In order to avoid the polarization by DC current, AC impedance was measured at various humidities between a pair of gold electrodes deposited on the grafted film in the frequency range of 100 Hz to 1 MHz.

In the case of PTFE-graft-poly(styrene sulphonate) Li, K, Na salts, as well as free acidic form, were prepared by the exchange of cations. Figure 3.21 shows the impedance as a function of humidity. The impedance of the alkali salt form is one order of magnitude higher than that of the acidic form. There are no significant differences among the alkali salt forms.

From the Cole-Cole plots (see Section 2.1, Figures 2.5 and 2.6) of the data obtained at several temperatures, the resistance component, R_b, was

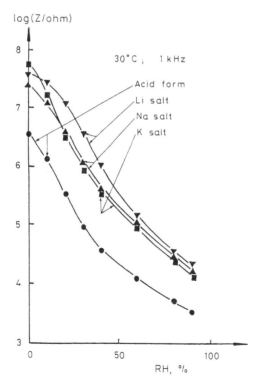

FIGURE 3.21. Impedance as a function of humidity for PTFE-graft-poly(styrene sulphonate) (temperature: 30°C; frequency: 1 kHz).

determined by intercepting the semicircle to the real axis. In this case, a simple equivalent circuit with parallel resistance and capacitance components was assumed. The value of R_b exponentially decreases as the humidity increases, while C_g remains constant over the whole humidity range. The relation between log R_b and the weight of sorbed water is shown in Figure 3.22. The figure shows that the resistivity does not depend on the species of cation but on the amount of sorbed water. The results suggest that it is not the effect of alkali ions but that of protons from the sorbed water that predominantly contributes to the charge transport.

Another graft polymer, that is, polyethylene-graft-poly-(2-acrylamido-2-methylpropane sulphonate) was also synthesized [59]. The humidity dependence of impedance was also measured for Li, K, Na salt as well as for the acidic form. The impedance of the latter is one order of magnitude

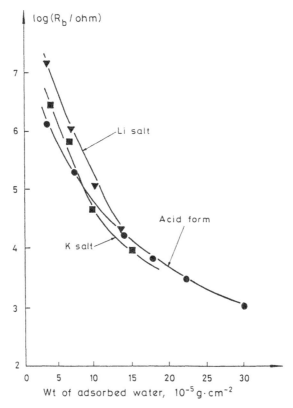

FIGURE 3.22. Plot of log R_b vs. the weight of sorbed water per unit area of the grafted films.

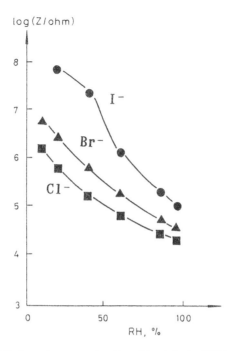

FIGURE 3.23. Impedance as a function of humidity for PTFE-grafted PVPy.

lower than that of the alkali salts. The results are similar to those obtained with PTFE-graft-poly(styrene sulphonate).

In the case of PTFE-graft-quarternized poly(vinyl pyridine) (PVPy) [58], the polymer chain has cationic sites. The counter-ion is a halide ion, which can be chosen by selecting the quaternization reagents alkyl halides. Sakai et al. have prepared three kinds of graft polymers having iodide, bromide, and chloride anion. In contrast to the polymers having a sulphonate group, in the case of a polymer having a pyridyl group, the anion seems to be the charge carrier, as shown in Figure 3.23. The impedance of PTFE-graft-quaternized PVPy having chloride ions is the lowest among the three halides. From the complex impedance plot (Cole-Cole plot), the parallel resistance, R_b, was determined as the intercept to the real axis. In Figure 3.24, log R_b was plotted as a function of the number of water molecules sorbed at each pyridyl group for the samples quaternized with ethyl halides or *n*-butyl halides. The figure shows that R_b depends on the amount of sorbed water, irrespective of the size of alkyl group bonded to the pyridyl group.

When the amount of sorbed water is small, the value of R_b depends on the

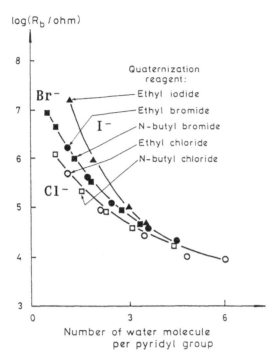

FIGURE 3.24. Plot of log R_b vs. the number of sorbed water molecules per each pyridyl group using various quarternization reagents.

species of anion, while in the range of high water content the difference between the halide anions becomes small. The activation energy was estimated from the Arrhenius plot of R_b. In Figure 3.25, the activation energy, *E,* was plotted against the number of sorbed water molecules. These figures also suggest that the majority carriers are the anions at low humidities and the protons at high humidities. It seems that in the low-humidity region, the anion jumps from the positive site of the pyridyl group to the nearest positive site, but the transport number of protons becomes predominant as the amount of sorbed water increases.

The other type of copolymers synthesized from ionic and nonionic monomers show the similar electrical properties to those in Figures 3.21 or 3.23. Poly(vinyl pyridine) cross-linked with α,ω-dihalogenoalkanes also has similar characteristics to those of PTFE-quaternized PVPy.

Poly(2-hydroxy-3-methacryloxypropyl trimethylammonium chloride) (PHMPTAC) has a quaternary ammonium group and is adequate to prepare humidity sensors. Since it is soluble in water, it must be grafted to a hydrophobic polymer [60,61], or cross-linking between HMPTAC polymer chains should be performed [62]. In the case of polyethylene-graft-PHMPTAC, the activation energy for R_b was estimated from the complex impedance analysis followed by Arrhenius plotting. In Figure 3.26, the activation energy was plotted against the humidity. There is a maximum around 40% relative humidity. In the low-humidity region, the carrier chloride ion obtains more hydrating water molecules as the humidity increases, so that the activation energy for mobility increases. Above 40% RH, the amount of sorbed water is sufficient to promote the transfer of hydrated chloride ion. In addition, the protons from the sorbed water may also become the carrier at high humidities.

Another type of polymeric material for humidity sensors is alkali salt–poly(ethylene oxide). It was reported [63] that the humidity sensor composed of poly(ethylene oxide) doped with a lithium salt has good impedance variation in the range of 10^7 to 10^3 Ω. Sorption of water into the film leads to an increase in the dielectric constant of the hybrid film, causing the decrease in the dissociation energy of the alkali salt, resulting in the decrease in the impedance. These hybrid films have good sensitivity to humidity, but they are not resistive to water and cannot be used at high humidities.

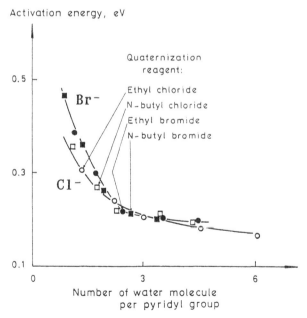

FIGURE 3.25. Plot of activation energy, *E*, vs. the number of sorbed water molecules per each pyridyl group using various quaternization reagents.

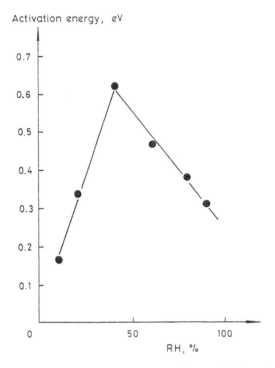

FIGURE 3.26. Plot of activation energy for R_b vs. relative humidity for polyethylene-graft-PHMPTAC.

Humidity sensors were also prepared from organopolysiloxanes having hydrophilic groups such as SO_3H and $N^+(CH_3)_3Cl^-$. The films were prepared either by grafting to a pressed silica gel disc or by cross-linking on a sintered alumina plate. It was demonstrated by the Arrhenius plots that the proton is the majority carrier in the case of sulphonic acid sidegroups, while it is the Cl^- anion in the case of quaternary ammonium types [64].

3.4 SORBENT POLYMER COATINGS FOR CHEMICAL SENSORS[2]

A chemical microsensor can be defined generally as the combination of

[2]This section has been contributed by Jay W. Grate, *Molecular Science Research Center, Pacific Northwest Laboratory, Richland, WA 99352;* Michael H. Abraham, *Chemistry Department, University College London, London WCIH OAJ, United Kingdom;* and R. Andrew McGill, *Geo-Centers, Inc., 10903 Indian Heah Highway, Fort Washington, MD*

a physical transducer with a chemically selective thin-film material. This concept is illustrated in Figure 3.27 with a polymer-coated sensor for the detection of vapours in the gas phase as described in Chapter 2. The physical transducer by itself cannot directly detect the analyte molecules in the gas phase. Instead, the transducer senses changes in the physical properties of the layer on its surface. It is the sorption of analyte molecules from the gas phase and the resulting modification of the sensed properties of the selective layer that results in the detection of the species in the gas phase. Thus, the sensor's response is dependent on two conceptually distinct processes, sorption and transduction.

Two types of sorption can be distinguished. Adsorption is the collection of a species on a surface. In absorption, adsorbed species go on to dissolve into the bulk of the material. Absorption is dependent on the strengths of various fundamental interactions between the absorbed species and the sorbent material. In sensor applications, bulk absorption can collect more vapour to a sensor surface than surface adsorption, thus offering higher sensitivity. In addition, sensors based on absorption have greater resistance to surface contamination effects that can degrade the performance of sensors relying entirely on surface adsorption for their selectivity.

This section will focus on the absorption of organic vapours by polymer materials. Although a variety of materials can be considered for use as the sorptive layer, polymer materials are particularly useful because of the ease with which thin, adherent films on sensors can be prepared and the wide range of chemical affinities that can be achieved by varying the polymer's chemical structure. In addition, rapid and reversible sensor responses can be obtained if the polymer is amorphous and its static glass-to-rubber transition temperature is below the sensor's operating temperature. The greater thermal motion of polymer chain segments of a rubbery (as opposed to glassy) polymer and the greater free volume allows rapid diffusion of vapours through the material and provides access to sites for interactions.

20744. The authors gratefully acknowledge the students and coworkers who participated in the development of sorbent polymers for chemical sensors and in the determination of LSER relationships, including Arthur Snow, Gary S. Whiting, Jeril Andonian-Haftvan, Jonathon W. Steed, Jan Hamerton, and Prina Sasson. Pacific Northwest Laboratory is operated for the U.S. Department of Energy by Battelle Memorial Institute under contract DE-AC06-76RLO 1830. The research activity of the authors has focused primarily on acoustic wave devices such as SAW vapour sensors, where the studies have addressed issues including vapour/polymer interactions, sorption equilibria, polymer design and synthesis, sensor array systems with pattern recognition, mechanisms of reactions, solvent effects, the hydrophobic effect, theoretical and experimental studies of electrolytes and of nonelectrolytes, hydrogen-bond acidity and hydrogen-bond basicity, and application of solubility theories to polymeric materials.

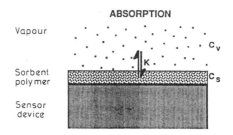

FIGURE 3.27. Illustration of vapour absorption by a sorbent polymer layer on a sensor device.

3.4.1 Sensor Responses and the Partition Coefficient

The equilibrium distribution of a vapour between the gas phase and a sorbent phase is given by the partition coefficient, K, defined as

$$K = C_s/C_v \qquad (3.32)$$

As shown in Figure 3.27, C_s represents the concentration of the vapour in the sorbent phase (the polymer in the case of a polymer-coated sensor), and C_v represents the concentration of the vapour in the gas phase. The partition coefficient provides a thermodynamic measure of the strength of sorption. It is related to the standard Gibb's free energy of solution of a gaseous solute, ΔG_s^o, by

$$\Delta G_s^o = - RT \ln K \qquad (3.33)$$

where the standard states are unit concentration in the gas phase and unit concentration in solution.

The response, S, of a gas phase chemical microsensor is empirically determined as a function of the gas phase vapour concentration:

$$S = f(C_v) \qquad (3.34)$$

Sensor sensitivity is defined as the incremental change in signal that occurs in response to an incremental change in analyte concentration in the analytical environment (i.e., the gas phase in this case). Mechanistically, however, sensor response is actually a function of the concentration of analyte in the sorbent layer on the sensor's surface.

$$S = f(C_s) \qquad (3.35)$$

$$S = f(KC_v) \qquad (3.36)$$

Equation (3.36) follows from Equation (3.35) and the relationship between the concentrations of vapour in the sorbent and gas phases given by Equation (3.32). Equation (3.36) provides a very general relationship that illustrates the key role of sorption and the partition coefficient in the response of a chemical sensor. The influence of the sorption step is given by KC_v, while the transduction step is accounted for the function operating on KC_v. Chemical selectivity can occur in both the sorption and transduction steps. This section will focus entirely on the selectivity that can be achieved in the sorption step. This is the primary factor influencing sensitivity and selectivity of polymer-coated acoustic wave vapour sensors such as the quartz crystal microbalance (QMB) and the surface acoustic wave (SAW) device [65,66].

3.4.2 Fundamental Interactions Contributing to Sorption

The process of absorption can be considered in terms of a solubility model where the vapour is the solute and the sorbent phase is the solvent [67]. Three conceptual steps are involved. First, a cavity must form in the solvent. This requires breaking solvent-solvent interactions and is endoergic. Second, the monomeric solute fills the cavity with concomitant reorganization of solvent around the cavity. Third, attractive interactions, which are by definition solubility interactions, form between the solute and the solvent. These interactions are exoergic and favour sorption.

The types of fundamental interactions that can form between neutral organic species include noncovalent interactions such as dispersion interactions (also known as London forces or induced-dipole/induced-dipole interactions), dipole/induced-dipole interactions (also known as induction interactions), dipole/dipole interactions (also known as orientation interactions), hydrogen-bonding interactions, π-π interactions, and charge-transfer interactions [68,69]. When noncovalent interactions are set up in a discrete geometry such as a cavity structure, host-guest inclusion complexes can be formed. The general term van der Waals interactions include dispersion, induction, and orientation interactions, as well as electrostatic interactions involving charges and polarizable or dipolar species.

Charge transfer interactions are not generally important between typical organic vapours and nonconducting organic polymers and will not be considered further in this chapter. π-π interactions can be modelled as a combination of dispersion interactions which provide the main force for association, combined with electrostatic interactions that govern the overall geometry [70]. Therefore, these interactions are not necessarily different from the more fundamental interactions noted above.

The strengths of the various solubility interactions and, hence, the strength of sorption depend on the respective solubility properties of the solute and the solvent. These properties, which include polarizability, dipolarity, hydrogen-bond basicity, and hydrogen-bond acidity, are determined by chemical structure. In all but the simplest processes, multiple interaction types occur simultaneously.

3.4.3 Linear Solvation Energy Relationships

A full understanding of the polymer/vapour interaction cannot be developed without some means of evaluating the properties of the sorbent polymer and the vapours to be detected. In addition, it is informative to be able to evaluate the relative contributions of each type of interaction to the overall sorption process. Solvation parameters and linear solvation energy relationships (LSERs) provide a useful approach to evaluating properties and interactions [71]. For the absorption of a vapour into a polymer, the relationship takes the form given in the following equation:

$$\log K = \text{constant} + rR_2 + s\pi_2^H + a\alpha_2^H + b\beta_2^H + l \log L^{16}$$

$$(3.37)$$

Sorption is modelled as a linear combination of a polarity term ($s\pi_2^H$), a polarizability term (rR_2), a hydrogen-bonding term in which the vapour is the hydrogen-bond acid ($a\alpha_2^H$), a hydrogen-bonding term in which the vapour is the hydrogen-bond base ($b\beta_2^H$), and a combined disperson interaction and cavity term ($l \log L^{16}$). The usefulness of these relationships is that they express the measure of absorption, $\log K$, in terms of solvation parameters R_2, π_2^H, α_2^H, β_2^H, and $\log L^{16}$, which describe solubility properties of the solute vapour and coefficients s, r, a, b, and l, which characterize the complementary properties of the polymer. The terms rR^2, $s\pi_2^H$, $a\alpha_2^H$, $b\beta_2^H$, and $l \cdot \log L^{16}$ indicate the contributions of particular interactions to the overall sorption process.

With the exception of R_2, the solvation parameters for the vapours are all derived from thermodynamic measurements of equilibrium phenomena. R_2 is a calculated excess molar refraction parameter that provides a quantitative indication of polarizable n and p electrons [72]. The parameter π_2^H measures the ability of a molecule to stabilize a neighbouring charge or dipole [73]. (Values of π_2^H are approximately proportional to molecular dipole moments for nonprotonic, aliphatic solutes with a single dominant dipole.) The hydrogen-bonding parameters α_2^H and β_2^H measure hydrogen-bond acidity and basicity, respectively [74,75].

Log L^{16} is the gas-liquid partition coefficient of the solute on hexadecane at 25 °C (determined by gas-liquid chromatography) [76]. [L is the symbol for the Ostwald solubility coefficient, which is defined identically to the partition coefficient as in Equation (3.32).] This parameter is combined measure of exoergic dispersion interactions (that increase log L^{16}) and the endoergic cost of creating a cavity in hexadecane (leading to a decrease in log L^{16}). These solvation parameters are available for hundreds of organic molecules and have recently been reviewed [71].

Table 3.6 provides the solvation parameters for a set of vapours selected as illustrative examples. All the solvation parameters except log L^{16} are scaled to have values of zero for aliphatic hydrocarbons. Aliphatic hydrocarbons can only interact by dispersion interactions, Aromatic hydrocarbons are more polarizable than aliphatic hydrocarbons, so their R_2 and π_2^H values are nonzero. Similarly, chlorinated aliphatic hydrocarbons are more polarizable and more dipolar than simple aliphatic hydrocarbons and, thus, have nonzero R_2 and π_2^H values. A great variety of organic compounds are basic and have positive β_2^H values, including esters, ethers, amines, nitriles, amides, and alcohols, to name just a few. Hydrogen-bond acidic compounds are generally those that contain hydroxyl groups such as alcohols and phenols, and these have positive α_2^H parameters. Dispersion interactions are ubiquitous among organic compounds. Log L^{16} values increase with solute size and molecular weight along any homologous series. Solvation parameters provide a means of evaluating the properties of analyte vapours to be detected.

The solubility properties of a sorbent polymer material can be determined by measuring the partition coefficients (K) of a diverse set of vapours into the material and regressing the log K values against the known solvation parameters of those vapours. This LSER method yields the regression coefficients $r, s, a, b,$ and l in Equation (3.37), as well as the constant. Gas-liquid chromatography is a well established method of determining the required partition coefficients [77] and has been used in the determination of the LSER equations characterizing the polymers described below.

The r-coefficient measures the ability of the sorbent phase to interact with solute n and p electrons and indicates polarizability. Usually r is slightly positive, but it can be negative if the phase is fluorinated. The s-coefficient measures the sorbent phase dipolarity/polarizability. The a-coefficient measures the sorbent phase hydrogen-bond basicity that interacts with the solute hydrogen-bond acidity. Similarly, the b-coefficient measures the sorbent phase hydrogen-bond acidity that interacts with the solute hydrogen-bond basicity. The l-coefficient is a combined measure of dispersion interactions that tend to increase l and cavity effects that tend to decrease l.

3.4.4 Selected Polymer Examples

The LSER approach is illustrated here with four specific polymers selected so that each emphasizes a different property. These include fluoro-polyol as a hydrogen-bond acidic polymer, poly(ethylenimine) as a

TABLE 3.6. Solvation Parameters for Various Organic Vapours.

Solute	Polarizability R_2	Dipolarity/ Polarizability π_2^H	Acidity* α_2^H	Basicity* β_2^H	Dispersion/Cavity $\log L^{16}$
Hexane	0	0	0	0	2.668
Benzene	0.610	0.052	0	0.14	2.786
Dichloromethane	0.387	0.57	0.10	0.05	2.019
Diethyl ether	0.041	0.25	0	0.45	2.015
Butanone	0.166	0.70	0.00	0.51	2.287
Ethylacetate	0.106	0.062	0	0.45	2.314
Acetonitrile	0.237	0.90	0.07	0.32	1.739
Triethylamine	0.101	0.15	0	0.79	3.040
N,N-dimethylformamide	0.367	1.31	0	0.74	3.173
Ethanol	0.286	0.42	0.37	0.48	1.485
Phenol	0.805	0.89	0.60	0.30	3.766
Hexafluoroisopropanol	−0.240	0.55	0.77	0.10	1.392

* These are for the solute surrounded by excess of solvent. Therefore, the α_2^H and β_2^H parameters, for example, are really "effective" or "summation" $\Sigma\alpha_2^H$ and $\Sigma\beta_2^H$ parameters, but we retain the simpler nomenclature here.

hydrogen-bond basic polymer, a siloxane with pendant cyanopropyl groups as a dipolar polymer, and poly(isobutylene) as a material limited to dispersion interactions. The chemical structures of these polymers are shown in Figure 3.28. All of them have glass-to-rubber transition temperatures below room temperature. Partition coefficients for vapours into these polymers were measured at 298 K. Their LSER coefficients are given in Table 3.7.

Fluoropolyol is expected to be hydrogen-bond acidic because it contains both hydroxyl groups and fluoro substitution. In addition, it has been proven to be useful in a number of chemical sensor studies [78–83]. In Table 3.7, fluoropolyol is the only polymer with a large *b*-coefficient (4.09), indicating that is indeed quite hydrogen-bond acidic. The other coefficients indicate that it is also somewhat basic ($a = 1.49$) and dipolar ($s = 1.45$). Therefore, this will be a very sensitive material for detecting hydrogen-bond bases, but it may not be as selective as a material with similar acidity but less basicity or dipolarity.

Poly(ethylenimine) was selected because the amine groups are expected to be quite basic. The large *a*-coefficient of 7.02 for this polymer indicates that this is indeed the case. The *s*-coefficient of 1.52 is somewhat surprising, considering the structure of this material (as shown in Figure 3.28) and the fact that simple amines are not particularly dipolar. However, real samples of poly(ethylenimine) usually contain some quaternized nitrogens, and these will lead to ion-dipole interactions that could contribute to the *s*-coefficient. Basic polymers are quite sensitive to hydrogen-bond acidic vapours, and generally sorb water quite well. Such polymers can be used to

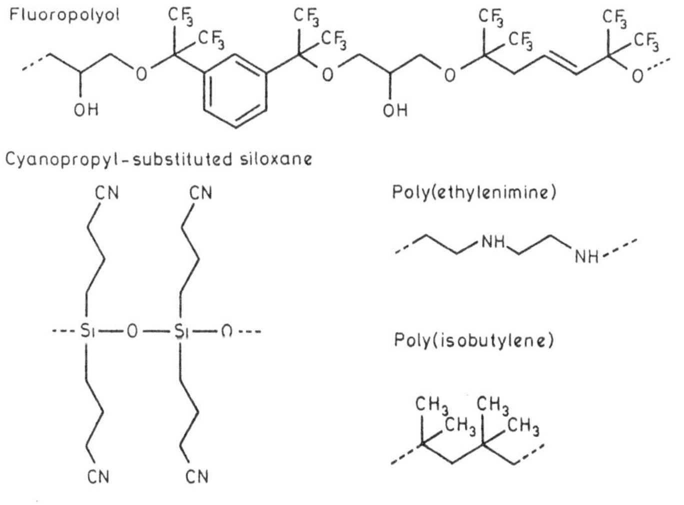

FIGURE 3.28. The chemical structures of the repeat units of four sorbent polymers.

TABLE 3.7. *LSER Coefficients for Sorbent Polymer Examples.*

LSER Coefficients	Polarizability (r)	Dipolarity (s)	Basicity (a)	Acidity (b)	Dispersion/Cavity (l)
Fluoropolyol	−0.67	1.45	1.49	4.09	0.81
Poly(ethylenimine)	0.50	1.52	7.02	(0)*	0.77
Cyanopropyl siloxane	0.00	2.28	3.03	0.52	0.77
Poly(isobutylene)	−0.08	0.37	0.18	(0)*	1.02

* Terms in parentheses were not statistically significant and were therefore dropped from the regression analysis.

make humidity sensors. Poly(ethylenimine) has been used in various chemical sensor studies [81,83,84].

The cyanopropyl-substituted siloxane was selected because of the high dipolarity of the nitrile groups. A variety of other functional groups such as amides and sulphoxides also have strong dipoles, but they are also strongly basic. Nitrile groups are moderately basic by comparison; therefore, this polymer represents an attempt to obtain maximum dipolarity with minimal basicity within the constraints of real materials. The s-coefficient of 2.28 for the cyanopropyl-substituted siloxane confirms that it is quite dipolar while the a-coefficient of 3.03 indicates that it is much less basic than poly(ethylenimine). Therefore, this material will be superior for the detection of dipolar vapours.

Poly(isobutylene) was selected as a material that is only capable of dispersion interactions, and all coefficients except the l-coefficient are small or zero. It has the largest l-coefficient of the group. This polymer can be useful in detecting simple aliphatic hydrocarbons and has been used in various chemical sensors applications [83,85–89].

The relative strengths of different solubility interactions between a vapour and polymer can be evaluated by multiplying the vapour's solvation-parameters by the complementary polymer LSER coefficients to produce the interaction terms rR_2, $s\pi_2^H$, $a\alpha_2^H$, $b\beta_2^H$, and $l \log L_{16}$. These terms (plus constant) add up to log of a partition coefficient, and therefore each term indicates the contribution of the interaction represented to the overall sorption process. The results of such calculations are presented in Table 3.8, pairing each of the four polymer examples with a vapour having complementary properties. Dispersion interactions are significant in all cases, although the magnitude of this interaction varies. It is the only interaction of significance between poly(isobutylene) and hexane. Strong hydrogen-bonding interactions are found in the fluoropolyol/triethylamine and poly(ethylenimine)/

TABLE 3.8. Calculated Interaction Terms.

Polymer/Solute	rR_2	$s\pi_2^H$	$a\alpha_2^H$	$b\beta_2^H$	$l\log L^{16}$
Fluoropolyol/ triethylamine	−0.07	0.22	0.00	3.23	2.46
Poly(ethylenimine) ethanol/	0.14	0.64	2.60	0.00	1.14
Cyanopropyl siloxane/ N,N-dimethylformamide	0.00	2.99	0.00	0.38	2.44
Poly(isobutylene)/ hexane	0.00	0.00	0.00	0.00	2.72

ethanol pairs. Dipole-dipole interactions are quite significant between the cyanopropyl-substituted siloxane and N,N-dimethylformamide.

3.4.5 Polymer Design

The design and selection of sorbent polymers for particular sensing applications require, first, that the solubility properties of the target analyte should be understood [67]. Solvation parameters for solute vapours are useful for this purpose. Then one can seek polymers with properties that are complementary to those to the target analyte in order to set up favourable interactions and promote sorption. The desired solubility properties are obtained by incorporating organic functional groups with those desired properties. At the same time, the material must be designed so that it has the desired physical properties: it is preferable that it be above its glass-to-rubber transition at the sensor's operating temperature in order to facilitate rapid reversible sorption, and it must adhere to the sensor's surface as a thin film.

The selected examples above illustrate that sorbent polymers do exhibit the solubility properties expected on the basis of chemical structure; therefore, materials with particular properties can be designed rationally, provided that the solubility properties of the organic functional groups to be incorporated are understood. In this regard, the solvation parameters of monofunctional organic compounds can be examined to determine which structures result in particular properties.

To sense a basic vapour, for example, one would want a hydrogen-bond acidic polymer. The properties are expected from materials with alcohol or phenolic moieties. From Table 3.6, it is apparent that phenolic solutes are substantially more hydrogen-bond acidic than simple alcoholic solutes. Even stronger hydrogen-bond acidity is observed if fluorine substituents are present to withdraw electron density from the hydroxyl group. This analysis leads to the selection of fluorine-substituted alcohols and phenols as candidates for sorbent materials for detection of basic analytes. It is also noteworthy that fluorine-substitution greatly decreases the basicity of an alcohol; therefore, a fluoroalcohol will also be more selective for bases than a simple aliphatic alcohol, since it will have much less affinity for the sorption of hydrogen-bond acidic vapours. This approach has been the basis of a number of studies into sorbent hydrogen-bond donating materials [82,90,91].

In designing a sorbent material for a sensor, some consideration must be given to whether the goal is to maximize sensitivity or selectivity [67]. To maximize sensitivity, sorption should be promoted by setting up all possible interactions with the vapour of interest and making these interactions as strong as possible. However, if selectivity is the primary goal, only one or two types of interactions should be maximized and all others minimized, thus limiting the types of vapours with which the polymer will interact strongly.

3.4.6 Sensors and Sensor Arrays

The application of LSER methods to chemical sensors has been developed primarily in connection with polymer-coated SAW vapour sensors. This approach was first suggested in 1988 and was set out in detail in a review in 1991 [67,78]. The solubility properties of various phenolic materials for chemical sensor applications were determined by LSER methods and reported in 1991, and preliminary LSER equations for five polymers of varying properties appeared in 1992 [90,92]. Currently, LSER equations for over a dozen polymers have been determined.

The application of the LSER approach to acoustic wave sensors is straightforward because the transduction step depends primarily on the amount of the vapour sorbed rather than on any particular property of the sorbed vapour. Therefore, the selectivity for particular vapour properties occurs in the sorption step. For acoustic wave sensors responding only to mass-loading effects, the sensors' response is related to the partition coefficient, according to Equation (3.38) [78,89]:

$$\Delta f_v = \Delta f_s C_v K / \varrho_s \qquad (3.38)$$

The response is a frequency shift, Δf_v, dependent on the amount of sorbent phase on the sensor surface expressed as a frequency shift, Δf_s, the density of the sorbent phase, ϱ_s, the concentration of the vapour in the vapour phase, and the partition coefficient. It has been found that polymer-coated SAW vapour sensors also respond to modulus changes of the polymer upon vapour sorption, substantially increasing sensitivity [89]. The modulus changes are related to the swelling of the polymer by the vapour. The responses of polymer-coated SAW vapour sensors can be estimated by Equation (3.39):

$$\Delta f_v = 4 \, \Delta f_s C_v K / \varrho_s \qquad (3.39)$$

The factor of 4 accounts for the extra sensitivity due to swelling-induced modulus effects. Because partition coefficients can be predicted for hundreds of vapours into polymers characterized by the LSER method and because SAW vapour sensor responses are directly related to those partition coefficients, it is possible to calculate estimates for a sensor's sensitivity before it is acutally made and tested. This capability can be useful in evaluating the likely success of an SAW sensor in meeting the requirements of a particular application. Alternatively, the relationship between SAW sensor responses and partition coefficients allows LSER relationships to be derived on the basis of polymer-coated SAW sensor responses to calibrated vapour streams instead of chromatographically determined partition coefficients [93].

It should be noted that estimations based on Equations (3.38) or (3.39) apply only to a sensor operating at the same temperature as the temperature at which the partition coefficient was determined. Vapour absorption and partition coefficeints decrease exponentially with temperature. Therefore, sensors dependent on vapour absorption will be more sensitive at lower temperatures, and they should be operated at constant temperatures. Varying temperature will result in varying sensitivity.

Sensors whose selectivity is based primarily on reversible sorption are not totally selective because all vapours will interact by dispersion interactions to some degree; however, selectivity can be greatly improved by incorporating sensors into sensor arrays. The array provides more chemical information than a single sensor, and this information can be processed by pattern recognition techiques in order to recognize and distinguish particular classes of vapours. The array provides the greatest information if the sensors in the array each provide unique and useful information. One approach to this goal is to have each sensor coating emphasize a different solubility interaction and to choose a set of coatings that span the entire range of solubility interactions. In this regard, the four coatings in Figure 3.28 could form the basis for a sensor array. (In fact, three of them have been included in published sensor array studies [79,81,83].) Detection systems using sensor array and pattern recognition are sometimes referred to as smart sensor systems or electronic noses [83,94–96]. Further discussion on the selection of polymers for sensor arrays and on achieving selectivity using sensor systems can be found in Reference [67].

The use of sorbent polymers is not limited to acoustic wave sensors and arrays, however. Fibre-optic sensors, for example, can be designed where the response is proportional to the amount of vapour sorbed by an applied polymer layer. Vapour sorption can alter properties such as film thickness, refractive index, or light absorption (see Section 2.6). In addition, sensors can be prepared using composites consisting of a sensing material dispersed in a sorbent polymer layer. In this case, designing the polymer matrix to promote sorption can concentrate the vapour at the surface of the incorporated sensing material, thus improving sensitivity and influencing selectivity. This approach has been demonstrated in the design of chemiresistors for organophosphorous detection where the selective layer consisted of crystallites of a substituted metallophthalocyanine in a fluoropolyol matrix [80]. The fluoropolyol served to promote the sorption of basic organophosphorous vapours, while the phthalocyanine material carried the current that was altered by vapour sorption. An examination of selectivity patterns showed that the primary determinant of selectivity was the strength of sorption into fluoropolyol.

In summary, the methods presented above serve to elucidate the roles of fundamental interactions in governing the sorption of vapours by polymers. They can provide guidance in the design and selection of polymers for use as chemically selective layers or as components of chemically selective layers on sensors. The methodology also suggests an approach to the design of sensor arrays.

3.5 SELECTIVE CHEMICAL SENSING: MOLECULAR RECOGNITION WITH POLYMERIC LAYERS AND CAGE COMPOUNDS[3]

Generally, chemical sensors convert a chemical state as input signal into an electrical output signal [97]. The input signal is determined by the concentration, partial pressure, or activity of different particles such as atoms, molecules, or ions in the gas phase or solution. The sensors consist of chemically sensitive, stable, and selective layers ("sss") and of transducers that generate the electrical signal (see Chapter 2). In addition, filters and membrane may be used to exclude interfering particles. The general structure is demonstrated in Figure 3.29.

The quantitative determination of molecules in the gas and liquid phase by means of chemical sensors requires their interaction with the chemically sensitive layer to be controlled thermodynamically. Under these conditions, the surface and bulk concentrations of molecules in the layers are adjusted unequivocally by their partial pressures or concentrations and by temperature.

Layers of polymeric and supramolecular organic compounds are gaining increasing interest to be used as materials for chemical sensors [67,98]. They show a large flexibility for tailoring recognition structures with the aim of a specific detection of ions and molecules. Of particular interest in

[3]This section has been contributed by Klaus-Dieter Schierbaum and Prof. Wolfgang Göpel, *Institute of Physical and Theoretical Chemistry, University of Tübingen, D-72076 Tübingen, Germany.* Dr. Schierbaum obtained his Ph.D. in chemistry in 1987 in the field of chemical sensors and interface analysis. Prof. Göpel has been director of the Institute since 1983, with research interests in interface properties of new materials for chemical sensors, catalysts and microelectronic devices.

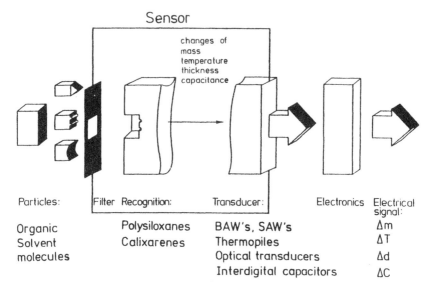

Particles:	Filter	Recognition:	Transducer:	Electronics	Electrical signal:
Organic		Polysiloxanes	BAW's, SAW's		Δm
Solvent		Calixarenes	Thermopiles		ΔT
molecules			Optical transducers		Δd
			Interdigital capacitors		ΔC

FIGURE 3.29. Key-lock arrangement of a typical chemical or biochemical sensor to detect atoms, molecules, or ions ("particles") in the gas or liquid phase. Additional components of complete sensor systems are also indicated.

the context is the detection of inorganic gases (like carbon dioxide, sulphur dioxide, and nitrogen dioxide) in air and organic solvent molecules (like aliphatic and aromatic organic compounds, chlorinated hydrocarbons, amines, alcohols, ketones, and ethers) in air or in water.

In the following, two different model systems will be discussed: *polymers* like polysiloxanes and polyetherurethanes (see Table 3.9) utilize the selective bulk absorption of molecules, which corresponds to their permeselective properties, and *supramolecular compounds* like calixarenes or högbergarenes (see Table 3.9), which are bound generally to polymeric surfaces and utilize the incorporation of molecules in molecular cages.

Different transducers may be applied to monitor the particle/layer interaction be means of mass changes, Δm (with bulk, surface, and plate acoustic wave devices, see Section 2.3); temperature changes, ΔT [with thermopile devices, see Section 2.5 and Figure 3.30(a)]; capacitance changes, ΔC [with interdigital electrode capacitors, see Section 2.1 and Figure 3.30(b)]; and thickness changes, Δd [with optical devices, see Section 2.6.4 and Figure 3.30(c)] [29,99–101].

This section is organized as follows: In Section 3.5.1, the physico-chemical fundamentals will be treated to understand the different types of interactions between polymeric or supramolecular compounds and the

TABLE 3.9. Polymeric and Supramolecular Compounds Investigated in Section 3.5.
"Me" and "Et" Denote the Methyl (CH_2) and the Ethyl Group (C_2H_5), Respectively.

	R	R'
Poly(dimethylsiloxane), PDMS	Me-	Me-
Poly(cyanopropylmethylsiloxane), PCMS	Me-	$CN(CH_2)_3$-
Poly(aminopropylmethylsiloxane), PAPMS	Me-	$NH_2(CH_2)_3$-
Poly-iso-propylanoic-acid methylsiloxane), PIPAMS	Me-	$HOOCCH_2C(Me)_2$-
Poly(phenylmethylsiloxane), PPMS	Me-	C_6H_5-
(γ-aminopropylethoxysilane/ propylmethoxysilane)-copolymer, PAPPS	NH_2-$(CH_2)_3$- Me-$(CH_2)_3$-	Me- Et-
Poly(diphenyl/phenylmethylsiloxane), PDPMPS	C_6H_5-	C_6H_5-
Polyetherurethane, PEU	Me-	C_6H_5-
5,11,17,23-Tetra-iso-propyl-25,26,27,28-tetrahydroxy-calix[4]arene = iso-propyl-calix[4]arene		
5,11,17,23-Tetra-tert-butyl-25,26,27,28-tetrahydroxy-calix[4]arene = tert-butyl-calix[4]arene		
1.21,23,25-Tetrakis[11-thiaheeicosyl]-2,20:3,19-dimetheno-1H,21H,25H-bis[1,3]dioxocino[5,4-i:5',4'-i']benzol[1,2-d:5,4-d']bis[1,3]benzodioxocin = Högbergarene		

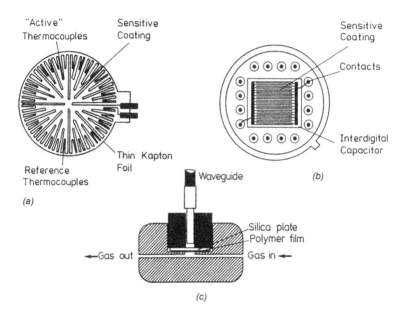

FIGURE 3.30. Schematic representation of various transducers to monitor particle/layer interactions: (a) planar thermopile with the thermocouples integrated on a thin polymer foil, (b) interdigital electrode capacitor, and (c) optical device connected via a waveguide to a light source and a spectrometer. Changes of optical thicknesses of the polymer film are determined interferometrically.

molecules to be detected. The specific detection of ions will not be treated in this part. Section 3.5.2 deals with sensor response signals resulting from mass-sensitive transducers, thermopile devices, interdigital electrode capacitors, and optical transducers, all of which are coated with the same or similar polymeric and supramolecular compounds. (Generally, sensor structures and responses are described in Chapter 4. However, the very special task of molecular recognition and the comparison of the same or similar materials applied on the surface of different sensor types mentioned above results in retaining this contributed section in its entirety—*The editor.*)

3.5.1 Fundamentals of Molecular Recognition

3.5.1.1 RECOGNITION SITES IN POLYMERIC AND SUPRAMOLECULAR COMPOUNDS

The weak molecular interaction between polymeric or supramolecular [see Figures 3.31(a) and 3.31(b)] compounds and solvent molecules results from short-range van der Waals "bonds" (due to dispersion forces between polarizable atoms or molecules, forces between polar molecules with permanent dipole moments and polarizable atoms or molecules, and forces between polar molecules) and long-range electrostatic "bonds" (due to point charges attributed to the differently electronegative atoms in the solvent molecule and in the polymeric or supramolecular compounds) [102]. To a first approximation, the different contributions of the overall interaction forces are additive, i.e., the total binding energy is given by the sum over individual binding energies from all interacting atoms. In this approximation, corresponding equilibrium distances between interacting molecules are determined by the sum of the van der Waals radii of the atoms, i.e., by "contact" interaction with the monitored molecules. Two extreme situations of recognition sites are realized in polymeric, as compared to supramolecular, compounds with their small molecular cages. The difference results from principal differences in their geometric and dynamic structures.

As a "model system" of a chemically very stable *polymer,* we consider polysiloxanes that consist of a polymeric [RR'Si−O−] "network" and different organic groups R and R' (see Section 3.1.4.2). Several types are given

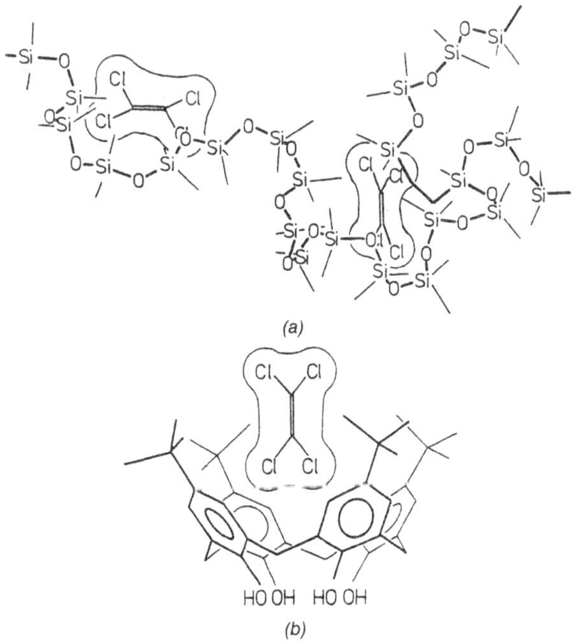

(a)

(b)

FIGURE 3.31. Typical recognition sites for organic solvent molecules (here: perchloroethylene, C_2Cl_4) in (a) poly(dimethylsiloxane) and in (b) *tert*-butyl-calix[4]arene.

in Table 3.9. The geometric structure of the simplest polysiloxane with $R = R' = CH_3$, i.e., poly(dimethylsiloxane) (PDMS), is well known [103]. At very low temperatures, short-range order forces lead to a helical conformation of the $Si-O$ chain over several atomic distances. However, the geometric structure on a larger scale and at higher temperatures may more adequately be described with a "worm-like" model of the individual polymer chains. This takes into consideration the high torsional flexibility of the $Si-O$ bond, which leads to very low glass and melting temperatures ($T_g = -125\,°C$ and $T_m = -40\,°C$) of PDMS [104]. The geometric structure of the polysiloxane matrix offers a statistical time-dependent structure and, hence, a distribution of recognition "sites" for organic molecules. The sites are formed by the short-range ordering of small intercepts of the $RR'Si-O$ chains around an individual solvent molecule.

As a "model system" of *supramolecular compounds,* we consider calix-arenes, which consist of several aromatic rings (in particular, four rings with *tert*-butyl and *iso*-propyl groups in para-position of the phenolic OH group are considered here; compare Table 3.9), condensed with methylene ($-CH_2-$) groups to form macrocycles of different sizes [105]. For the relatively small calix[4]arene molecule, the geometry of the recognition site does not change significantly upon interaction with guest molecules. To a good approximation, this compound has a time-independent structure. Since the "calyx" of calix[4]arenes has a well defined, but small, size and a limited torsional flexibility of the dihedral angle around the methylene groups, these molecules are expected to exhibit a pronounced shape selectivity for the incorporation of different solvent molecules.

The larger sizes of calix[6]arenes and calix[8]arenes, in contrast, are expected to form complexes with more than one of the small solvent molecules considered in the following and, hence, are expected to exhibit less selectivity. This situation may be considered as an "in between case" between the well-defined "static" structure of recognition sites in calix[4]arenes and the "dynamic" structure of recognition sites in polysiloxanes.

As mentioned previously, the interaction between recognition sites of the sensitive layer and the molecules to be detected adjusts the concentration at surface and bulk sites. Surface and bulk contributions may be separated quantitatively if the sheet concentrations, c_{sh}, of the molecules are determined for different thicknesses of the sensitive layers. If a number N of adsorbed and/or absorbed molecules is determined experimentally, the value of c_{sh} may be calculated from:

$$c_{sh} = N/A \qquad (3.40)$$

Here A denotes the area of the sensitive layer. Equation (3.40) yields values of c_{sh} with the unit particle concentration per area. The number $N = \Delta m \cdot N_L/M$ (where N_L is the Loschmid number and M is the molecular weight of the molecules to be detected) can be determined with measurements of the corresponding increase of mass, Δm, of the sensitive layer upon adsorption or absorption, e.g., by means of quartz microbalance oscillators.

For combined adsorption-absorption processes, the sheet concentration, c_{sh}, is a linear function of the thickness, d, of the layer [106]:

$$c_{sh} = c_{(s)} + dc_{(b)} \qquad (3.41)$$

Then, surface concentrations $c_{(s)}$ and bulk concentrations $c_{(b)}$ may be determined from the intercept and the slope in a $c_{sh} - d$ plot, respectively. (For nonreactive transducer surfaces, the enrichment of molecules at the layer/transducer interface may be neglected.) This approach will be applied to polysiloxane and calixarene layers next.

3.5.1.2 BULK ABSORPTION OF MOLECULES IN POLYSILOXANE LAYERS

A typical result of a $c_{sh} - d$ plot of poly(dimethylsiloxane) (PDMS) interacting with organic solvent molecules (here, perchloroethylene, C_2Cl_4) is shown in Figure 3.32 [102]. The straight line with an intercept of zero indicates, first, that the interaction of C_2Cl_4 with PDMS layers leads to their bulk absorption (i.e., the number of C_2Cl_4 molecules increases linearly with the PDMS thickness, and hence the bulk concentration $c_{(b)}$ is constant) and, secondly, that surface adsorption may be neglected (i.e., the surface concentration $c_{(s)}$ of C_2Cl_4 is zero).

The bulk absorption of organic solvent molecules leads to their enrichment in the polymeric phase with respect to the gas phase. This may be described by a temperature-dependent equilibrium constant, which is denoted as a partition coefficient [see Equation (3.32)]:

$$K = c_{(b)}/c_g \qquad (3.42)$$

i.e., the ratio of the concentration of molecules in the polymer $c_{(b)}$ and the gas phase c_g.

This effect can be understood in the framework of classical equilibrium thermodynamics if equilibrium bulk concentrations of the molecules "i" are adjusted in the polymer layers at constant partial pressures, p, and constant temperature, T [102]. Under these conditions, K is determined by the changes in the standard Gibbs enthalpy ΔG_S^o (here and in the following,

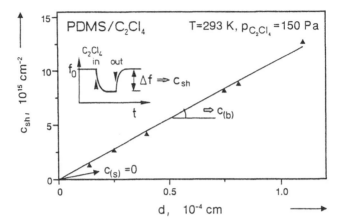

FIGURE 3.32. Sheet concentrations, c_{sh}, of dissolved C_2Cl_4 molecules (molecules per area) in PDMS layers as a function of thickness, d. The values are obtained from the frequency change, Δf, of a quartz microbalance oscillator coated with a PDMS layer of the thickness, d, in a gas exposure experiment with a constant partial pressure $p_{C_2Cl_4} = 150$ Pa and temperature $T = 293$ K (see inserted figure). $c_{(b)}$ and $c_{(s)}$ denote bulk and surface concentrations, respectively.

only molar quantities are considered) during the bulk absorption (i.e., the dissolution) of molecules "i" in the polymeric matrix.

Since changes in the Gibbs enthalpy ΔG_S^o may be split into changes in the enthalpy ΔH_S^o and entropy ΔS_S^o, which result from the different molecular interactions between dissolved molecules and polymeric macromolecules,

$$\ln K = -\Delta G_S^o/RT = -\Delta H_S^o/RT + \Delta S_S^o/R \qquad (3.43)$$

holds. In addition, ΔH_S^o can be split into changes of the inner energy ΔU_S^o and the volume work $p\Delta V/RT$ (Here it is assumed that the ideal gas law holds for small concentrations of organic molecules):

$$\ln K = -\Delta U_S^o/RT - p\Delta V/RT + \Delta S_S^o/R \qquad (3.44)$$

The dissolution includes different steps, i.e., condensation, mixing, and, if polymers swell simultaneously, elastic volume expansion. Each step leads to specific contributions to the overall changes ΔU_S^o and ΔS_S^o.

Equation (3.44) may be simplified for organic solvents if interaction energies between the solvent molecule and the polymeric macromolecule are, to a first approximation, considered to be identical with interaction energies between solvent molecules in the liquid state. It this case, ΔU_S^o is equal to the energy of condensation of the solvent. If, in addition, changes of ΔS_s^o are assumed to be identical for different solvents, then

$$\ln K = \alpha T_b/T \qquad (3.45)$$

holds [102]. Here, T_b and T denote the boiling and the absolute temperature, respectively. Particle/polymer systems that obey Equation (3.45) will be denoted at "T_b-controlled" in the following. Characteristic deviations from this simple logarithmic relationship result, for example, from characteristic functional groups in polymers. The latter may lead to different molecular interactions such as induced dipole/induced dipole, dipole/induced dipole, and dipole/dipole interactions, as well as interactions due to basicity and acidity (see Section 3.4).

3.5.1.3 SURFACE ADSORPTION AND BULK ABSORPTION OF MOLECULES IN CALIXARENE LAYERS

In contrast to polysiloxane layers interacting with organic solvent molecules, significant surface effects are found in calixarene layers. This follows quantitatively from an evaluation of the thickness-dependent sheet concentrations, c_{sh}, for different thicknesses, d, of the calixarene layers. A typical result obtained for *iso*-propyl-calix[4]arene is shown in Figure 3.33. Bulk and surface concentrations, $c_{(b)}$ and $c_{(s)}$, of C_2Cl_4 molecules are determined from the slope and the intercept in this figure. The value $c_{(s)}$ is then the layer concentration in the bulk as calculated from $c_{(b)}{}^{2/3}$. This effect indicates the favoured accessibility of calixarenes at surface sites if compared with bulk sites. However, at larger thicknesses the overall sensor signal is

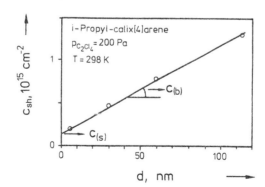

FIGURE 3.33. Sheet concentration, c_{sh}, of C_2Cl_4 molecules (molecules per cm²) in *iso*-propyl-callix[4]arene as a function of thickness, d, at $p_{C_2Cl_4} = 200$ Pa and $T = 298$ K.

mainly determined by the bulk since the number of bulk sites increases linearly with the film thickness.

Theoretical approaches to determine the calixarene/molecule interactions and, hence, selectivities and sensitivities of chemical sensors make use of force-field calculations. Typical results of binding energies, which are obtained with the energy routine of the SYBYL software [107], will be given in Section 3.5.2.2. A typical result of a geometric structure of an individual calixarene/molecule complex by one C_2Cl_4 molecule and the molecular cavity of *tert*-butyl-calix[4]arene has already been shown in Figure 3.31(b). Here, the different atoms are represented by their van der Waals spheres. This example clearly demonstrates the "key-lock" interaction, which is a prerequisite for the development of highly selective chemical sensors based upon cage compounds.

In contrast to calix[4]arenes, force-field calculations on calix[6]arene and calix[8]arene show that the larger molecular cavities of these compounds can take up more than one solvent molecule. This leads to more complicated adsorption behaviour because of coverage-dependent solvent/solvent interactions and pronounced dynamic changes of geometric cage structures.

3.5.1.4 SURFACE ADSORPTION OF ORGANIC SOLVENT MOLECULES IN HÖGBERGARENE LAYERS

A recent approach to exclusively utilize surface effects for the detection of molecules is based on self-organized monolayers of högbergarenes. This is schematically shown in Figure 3.34. Here, the högbergarene macrocycles provide molecular recognition sites for C_2Cl_4 molecules. The covalent coupling to the gold surface (of, for example, the electrode of a quartz microbalance sensor) results from strong S-Au bonds between the sulphide groups, which terminate the aliphatic chains. The latter may be considered as a "spacer" between the gold surface and recognition sites. Alternatively, högbergarenes with thiole groups may be used [108].

3.5.2 Specific Sensor Responses of Polymer-Based Chemical Sensors

3.5.2.1 SURVEY OF THE SENSING PRINCIPLES

The partition coefficients, K, determine to a certain extent sensitivities to monitor different solvents by means of Δm, ΔC, ΔT, and Δd effects. However, for many transducers additional parameters are required to adequately describe their sensor responses as a function of the concentration in the gas phase.

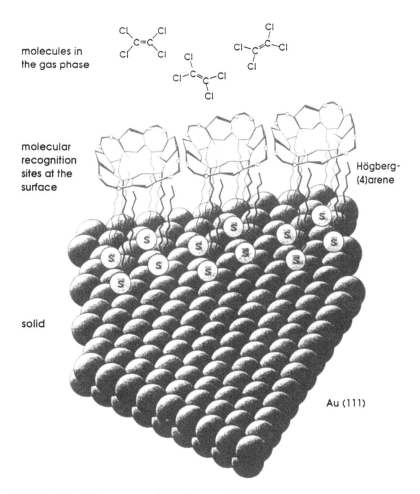

FIGURE 3.34. Högbergarene-alkylsulfide monolayer on a gold surface [here, the close-packed (111)-surface was indicated]. Due to the geometric "architecture" of the monolayer, the högbergarene macrocycles are very accessible for molecules (here: C_2Cl_4 from the gas phase.

For measurements of Δm with *quartz microbalance oscillators,*

$$\Delta m = (KMV/N_L) \cdot c_g \qquad (3.46)$$

holds to a first approximation with M as the molar weight of the solvent, V as volume of the polysiloxane layer, and N_L as the Avogadro number. Here, Δm denotes the mass change adjusted under equilibrium conditions. For this transducer principle, Δm is proportional to the concentration, c_g, of solvent molecules in the gas phase with their molecular weights, $M,$ and partition coefficients, K, as a parameter that desrcibes molecular properties. Quartz microbalance sensors convert the Δm effects to a first approximation linearly into frequency changes, Δf, according to $\Delta m = -\Delta f A C_f^{-1} f_0^2$ (here A denotes the coated area, C_f the mass sensitivity, and f_0 the fundamental frequency; see Section 2.3.1). However, the simple Equation (3.46) has to be modified to describe sensor responses of other mass-sensitive transducers (see Sections 2.3.2 and 2.3.3).

For measurements of ΔT with *thermopiles* [see Figure 3.30(a)], kinetic treatments of sensor gas/interactions are adequate to describe sensor responses. If thermopiles are exposed to rectangular-shaped concentration variations from 0 to c_g,

$$\Delta T(t) = \frac{1}{m \cdot c_{mass}} \cdot \int_0^t \left[\frac{dq^+}{dt} - \frac{dq^-}{dt} \right] dt \qquad (3.47)$$

holds with c_{mass} as the specific heat coefficient and m as the mass of the thermopile (in a first approximation, the contribution from the relatively thin polymer layer to both c_{mass} and m may be neglected [100]). The term $\Delta T(t)$ denotes the temperature change at time t ("transient sensor response") of thermopile sensors. In particular, maximum temperature changes, Δt_{max}, result from the compromise between the time-dependent heat production dq^+/dt and heat loss dq^-/dt (per second). Under chopped-flow conditions, the heat production per second may be calculated from

$$dq^+/dt = \Delta H_S^o \cdot KV \cdot 2 \cdot Dd^2 \sum_{n=0}^{\infty} \exp[-(2n + 1)^2 \pi^2/4 \cdot Dt/d^2]$$

$$(3.48)$$

For the heat loss per second,

$$dq^-/dt = \alpha_{ht} A \Delta T \qquad (3.49)$$

holds. In Equation (3.48), ΔH_S^o denotes the molar heat of absorption and D the bulk diffusion coefficient of the solvent in the polysiloxane layer with a thickness d. The heat transfer to the environment is assumed, in Equation (3.49), to be proportional to the temperature difference, ΔT, with α_{ht} as specific heat transfer (with the unit $Jm^{-2}K^{-1}s^{-1}$) and A as area. Equation (3.49) holds for "ideal" thermopiles with negligible heat transport along the thermocouples. However, ΔT-effects of thermopiles shown in Figure 3.30(a) may be significantly smaller. The molecular parameters for measurements of ΔT are molar heats of absorption, ΔH_s^o, diffusion coefficients, D, and partition coefficients, K, of solvent molecules. Thermopile transducers convert the ΔT effects to a first approximation linearly into thermo voltages, V_{thermo}.

For measurements of ΔC with *interdigital capacitors* [see Figure 3.30(b)], corresponding relationships contain contributions from changes in the relative dielectric coefficient, ϵ, of the polymer and/or from changes in the effective thickness, d, of the polymeric film [29] (see Section 3.1.4.2). The latter results from a temperature-and partial pressure–dependent swelling of the polymer film. This swelling can be made use of to design optical sensors for the determination of solvent molecules in the gas and liquid phase [101]. As a result, ΔC depends on electrode and layer thicknesses, on the permittivity of the polysiloxane, dipole moments μ, and polarizabilities α, of dissolved solvent molecules, and frequencies ν that are used to determine ΔC and, in addition, partition coefficients K. Corresponding relationships to describe sensor responses with molecular parameters are complex and will not be treated here.

For measurements of Δd with *optical transducers* [see Section 2.6.4 and Figure 3.30(c)] as they are reported [101],

$$\Delta d = sKc_g \tag{3.50}$$

holds with s as an empirical parameter, which is determined by the volume expansion upon dissolution of solvent molecules. The value of s depends mainly on the number of cross-links of the polymeric network structure.

3.5.2.2 QUARTZ MICROBALANCE TRANSDUCERS: MASS CHANGES AS SENSOR RESPONSES

The bulk dissolution of organic molecules into the polysiloxane layers leads to changes of Δf in the fundamental oscillation frequency f_0 of the quartz microbalance sensors. Typical sensor signals, Δf, upon exposure to different organic solvent molecules like n-octane (C_8H_{18}), methylcyclohexane (Me-C_6H_{11}), perchloroethylene (C_2Cl_4), chloroform ($CHCl_3$), tetrachloromethane (CCl_4), toluene (Me-C_6H_5), *iso*-propylether (i-Pr$_2$O), ethanol (EtOH), n-propanol (n-PrOH), *iso*-propanol (i-PrOH), n-butylamine (n-BuNH$_2$), and acetone (Me$_2$CO) coated with different modified polysiloxanes (for abbreviations, see Table 3.9) are shown in Figure 3.35 [109]. Here, the time-dependence of Δf for identical concentrations of different organic molecules is plotted at a constant temperature $T = 303$ K. Evidently, most of the molecule/polysiloxane interactions are completely reversibe with response and decay times in the order of minutes at $T = 303$

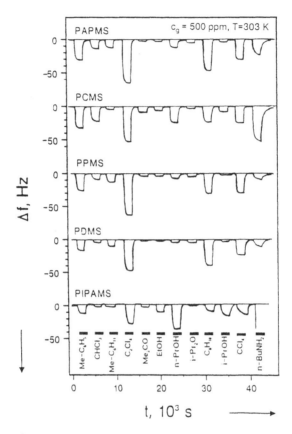

FIGURE 3.35. Typical frequency changes of quartz microbalance sensors coated with modified polysiloxanes (see Table 3.9) as a function of time t for an exposure to $c = 500$ ppm concentration of different vapours in synthetic air.

K. One exception is n-butylamine, which leads to partly or completely irreversible [for poly(iso-propylanoic-acid methylsiloxane)] sensor responses.

Bulk concentrations, $c_{(b)}$, of dissolved molecules and, hence, partition coefficients, K, are approximately independent of concentrations, c_g, at constant temperature, i.e., Henry's law holds. Even if two solvent components are present, the individual values of K remain constant in most cases.

As shown in Figure 3.36, a T_b-controlled dissolution with a linear relationship $\ln K \propto T_b/T$ is found for a variety of polymer/solvent systems. One typical example is PDMS interacting with nonpolar molecules. However, deviations occur for strong polar alcohols, ethers, and ketones with characteristic exceptions: if polar bonds are completely shielded by bulky groups, e.g., in di-*iso*-propylether (here the polar $C-O$ bonds are shielded by two *iso*-propyl groups), polar molecules behave like nonpolar molecules, and their dissolution in PDMS is T_b-controlled.

The influence of the different functional groups in modified polysiloxanes is obvious. In order to characterize unequivocally the influence of different functional groups on solubilities of different organic molecules in the poly-siloxane matrix, we normalize all partition coefficients and represent the data of poly(dimethylsiloxane) (PDMS), poly(cyanopropylmethylsiloxane) (PCMS), and poly(phenylmethylsiloxane) (PPMS) in a three-dimensional plot. This also makes it possible to describe quantitatively the selectives of sensor arrays. For example, the components, K_O, of corresponding three-dimensional normalized vectors are calculated from the partition coefficients, K, in PDMS, PCMS, and PPMS according to

$$K_O(PCMS) = K(PCMS)/\sqrt{K(PCMS)^2 + K(PDMS)^2 + K(PPMS)^2}$$
$$(3.51a)$$

$$K_O(PPMS) = K(PPMS)/\sqrt{K(PCMS)^2 + K(PDMS)^2 + K(PPMS)^2}$$
$$(3.51b)$$

$$K_O(PDMS) = K(PDMS)/\sqrt{K(PCMS)^2 + K(PDMS)^2 + K(PPMS)^2}$$
$$(3.51c)$$

Results are shown in Figure 3.37 for the different organic molecules.

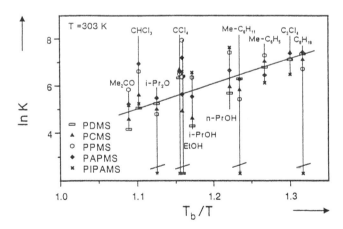

FIGURE 3.36. Logarithm of the partition coefficient, K, of different solvent molecules in PDMS, PCMS, PPMS, PAPMS, and PIPAMS as a function of T_b/T where T_b is the boiling temperature of solvents and $T = 303$ K as a measuring temperature. The solid line represents the T_b-controlled dissolution.

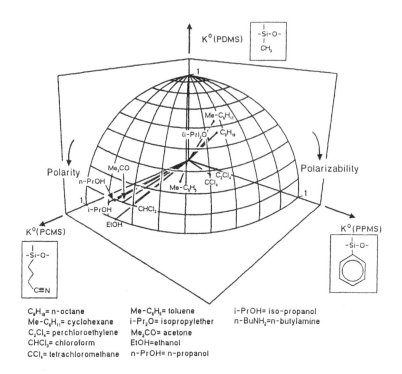

C$_8$H$_{18}$= n-octane   Me-C$_6$H$_5$= toluene   i-PrOH= iso-propanol
Me-C$_6$H$_{11}$= cyclohexane   i-Pr$_2$O= isopropylether   n-BuNH$_2$=n-butylamine
C$_2$Cl$_4$= perchloroethylene   Me$_2$CO= acetone
CHCl$_3$= chloroform   EtOH=ethanol
CCl$_4$= tetrachloromethane   n-PrOH= n-propanol

FIGURE 3.37. Three-dimensional plot of normalized vectors that correspond to the partition coefficients in PCMS, PPMS, and PDMS. The end points of each vector are positioned at a spherical surface, which is represented by azimuth and longitudinal lines and may be characterized by their polar coordinates. The small figures indicate monomeric units of the different polysiloxanes with the polar cyano groups in PCMS, the polarizable aromatic ring in PPMS, and the methyl group in PDMS. (Further explanations are given in the text.)

Polar molecules like ethanol (with a dipole moment $\mu = 1.66$ D), n-propanol ($\mu = 3.09$ D), iso-propanol ($\mu = 1.66$ D), acetone ($\mu = 2.69$ D), and chloroform ($\mu = 1.15$ D) are detected more selectively with PCMS [110]. (D is the Debye unit of dipole moment, 1 D $= 3,3456 \cdot 10^{-30}$ Asm.) This results from the interaction with the polar cyano groups.

Nonpolar or weakly polar molecules with polarizable atoms or groups like perchloroethylene, tetrachloromethane, and toluene are more selectively detected with both PCMS and PPMS, due to dipole/dipole and dipole/induced dipole interactions with the polar cyano group and the polarizable phenyl group. Here, the polarizability of the aromatic ring arises from π-electrons. Non-polar molecules like n-octane and methylcyclohexane are detected more selectively with PDMS. Evidently, di-iso-propylether behaves like nonpolar molecules. Polysiloxanes may also be modified with "basic" and "acidic" functional groups like in poly(amino-propylmethylsiloxane) (PAPMS) with *iso*-propyanoic acid groups and poly(*iso*-propylanoic-acid methylsiloxane) (PIPAMS) with propylamino groups. For a detailed discussion, see Reference [109].

These results demonstrate that modified polysiloxanes with different functional groups may be used to advantage in sensor arrays based upon quartz microbalance oscillators to identify organic molecules. Since at low concentrations these partition coefficients are independent for different organic molecules, even solvent mixtures may be identified with these sensor arrays.

Modified polysiloxanes may also be applied as layers of quartz microbalance sensors for the detection of inorganic gases like CO_2 (with γ-aminopropylethoxysilane/propylmethoxysilane-copolymer, PAPPS, see Figure 3.38 [111]) or for the detection of hydrocarbons in water [with poly-etherurethane (PEU), see Figure 3.39] [112].

The specific "key-lock" interactions between calix[4]arenes and solvent molecules may be monitored with quartz microbalance sensors. Typical results obtained for different thicknesses, $6 \leq d \leq 114$ nm, of *iso*-propyl-calix[4]arene layers are shown in Figure 3.40 [113]. The time dependence indicates fast initial adsorption at the surface and a subsequent slow diffusion of C_2Cl_4 molecules in the bulk. The incorporation of C_2Cl_4 into calix-

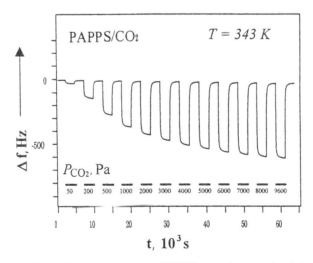

FIGURE 3.38. Typical frequency changes of PAPPS-coated quartz microbalance sensors as a function of time for stepwise exposures to various concentrations of CO_2 in air. (This sensor is operated at an elevated temperature $T = 334$ K in order to achieve short response and decay times.)

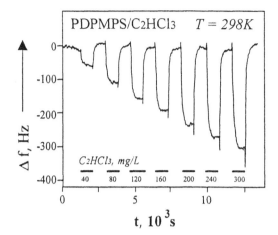

FIGURE 3.39. Typical frequency changes of PDPMPS-coated quartz microbalance sensors as a function of time for stepwise exposures to different concentrations of trichloroethylene (C_2HCl_3) dissolved in water. Here, the frequency is determined at the maximum of the resonance curve by means of an impedance analyzer. (For further details, see Reference [112].)

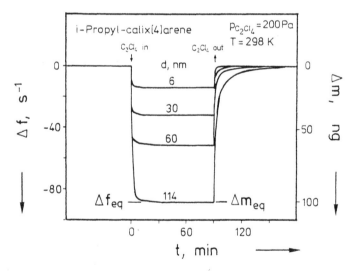

FIGURE 3.40. Typical responses of quartz microbalance sensors coated with different thicknesses, d, of iso-propyl-calix[4]arene as a function of time after exposure with $p_{C_2Cl_4} = 200$ Pa in air at $T = 298$ K.

166

[4]arene macromolecules at bulk sites leads to the pronounced increase of sensor signals, Δf, for larger layer thicknesses, d.

A correlation between theoretical binding energies may be obtained from force-field calculations and the experimental partition coefficients, K (see Figure 3.41) [113]. Here, binding energies are determined as differences of total energies (i.e., the sum over bond stretching energies, E_{str}; angle bending energies, E_{bend}; out-of-plane bending energies, E_{oop}; torsional energies, E_{tors}; van der Waals energies, E_{vdw}; and electrostatic energies, E_{elec}) of the calixarene/molecule complex and the "empty" calixarene. Electrostatic energies are calculated from partial charges, as determined at the different atomic positions on a semi-empirical level using, for example, the PM3 method of the MOPAC package [114]. The latter were determined from the experimental bulk concentrations $C_{(b)}$ (compare Figure 3.33). The partition coefficients are given by the Gibbs enthalpy with its enthalpy and entropy contributions [see Equation (3.44), which can also be applied here].

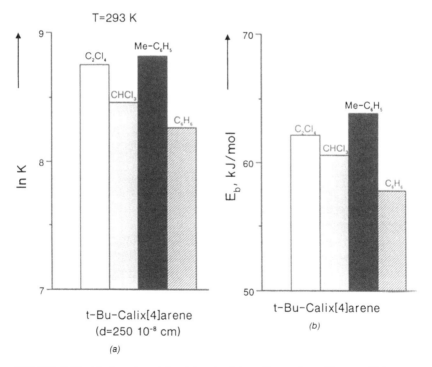

FIGURE 3.41. (a) Experimentally determined logarithms of partition coefficients at $T = 293$ K for different solvent molecules in *tert*-butyl-calix[4]arene. (b) Theoretically determined binding energies E_b of the same solvents in *tert*-butyl-calix[4]arene.

3.5.2.3 CALORIMETRIC TRANSDUCERS: TEMPERATURE CHANGES AS SENSOR RESPONSES

Temperature changes during the dissolution of solvent molecules in polysiloxanes are utilized in calorimetric sensors like thermopiles. However, since the formation of heat is zero under equilibrium conditions, ΔT effects are only monitored during pressure variations with $dc_g/dt \neq 0$ [100]. With an experimental set-up that produces fast and rectangular-shaped partial pressure variations, an unequivocal determination of sensor signals (such as the maximum temperature change) is possible. A typical result obtained with PDMS-coated thermopiles exposed to C_2Cl_4 is shown in Figure 3.42. The temperature changes, ΔT_{max}, depend linearly on partial pressures, $p_{C_2Cl_4}$, during a constant absorption "bath" temperature of the outer thermopiles. The response times, t_{max}, to reach ΔT_{max} depend only on the diffusion coefficients of the solvent molecules.

3.5.2.4 CAPACITANCE TRANSDUCERS: CHANGES OF DIELECTRIC PROPERTIES AND THICKNESSES AS SENSOR RESPONSES

Capacitance sensors for detecting molecules may be used if the interaction between sensitive layers and molecules changes the dielectric properties and/or the thickness of the layers [29]. As shown in Figure 3.43 [and according to Equations (2.1) and (2.2)], the total capacitance of an interdigital capacitor (IDC) can be separated into contributions from the gas phase (index g), the substrate (index s), and the polymer (index p) with

$$C = C_g + C_s + C_p \qquad (3.52)$$

The three capacitances have three different permittivities $\epsilon_i (i = g, s, p)$ characteristic of the gas phase, the substrate, and the polymer. The most pronounced response of capacitance sensors is expected for strong polar molecules like SO_2 with large dipole moments that interact with the polymer films, which leads to a change in the permittivity, ϵ_p, of the polymer and, hence, to a sensor response, $\Delta C = \Delta C_p$. In contrast, ϵ_g in the gas phase and, hence, the contribution of C_g is practically constant, even for large variations of the partial pressure of molecules with high dipole moments. In addition, the capacitance, C_s remains constant for inert substrates like glass.

As shown in Figure 3.43 and Equation (2.2), capacitance changes determined with interdigital electrode devices contain contributions from changes in the relative permittivity of the polymer and/or from changes in

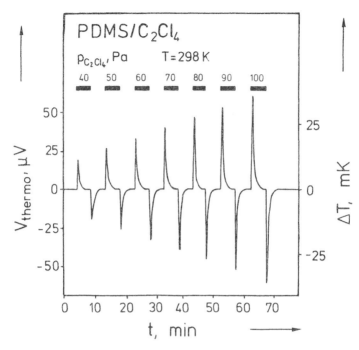

FIGURE 3.42. Typical responses (V_{thermo} α ΔT) of the thermocouples of calorimetric thermopile sensors coated with PDMS as a function of time for stepwise exposures to different partial pressures of C_2Cl_4 in air. (V_{thermo} is the voltage signal of the thermopile.)

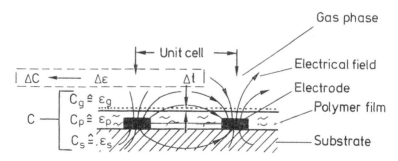

FIGURE 3.43. Schematic representation of a "unit cell" of polymer-coated interdigital structures with the three permittivities.

169

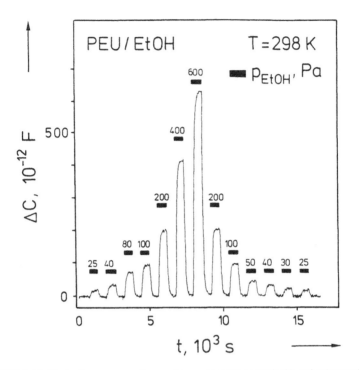

FIGURE 3.44. Typical responses of capacitance sensors coated with polyetherurethane as a function of time for stepwise exposures to different partial pressures of ethanol in air.

the effective thickness, here t [according to Equation (2.2)], of the polymeric film [29]. Molecules with lower or even with zero permanent dipole moment may also be detected if a significant "swelling" of the polymer occurs during the absorption of these molecules and if permittivities, ϵ_p of the polymer are large. As typical examples, polysiloxane with polar groups show large permittivity values (see also Section 3.1.4.2).

However, this sensor principle may be applied favourably for the detection of polar molecules. A typical result for the sensitive detection of n-propanol with polyetherurethane is shown in Figure 3.44. Equilibrium values are obtained after response times in the order of minutes. The ΔC values are completely reversible for these systems. A linear relation between ΔC and the partial pressure, p_{EtOH}, has been found within the experimental error.

3.5.2.5 OPTICAL TRANSDUCERS: CHANGES OF THICKNESSES AS SENSOR RESPONSES

As previously has been reported by Gauglitz et al., thickness changes,

Δd, result from a temperature- and partial pressure–dependent swelling of polymeric films [101]. This swelling can be made use of designing optical sensors for the sensitive determination of solvent molecules in the gas and liquid phase. Linear calibration curves ($\Delta d \alpha C_g$) have been found for many solvent molecules.

From the spectrally dependent optical reflection monitored with a fast diode–array spectrometer, changes in the optical thickness, nd, of the transparent thin polymer films are determined (see Section 2.6.4 and Figure 2.28). Hence, n and d denote the refraction index and the physical film thickness, respectively. From the wavelength dependence of the interferences, changes in the refractive index of the polysiloxane films can be excluded from contributing to the sensor signal for the specific gases [101].

Changes in the thickness, i.e., swelling of polymer membranes in contact with organic solvents is a well known phenomenon observed, e.g., during the transport of neutral molecules through membranes. For certain polymer/solvent systems, e.g., polydimethylsiloxane/chlorinated hydrocarbons, the swelling is reversible if only low concentrations of organic molecules are adjusted in the polymer matrix. By measuring the shift in the intensity maxima formed by constructive interference of both the incident and the reflected light beams in the polysiloxane film, the thickness changes, Δd, can be determined sensitively. Hence, measurements of the relative thickness changes, $\Delta d/d$, may lead to very low detection limits in the ppm range in the gas phase for certain organic solvents such as C_2Cl_4. Typical results are shown in Figure 3.45 for PDMS-coated optical transducers. A

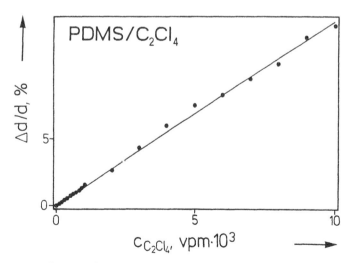

FIGURE 3.45. Relative thickness change, $\Delta d/d$, vs. a concentration of C_2Cl_4 in the range of 10–10^4 vpm.

remarkable linearity and a complete reversibility of the sensor signal during repeated test cycles are found. These optical transducers may also be applied to determine organic hydrocarbon compounds in water [115].

In conclusion, one can see that the experimental and theoretical work on polymeric and supramolecular compounds indicates recent approaches to design highly selective chemical sensors. The latter is based on different transducer principles to monitor the change of physical-chemical properties of the selective layer upon interactions with molecules to be detected in air and water. In particular, changes of mass (Δm), temperature (ΔT), capacitances (ΔC), and thicknesses (Δd) may be monitored—each of them providing specific advantages and disadvantages in practical applications. A comparison of the different transducer principles in view of practical applications is given in Reference [116].

Since supramolecular chemistry aims at the preparation of molecules with specific binding sites inside their "cavities," even very complicated supramolecular recognition structures may be available in the future. In addition, techniques for the preparation of self-organized layers with a well-defined "architecture" have been developed (see Figure 3.34). This will make it possible to design completely new highly specific chemical sensors, which exclusively utilize very fast adsorption processes between recognition sites at the surface of these monolayers and the molecules to be detected.

3.6 POLYMERIC SEPARATION MEMBRANES

During the first century (1855–1955) of their existence, polymeric membranes performed a useful, but limited, role as research tools of microbiologists, analytical chemists, and physicians. Principal concerns were with filtration of colloids, proteins, blood sera, enzymes, toxins, bacteriophages, and other viruses. The 1960s and 1970s witnessed the renaissance of interest in membrane separation processes to the extent that their large-scale application in the water treatment, food, and medical industries appeared imminent. In the last two decades, the rapid development of chemical and biosensors have given another driving force to the investigations.

Polymer membranes can be applied to gaseous and liquid-state separation of different constituents, which can be used not only in industrial and medical cleaning processes, but in analytical chemistry and even in the field of sensors with success. A few examples from the practical utilization are as follows:

- gas separation membranes in gas sensors to improve their selectivity to certain gas constituents

- ion separation membranes in ISEs and ISFETs to improve their selectivity to a certain ion type
- separation of dissolved gas components from solvents for measuring the partial pressure of dissolved gases
- diffusion membranes in amperometric electrochemical cells

3.6.1 Permeability and Permselectivity

Since the principal function of a polymeric membrane is to act as a variable resistance to the passage of permeating species, its most important characteristics, to which all others must be considered secondary, are permeability and permselectivity. The former is a measure of the rate at which a given species permeates a polymeric barrier, and the latter is a measure of the rates of two or more species relative to one another. Different techniques have evolved for measuring these quantities in the various membrane separation processes, which are described in this section. The following considerations are valid mainly for the gas permeation, but they can be easily applied for the other process types.

Gas permeation is a function of two factors: diffusion and solubility. Because the solubility of various gases within a polymeric membrane generally varies by less than two orders of magnitude whereas the permeability can vary by as many as five orders of magnitude, it is apparent that gas permeation is primarily a diffusion-controlled process.

The permeability constant, P_m, which is characteristic for a given polymer-permeant system, can be defined as the ratio of the flux of the gas or vapour to the pressure gradient across the thickness of the membrane:

$$J/A = P_m \Delta p / t \tag{3.53}$$

where

Δp = the pressure or partial pressure difference
t = the thickness of the membrane
J = the material current
A = the area of the membrane's surface

P_m can be separated into the diffusion and partition coefficients according to the model shown in Figure 3.46. The two compartments with pressures (or partial pressures) p_1 and p_2 are separated by the membrane. The equilibrium distribution of a vapour between gas phase and a sorbent polymer is given by the partition coefficient defined by Equation (3.32). Accordingly

$$p_{1M} = Kp_1 \qquad \text{and} \qquad p_{2M} = Kp_2 \tag{3.54}$$

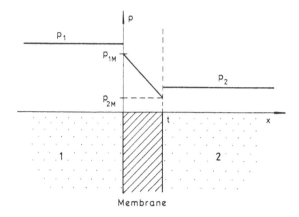

FIGURE 3.46. Schematic model of gas permeation through polymeric membranes.

where p_{iM}'s are the (partial) pressures in the boundary regions of the membrane. The particles are transported through the membrane because of the diffusion caused by the pressure (concentration) difference. According to the Fick's law [117]

$$J/A = D(p_{1M} - p_{2M})/t \qquad (3.55)$$

where D is the diffusion coefficient.

Taking the expressions of Equations (3.54) into Equation (3.55)

$$J/A = DK(p_1 - p_2)/t \qquad (3.56)$$

thus, the permeability [118]

$$P_m = DK \qquad (3.57)$$

Permselectivity can be expressed as a separation factor, α_{AB}, that is, the ratio of permeabilities for components A and B [119]

$$\alpha_{AB} = P_{mA}/P_{mB}$$

$$(3.58)$$

The unit of the permeability coefficient is [mol m/s m² Pa] = $2.99 \cdot 10^5$ [(cm³, STP)/(cm²s Hgcm)]. The latter is often used in the literature.

As for all membrane separations, gas and vapour permeability is a function of membrane properties, the nature of the permeant species, and the interaction between membrane and permeating species. Expressed symbolically, the permeability constant [119]

$$P_m = X \, (\text{membrane}) \, Y \, (\text{gas}) \, \Theta \, (\text{membrane-gas}) \qquad (3.59)$$

where X is a function of the membrane's chemical and physical properties, i.e., structure on the molecular, microcrystalline, and colloidal levels, and Y is a function of gas properties, such as critical temperature, polarity, and steric factors. Θ is a correction function due to the interaction of membrane with the permeant species.

Gas and vapour permeation processes can be divided into several categories, depending upon the extent of membrane-gas (or membrane-vapour) interaction. The simplest case is that of the permanent gases, which have virtually no interaction with the membrane. More complicated are the condensable vapours, which by their very nature tend to interact more strongly. A distinction may be made between those vapours that interact with (swell) the membrane more strongly than with themselves and those vapours that do not have much affinity for the membrane but tend to form clusters.

Permeant polymeric membranes can be grouped into two main categories:

- dense or homogeneous film-type membranes through which gases can permeate by segregation, absorption, and diffusion (see the former model of Figure 3.46), for example, silicone rubber, PTFE, polyethylene, Mylar® (DuPont), etc.
- porous or heterogeneous membranes through which the gases can permeate in the gas phase by ultrafiltaration (see Section 3.6.2), for example, capillary membranes: cellulose acetate, PTFE, and PVC.

Variations in temperature have a great many largely interrelated effects upon membrane separation processes. In addition to the increased motion of gas or solvent and solute molecules and decreasing interactions between them, the membrane can undergo temperature-dependent changes on the molecular, microcrystalline, and colloidal levels. Membrane-solvent and membrane-solute interactions are also affected. It should be noted, however, that the partition coefficient can either increase or decrease with increasing temperature. Generally, the temperature dependence of the permeability constant can be expressed as an Arrhenius relationship [12]:

$$P_m = P_{mo} \exp[\, - \, E_p/RT] \qquad (3.60)$$

where E_p is the activation energy of permeation. The diffusion and partition coefficients can also be expressed as Arrhenius functions [see also Equation (3.33)]. Thus,

$$E_p = \Delta G_S^o + E_D \tag{3.61}$$

where E_D is the activation energy for the diffusion process.

The combination of two or more membranes in series results in a composite membrane. The permeability constant, P_m, for a composite membrane may be expressed [120]

$$P_m^{-1} = \sum_i x_i P_{mi}^{-1} t^{-1} \tag{3.62}$$

where

P_m = the overall permeability constant
x_i = the thickness of the individual membranes
P_{mi} = the permeability coefficient of the films
t = the total thickness of the composite structure

The mechanism of permeation is different at the dense and porous membranes. All membranes contain some free space or void volume; however, it is usually not until the fraction of void volume becomes comparable to that occupied by the polymer so that the various microvoids interconnect that the resultant membranes are considered porous.

3.6.2 Separation Process

Investigations of the present section are restricted to the electrically neutral separation processes that are as follows: gas permeation, dialysis, osmosis, reverse osmosis, and ultrafiltration.

Gas permeation can be described by the ideal model (see Section 3.61) in the case of noncondensable gases, when the membrane structure is not perturbed by the dissolved molecules [119]. The low degree of membrane-solute and solute-solute interaction characteristics of permeant gases supports separation primarily on the basis of permeant size.

The permselectivity of polymeric membranes to condensable gases and vapours is more complex, owing to increasing membrane-solute and solute-solute interactions. For gases that tend to interact strongly with the membrane and swell it, their sorption by the membrane is energetically favourable. The solubility factor in permeation becomes of much greater importance than for permeant gases. Depending on the closeness between the solubility parameters of the permeant and the polymeric membrane [119–122]:

- An amorphous membrane will swell (or dissolve if the permeant concentration is high enough).
- A covalently cross-linked polymer will swell to varying extents.
- A semicrystalline polymer (whose crystallites function as virtual cross-links) will swell to varying extents.

The relationship between gas sorption and its partial pressure becomes more complicated since macromolecular displacement in the membrane will be affected by permeant concentration. Mixtures of swelling permeants are even more complex since solute$_a$-solute$_b$ and solute$_b$-membrane interactions can influence solute$_a$-membrane interactions. The permeability of a given membrane to a swelling permeant will be strongly concentration-dependent. The overall permeabilities of membranes to swelling permeants are generally several orders of magnitude higher than to permeant gases. The permselectivity of swollen membranes decreases with an increasing degree of swelling because the increasing interchain displacement lessens the possibility of separations on the basis of size. For practical applications, however, the accompanying gain in permeability may be so large that the loss of selectivity is acceptable (for instance, at PHEMA hydrogels).

A special case of ordinary gas permeation by diffusion is the process known as *liquid permeation,* or *pervaporation,* in which two compartments are separated by the membrane [123]. The first one contains the feed in the liquid state (for example, dissolved gases). The other compartment is maintained at a lower pressure to ensure the absence of liquid.

Dialysis is one of the oldest among the membrane separation processes, having been used to separate colloids from "crystalloids" i.e., low-formula-weight electrolytes [124]. In dialysis, solute from a more concentrated solution permeates a membrane to enter a less concentrated solution (dialysate). It is mainly a diffusion-controlled process; i.e., penetration of the membrane by the solute occurs as a result of molecular motion occasioned by a concentration gradient.

The rate-determining step in dialysis can be either transport across the membrane or transport across the liquid film at the membrane/solution interface. Although liquid-film control of dialysis can be a result of high membrane permeability, it is also sometimes caused by the accumulation of solute of the product side of the membrane.

Osmosis is the phenomenon whereby solvent permeates a semipermeable membrane separating two solutions of different concentrations. A net flow of solvent occurs from the less concentrated to more concentrated solution. In equilibrium state, a pressure difference (the osmotic pressure) tends to stop this flow.

Dialysis differs from osmosis with respect to the nature of the permeant (there is a net flow of solvent rather than solute in the latter) and direction of flow (permeation is from the less concentrated to the more concentrated solution). Whereas semipermeability in osmosis refers to solute retention with solvent permeation, in dialysis it refers to the permeation of some solutes and the retention of others.

The osmotic pressure, π, in dilute solutions can be expressed similarly to the ideal gas law:

$$\pi = \Delta CRT/M \qquad (3.63)$$

where

ΔC = the concentration difference
R = the universal gas constant
T = the absolute temperature
M = the mass of the molecules

When pressure, Δp, in excess of the osmotic pressure is applied to the more concentrated solution, the net flow of solvent is into the more dilute solution, and the process is known as *reverse osmosis*. The flux of water flow through the membrane in reverse osmosis, as in other membrane processes, is a function of both diffusion and solubility factors [125]:

$$J/A = D_w c_w V_w (\Delta p - \pi)/RTt \qquad (3.64)$$

where the index w refers to water. The new symbols are as follows:

c_w = the average concentration of water solvent in polymer
V_w = the partial molar volume of water in polymer

Ultrafiltration is closely related to reverse osmosis. Both processes utilize pressure to force solvent to permeate a membrane while the solute is retained to a greater or lesser extent [126]. The difference between the two processes is ill-defined, and, in fact, they overlap to some extent. Nevertheless, reverse osmosis refers to separations of smaller solute particles from solutions of such appreciable molar concentration that the opposing movement of solvent by osmosis is not negligible. Ultrafiltration, on the other hand, implies the separation of large solute particles from the solvent. The particles may be so large that a suspension, rather than a true solution, is present.

In ultrafiltration, the dispersed phase ("solute" in its most general sense) passes through the membrane less readily than the "solvent" for one of several reasons:

- It is adsorbed on the surface of the filter and in its pores (primary adsorption).
- It is either retained within the pores or excluded from them (blocking).
- It is mechanically retained on top of the filter (sieving).

The effect of surfactants in increasing permeability in ultrafiltration is to be contrasted with their opposite effect in reverse osmosis. Whereas they coat pore walls in ultrafiltration, thereby increasing lubricity, they fail to permeate the tighter reverse-osmosis membranes, and by remaining at the membrane/solution interface, they operate as liquid membranes in series with the underlying solid (gelatinous) membranes.

3.6.3 Behaviour of Polymers

The interactions of the various permeant species with a polymeric membrane may vary from weak to strong. An example of the former is gas permeation, which occurs primarily by a diffusion-controlled mechanism. The slight differences between the permselectivity of the various gases through closely related polymeric membranes (see Table 3.10) are to be contrasted with the 10^5 difference in permeability that exists between PTFE and PDMS (see Table 3.11) [119,127].

Such large differences in permeability are attributable to differences in interchain displacement and flexibility, which, in turn, are causally related to polar and steric effects. Everything else being equal, polar macromolecules such as PTFE will have a stronger tendency to form the rigid associations leading to crystallite formation than nonpolar types. Because of the weak cohesive forces, nonpolar molecules will generally be more flexible and widely spaced.

A crystalline or even a paracrystalline domain is a region of high mole-

TABLE 3.10. *Permselectivities of PDMS and PDMS-Copolycarbonate Related to N_2 Permeability [127].*

	Permselectivity, α_{Gas/N_2}	
Gas	PDMS	PDMS-Copolycarbonate
H_2	2.2	3.0
O_2	2.0	2.3
CO_2	10.8	13.9

TABLE 3.11. Oxygen Permeabilities in Various Polymers [127].

Polymer Membrane	O_2 Permeability 10^{-15} [mol m/s m^2 Pa]
PTFE	0.0013
Mylar®	0.0063
Nylon 6	0.013
PVA	0.033
PVC	0.047
Methyl cellulose	0.233
Cellulose acetate	0.267
PE, high density	0.33
Polystyrene	0.4
Natural rubber	8
Fluorosilicone	36.6
PDMS	167

cular order and dense packing. Permeant molecules are insoluble in polymer crystallites and poorly soluble in paracrystalline domains so that they tend to permeate through disordered regions. The permeation sites may either be amorphous material or other interstices between crystallites. Crystallites and other ordered regions reduce permeability in two ways: by reducing the volume fraction of the membrane available for solution of the permeant species and by forcing diffusion to occur in tortuous paths between and around crystallites.

Where membrane-permeant interaction occurs, the effects of crystalline morphology on permeability are even more pronounced than they are in noncondensable gases, the reason being that crystallites act as virtual crosslinks to impose a swelling limit upon the amorphous regions. Diffusivity, of course, increases greatly with an increasing degree of swelling. Permselectivity is higher in crystalline than in amorphous membranes because of the more limited swelling. A well-known example is polyethylene: permeability values for xylenes, for example, are different by almost one order of magnitude in low-density ($\varrho = 0.918$ g/cm^3) and in high-density ($\varrho = 0.957$ g/cm^3) polyethylene [122].

Annealing a membrane while it is swollen with a permeant can effect up to an order of magnitude increase in permeability with little loss in permselectivity. Moreover, uniaxial stretching below the melting point can result in a significant reduction of swelling with a negligible increase in crystallinity. Although permeability decreases by at least two orders of magnitude, permselectivity increasers dramatically [119].

Plasticization increases permeability because, by reducing cohesive forces between polymer chains, it leads to increased diffusivity. Permeability of PCTFE for N_2, O_2 and CO_2 can be increased by more than one order of magnitude by the application of a plasticizer from the low molecular weight version of the same material [128]. Since copolymerization provides internal plasticization, it also results in a decrease in cohesive energy density (CED) and, hence, increased permeability.

Permeation depends on compatibility of the liquid or gas with the polymer. Permeation of moisture through the hydrophobic polyolefins is extremely slow. Cellulose, however, is compatible with water vapour, and the permeability constant is quite high. Permeability of various polymers to water vapour is illustrated by Table 3.12 [129]. Water vapour can permeate through hydrophilic polymeric membranes as a swelling permeant and through hydrophobic membranes as a clustering permeant. In the former, it leads to increasing permeability with increasing concentration and, in the latter, to permeability constants that are independent of pressure.

The water content of the polymer has a decisive effect on the swelling and deswelling behaviour of hydrogels. The oxygen permeability change is more than one order of magnitude when the water content is changed from 35% to 60% in PHEMA hydrogels [130].

Fabrication processes of semipermeable polymeric membranes vary from casting or stretching of dense membranes to sintering (see Chapter 1) or in situ formation of porous membranes [119]. In situ formation means the formation of membranes directly upon their porous supports, which permits the simplification of the handling and mounting procedures. In sensors, films are generally deposited directly onto the substrates. Electrochemical deposition, combined with doping-dedoping processes, gives a possibility to synthesize polymers with controlled pore size in the ionic dimension ranges. That will be described in connection with ECPs.

Chemical separation processes and technologies of semipermeable polymeric membranes are a special area of chemistry. Only a short survey

TABLE 3.12. Relative Permeability of Various Polymers to Water Vapour (at 20°C) [129].

Polymer Membrane	Permeability (Relative to Mylar®)
PU	32
Neoprene	3.2
Mylar®	1.0
Low-density PE	0.3
PP	0.12
High-density PE	0.072

was possible in the framework of this book in order to explain their possibilities in sensors.

3.7 POLYMERS IN ION SENSITIVE MEMBRANES[4]

The properties of plasticized polymeric, mainly PVC membranes as used in ion-selective electrodes are reviewed in this section. The kinetic processes at the water/polymer interface are also described with particular reference to membranes.

3.7.1 Background

Ion-selective electrodes (ISEs) based on plasticized polymers (e.g., PVC), in which a neutral carrier such as valinomycin (an ionophore) is dossolved, are now widely used [131–135]. A typical membrane used to sense K^+ ions in water consists of 66% (by weight) of di-octyl sebacate (a plasticizer), 33% PVC, and 1% valinomycin, plus a small amount of potassium tetraphenylborate, $KBPh_4$. In operation, this membrane (0.2 mm thick) is placed between the test solution containing K^+ and a solution of 0.1 M KCl, and the potential between the two aqueous solutions is measured [The general set-up is shown in Figure 2.19(a)]. From this measurement, the K^+ concentration in the test solution is inferred. In this section, we relate the operation of such ion-selective electrodes to the physical chemistry of the membranes and to general electrochemical principles.

The role of an ionophore (e.g., valinomycin) is to bond (i.e., form a complex) to the ion being sensed (e.g., K^+), known as the primary ion, rather than to any other similar ions (e.g., Na^+) within the material of the membrane. This ion is known as the ion to which the ionophore is selective. Figure 3.47 shows some typical ionophores that are currently used.

In some cases more than one ionophore molecule bonds to a single ion. The number of molecules of the ionopbore in the membrane generally exceeds the number of primary ions in the membrane by at least a factor of

[4]This section has been contributed by Prof. Ronald D. Armstrong, *Department of Chemistry, Bedson Building, University of Newcastle upon Tyne, Newcastle upon Tyne NE1 7RU, United Kingdom* and Prof. George Horvai, *Institute for General and Analytical Chemistry, Technical University of Budapest, 1521 Budapest, Hungary.* The authors would like to thank Prof. E. Pungor for helpful discussions. Prof. George Horvai also acknowledges the support from PECO (1079). Both authors are professors. Their main research interests are the physicochemical investigation of ion-selective membranes, reliability of potentiometric measurements, electrochemical detectors in the chromatography, and analysis of liquid-liquid boundaries.

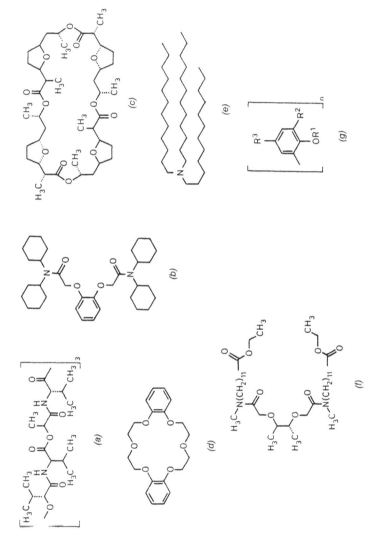

FIGURE 3.47. Structure of the often used ionophores: (a) valinomycin (for K⁺); (b) ETH2120 (for Na⁺); (c) nonactin (for NH₄⁺); (d) dibenzo-18-crown-6; (e) tridodecylamine (for H⁺); (f) ETH1001 (for Ca²⁺); and (g) a calixarene structure (see also Figure 3.31).

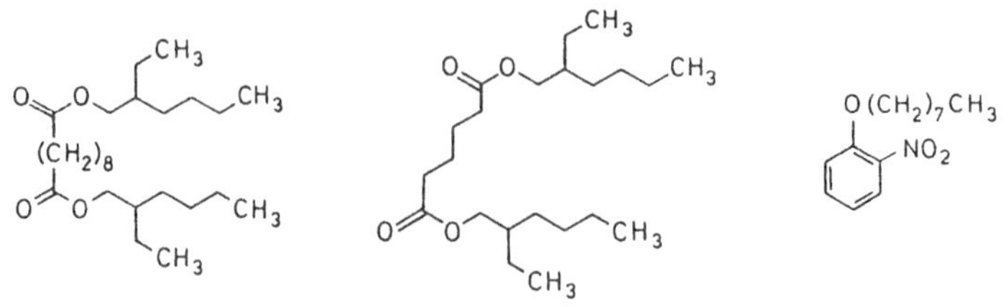

FIGURE 3.48. Structure of typical plasticizers used in ISE membranes: (a) bis(2-ethylhexyl)sebacate (DOS); (b) bis(2-ethylhexyl)adipate (DOA): and (c) 2-nitrophenyl-octly-ether (NPOE).

ten, so that only a small proportion of ionophore molecules in a membrane are bonded to ions at the same time. It is because the ionophore bonds strongly to the primary ion that this ion is able to exchange easily between the water phase and the polymer phase. Thus, for an ion such as K^+ when there is no potassium ionophore in the membrane phase, the free energy of transfer of K^+ from water to the polymer may be as high as 40 kJmol^{-1}, so that no significant concentrations of potassium ions will normally be present in the polymer under these conditions. However, this value may change to -2 kJmol^{-1} with a potassium ionophore present. Ionophores are, in most cases, small molecules; however, they must be lipophilic so that they remain in the polymer phase and do not enter the contacting aqueous phases to any significant extent.

Figure 3.48 shows some of the plasticizers used in ISE membranes, while Table 3.13 lists some of the polymers used. (A large number of membrane components is marketed by Fluka Chemie AG, Switzerland [136].) It is generally found that a particular combination of ionophore, plasticizer, and polymer is the most effective way for sensing the concentration of a particular ion.

It has been found that ions must be present in an ISE membrane if it is to work properly. The presence of ions in the membrane can be achieved by adding a substance such as KBPh$_4$ to a membrane (which is to sense K^+).

TABLE 3.13. Some Polymers Used in ISE Membranes.

PVC	(usually high molecular weight)
PVC-COOH	(Carboxilated, 1.8% COOH groups)
PV(C/Ac/A)	[copolymer of vinyl chloride, vinyl acetate (3%) and vinyl alcohol (6%)]
PVC-NH$_2$	(aminated)
Silicone rubber	(Silopren® (Bayer) K 1000 and cross-linking agent)

3.7.1.1 BEHAVIOUR REQUIRED OF MEMBRANE FOR NERNST RESPONSE

Figure 3.49 shows schematically an electrochemical cell used often in practice to determine ion concentrations in an aqueous solution (described in Section 2.4.1 in general).

ISE response is called Nernstian if the ISE potential changes by [see Equation (2.16)]

$$(RT/zF) \ln 10$$

for every tenfold activity change of the primary ion. Here R is the gas constant, T the temperature in Kelvins, z the valency of the ion, and F the Faraday constant. Since the potential of a single electrode cannot be measured, in a more correct definition, ISE potential should be replaced by the *emf* of the galvanic cell consisting of the ISE, the sample, and the reference electrode. This latter definition raises, however, the problems of liquid junction potentials at the boundary sample/reference electrode. Both definitions lead to difficulties with solutions of high ionic strength (high concentration of ionic species), where individual ionic activities are not uniquely defined. All these problems lead to the practical difficulty of checking whether an electrode behaves in the Nerstian way. The test with the safest theoretical background appears to be in a so-called cell without transference, against a well-characterized reference electrode, e.g., silver/silver chloride. In many practical situations, the theoretical difficulties are less severe, e.g., at constant ionic strength and when a concentrated KCl salt

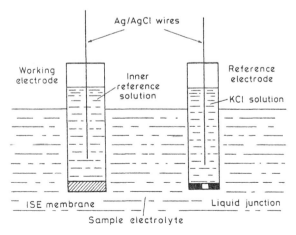

FIGURE 3.49. A typical potentiometric cell arrangement with membrane ISEs using liquid electrolytes.

bridge can be used. In such cases both definitions can be used. We shall only consider these cases further on. A further restriction of Nernstian response has to be made because of the lower detection limit of the electrodes. Below a certain concentration of the primary ion, all electrodes exhibit a non-Nernstian, gradually vanishing response.

The membrane potential is the sum of three contributions: two interfacial potentials arising between the membrane and the neighbouring phases and a potential drop within the membrane. The interfacial potentials can be dealt with more easily when the adjoining phases are both aqueous solutions. With some practical sensors like coated wire electrodes of ISFETs (see Sections 2.2.1 and 2.4.1), this is not the case. In this section, we shall restrict our theoretical discussion to aqueous/membrane/aqueous systems.

Sufficient conditions for a Nernstian transmembrane potential drop between the bulks of the two aqueous phases are that: (1) one and only one kind of ion can pass across the membrane and the two phase boundaries, (2) the concentration of this ion is appreciable in all three phases, and (3) there are no concentration gradients within the phases. The thermodynamic prediction for the transmembrane potential, ΔE, in this case is

$$\Delta E = (RT/zF) \ln (a'/a'') \qquad (3.65)$$

where a' and a'' are the activities of the primary ion in the two aqueous phases.

In principle, since what is required is the constancy of the electrochemical potential of the primary ion, i, throughout the system, it would be possible for a Nernstian response to be found if electroneutrality was not maintained within the membrane phase, i.e., if only the ion, i (e.g., K^+), was present within it. However, as we shall show later for the PVC-based membranes that are normally used, electroneutrality does, in fact, hold within the membrane except for space charge regions at the membrane/water interfaces. Within the bulk of the membrane, the charge due to the presence of the primary ion, i, is balanced by an equal number of opposite charges, which may be mobile or immobile in the polymer phase. If they are mobile, it is a basic requirement that, for either thermodynamic or kinetic reasons, they should be retained in the membrane if Nernstian behaviour is to be found.

The above conditions are nearly met, for instance, when a plasticized PVC membrane with the neutral ligand valinomycin separates two aqueous KCl solutions. Many experiments indicate that, in this system, only K^+ can pass from one aqueous phase into the other, whereas Cl^- ions cannot. The bulk potassium concentration in the membrane is either sustained by the ion-exchange properties of the membrane [137–141] or is present as a result of the deliberate addition of a K^+ salt to the membrane. The negative charges within the membrane do not cross into the aqueous phase. The above conditions for Nernstian response are not met when the same electrode contacts a solution of K^+ mixed with interfering ions, e.g., NH_4^+, Cs^+, or solutions with lipophilic anions, e.g., SCN^-. These ions are capable of passing into or across the membrane.

3.7.1.2 BEHAVIOUR REQUIRED FOR RAPID RESPONSE

The rate of ISE response following an abrupt change in solution concentration has been thoroughly discussed in a recent book [142]. Intuitively, one might expect that the *emf* will be stabilized after a concentration jump only when all the possible processes (diffusion, equalization at the interface) have achieved equilibrium or steady state. It has been shown [143], however, that under certain conditions the *emf* may be time-invariant already when these processes are still far from the final state. It is not always necessary, therefore, that a steady state or equilibrium be reached as soon as possible, yet a fast establishment of equilibrium or a steady state is certainly a sufficient condition for fast potential response. The detailed mechanism of the response to an abrupt concentration change is dealt with in Section 3.7.3.2.

3.7.1.3 BEHAVIOUR REQUIRED FOR LOW IMPEDANCE SYSTEM

In a first approximation, the resistance of an ISE should be immaterial for the *emf* response. The noise of the *emf* reading may be seriously increased by high electrode impedance, however.

Microelectrodes with "membranes" of only several microns diameter but more than 100-μm thickness are particularly affected. The specific resistance of the membrane can be decreased by lipophilic salts, which dissociate substantially in the membrane medium. The lipophilic salt may dissociate into two lipophilic ions, as in the case of tetraalkylammonium tetraphenylborates, or into one lipophilic and one hydrophilic ion (e.g., potassium tetraphenylborate), where the hydrophilic ion is the ion being sensed. In the latter case, the molar concentration of the lipophilic salt should not exceed that of the neutral carrier (1:1 complexation between the carrier and the hydrophilic ion being assumed).

3.7.1.4 BEHAVIOUR REQUIRED FOR LONG-TERM STABILITY

The shelf lifetime and service lifetime of ISE membranes are of utmost interest in practical applications. Whereas the shelf lifetime of neutral carrier ISE membranes made from plasticized PVC appears to extend to years, the service lifetime in flow-through cells is typically reported to be only a few months or even less. Oesch and Simon [144] have shown that the deterioration of membrane response is, first of all, due to loss of membrane components to the aqueous phase. Dinten et al. [145] stressed the importance of the lipophilicity of the membrane components in relation to a lifetime. Chemical reasoning would say that the loss of membrane components should be related only to distribution equilibria and diffusion rates. However, it has often been found [146] that exudation of components from plasticized PVC is a more involved and almost unpredictable process; therefore, predicitons should be used with care.

Lindner et al [147] quote the following possibilities to slow down the leaching of membrane constituents:

- use of plasticizer-free membranes and/or membrane polymers with inherent ionophore properties [148–151]
- covalent coupling of known ionophores to the membrane matrix [152]
- making the distribution of components into the membrane more favourable [153]
- using improved lipophilic salt additives [154,155]

3.7.2 Bulk Properties of Membranes

Liquid membranes were originally made by impregnating a porous disc, e.g., filter material, with a solution of the electroactive material in a water-immiscible solvent. This practice has now almost completely been replaced by incorporating the electroactive material(s) into plasticized PVC [156]. PVC menbranes are most easily handled and prevent the rapid loss of expensive electroactive materials.

Alternative polymeric membranes have been tested mainly to improve adherence to solid-state devices or to reduce adsorption of macromolecules from the analytical samples. Notable in this respect are silicone rubber, epoxies, polyurethanes, and modified PVCs [149,157–163] (see Section 4.7).

Plasticized PVC is composed of PVC and a plasticizer. The latter is typically a water-immiscible, high-boiling organic solvent. Commercial plasticized PVC products usually contain less than 33 wt% plasticizer. ISE membranes, on the other hand, are typically made with 66% plasticizer, and electrode performance deteriorates with decreasing plasticizer content [144].

For many years after the introduction of plasticized PVC ISE membranes, little attention was paid to the physical and chemical structure of these membranes. A rare exception was the outstanding study of Fiedler and Ruzicka [64], showing that electrodes are functional only above the glass transition temperature, T_g, a transition temperature similar to the melting point of solids. This clearly demonstrates that the plasticized PVC has to be in a liquid-like state, with the plasticizer molecules and PVC chains forming a true (thermodynamically stable) solution (with mixing at a molecular level) in which ionophores and salts are dissolved. In addition, since there is a single T_g, we are dealing with a one-phase system. Notwithstanding this study, most researchers appear to have considered the PVC component as an inert support and the plasticizer as a free liquid. Moreover, the plasticized PVC was regarded as an inert and sufficiently pure matrix with the only role of being an immobilized, water-insoluble solvent.

Recent studies [65–67, 137–141] show that this picture is not correct. Both the physical structure and the chemical contaminations of plasticized PVC are important for the explanation of a variety of experimental observations. It should be stressed here, before going into detail, that the particular features that explain the behaviour of plasticized PVC membranes. It is possible to construct ISEs without using a polymer framework [65], and such ISEs may operate in a different manner. The role of impurities in the PVC matrix appears to be somewhat diminished when deliberately added salts, e.g., tetraphenylborates, are used in the membrane.

3.7.2.1 DIELECTRIC CONSTANT OF PLASTICIZED PVC

The dielectric constant of plasticized PVC is obviously important when considering the material for use in ion-selective electrode membranes. For example, a low dielectric constant material is desirable when monovalent ions are to be selected against divalent ions. The dielectric constant of the polymer is a function both of the proportion of PVC to plasticizer and the nature of the plasticizer. Since most membranes used in ion-selective electrodes contain between 60 and 70% by weight of plasticizer, we must, in general, expect the ϵ_r values of such membranes to be similar to those of the pure liquid plasticizer, especially since the plasticized PVC will generally be above its T_g at room temperature and essentially a viscous liquid. The situation has been recently studied [68] in detail for PVC plasticized by dioctyl sebacate (DOS), where it was found that for pure liquid DOS, $\epsilon_r = 4.2$, whereas for PVC (33%)–DOS, $\epsilon_r = 4.8$. The increase in dielectric constant in this case is probably due to the movement of the $C-Cl$ bonds of the PVC in the electrical field. It is probable that a similar situation will be found for other low dielectric constant plasticizers.

For the selection of divalent cations such as Ca^{2+} over monovalent cations, a high dielectric constant plasticizer, e.g., o-nitrophenyl octyl ether (NPOE), is generally used. Smaller molecules such as nitrobenzene are not suitable since, although they plasticize PVC, they are readily lost from the membrane when it is contacted with an aqueous solution. Recent measurements show that for PVC (33%)–NPOE, $\epsilon_r = 14$, which is considerably lower than for pure liquid NPOE ($\epsilon_r = 21$). It seems probable that such a lowering of ϵ_r by the addition of PVC will be found for other high dielectric constant plasticizers. Thus, it appears that it is difficult to obtain high dielectric constant membranes based on PVC since: (1) they must contain at least 20% PVC to be mechanically stable and (2) the plasticizer molecule must be sufficiently large not to be lost to the aqueous solution. A dielectric constant for the membrane higher than twenty is difficult to envisage.

3.7.2.2 IONIC CONDUCTIVITY OF PLASTICIZED PVC

The ionic conductivity of plasticized PVC, as with other ionic conductors, is given by Equation (3.25). The mobilities of a number of ions have recently been measured [169] in PVC (33%)–DOS. These are shown in Table 3.14, from which it can be seen that the ions have a mobility in the polymer, which is typically about 1000 times lower than that in water. These mobilities are very sensitive to the exact level of PVC in the membrane. Thus, if we go from a 33% PVC membrane to a 40% PVC membrane, the mobilities of the ions are decreased by a factor of about ten. As it is discussed more fully in the next section, the membranes on which these measurements have been made have all been in contact either with moist air or with liquid water. We must, therefore, expect that ions such as Li^+ will be

TABLE 3.14. Ionic Conductivities (Mobilities) in PVC (33%)-DOS Membranes at 25°C.

Ion	$\lambda^*(cm^2\ mol^{-1}\ \Omega^{-1})$	Mobility Relative to K^+
K^+	0.11	1.0
Na^+	0.079	0.71
K^+val	0.071	0.65
Na^+val	0.015	0.14
Na^+DB18C6	0.025	0.23
$AsPh_4^+$	0.011	0.1
BPh_4^-	0.011	0.1

* $\lambda = \mu e z$, see Equation (3.25).

TABLE 3.15. Dissociation Constants for MBPh$_4 \leftrightarrow M^+ + BPh_4^-$ in PVC(33%)-DOS at 25°C.

Ion	Ligand	Ligand/M$^+$	K (mol dm^{-3})
Na$^+$	–	0	$(2.3 \pm 0.6) \times 10^{-3}$
Na$^+$	val	2.0	$(2.4 \pm 0.3) \times 10^{-3}$
Na$^+$	DB18C6	2.0	$(1.7 \pm 0.4) \times 10^{-3}$
K$^+$	–	0	$(5 \pm 2) \times 10^{-4}$
K$^+$	val	2.0	$(1.4 \pm 0.4) \times 10^{-4}$
NBu$_4^+$	–	0	$(3.8 \pm 0.4) \times 10^{-4}$

present in the membrane as the hydrated ion, perhaps with six water molecules around it.

Most salts, when dissolved in plasticized PVC, behave as weak electrolytes with a large proportion of ion pairs being formed. For a number of salts, the ion pair dissociation constant has been measured [170] in PVC (33%)–DOS. These are shown in Table 3.15. The value of the ion pair dissociation constant generally lies between 10^{-4} and 3×10^{-3} mol/dm^3. This means that if a salt is dissolved in one of these membranes to a millimolar level, the ion concentration will probably be at least 3×10^{-4} M. When the plasticizer is NPOE (at 2/3 by weight), 1 mM concentration salts are likely to be completely dissociated because of the higher dielectric constant.

Thus, for any particular plasticized PVC with salt (which may be an impurity) dissolved in it, as recognized by Wartman [171], the conductivity will depend on: (1) the extent to which the salt is dissociated into ions and (2) the mobilities of the ions formed. The degree of dissociation, as well as the ionic mobilities, may be a sensitive function of the PVC/plasticizer ratio. Thus, for PVC-DOS polymers containing 1 mM NaBPh$_4$, the conductivity goes through a maximum as the PVC content is increased, in spite of the fact that the mobilities of Na$^+$ and BPh$_4^-$ decrease monotonically with increasing PVC level [168]. This is because the degree of dissociation of Na$^+$-BPh$_4^-$ ion pairs is very small in pure DOS but increases sharply with the addition of PVC and is probably due to the solvation of Na$^+$ by Cl$^-$.

The ionic conductivity of plasticized PVC to which no ions have been deliberately added is a function of the level of ionic impurities in the PVC and the plasticizer. In many cases, the impurity levels of ions may be about 10^{-4} mol/dm^3, though this can be reduced by recrystallization of the PVC. For a membrane based on impure PVC + plasticizer, the bulk resistance for an area of 1 cm^2 and thickness of 0.1 mm will be (for 33% PVC) between 10^6 and 10^7 Ω. Obviously, with deliberately added salt, the resistance could be much lower.

3.7.2.3 DIFFUSION OF NEUTRAL MOLECULES IN PLASTICIZED PVC

It is important to know the diffusion coefficients of neutral ionophores, which are incorporated in PVC membranes. Buck and coworkers [172] have recently envolved a technique from which these diffusion coefficients can be evaluated, based on measurements of current voltage curves. Thus, for ETH1001 in PVC (33%)–NPOE a diffusion coefficient of 5×10^{-8} cm²/s was found, while for valinomycin in PVC (33%)–DOS, the value was 1.51×10^{-8} cm²/s, and it seems likely that the diffusion coefficients of most ionophores will fall in this range for membranes with 33% PVC. Of course, with increasing PVC content the diffusion coefficients will sharply decrease. For example, at 40% PVC we would expect diffusion coefficients near 10^{-9} cm²/s.

When PVC makes contact with water, water molecules must enter the polymer. There is little information in the literature on the equilibrium concentration of water molecules in plasticized PVC, which is in contact with liquid water. Measurements in Reference [173] suggest that in PVC (33%)–DOS there is an uptake of water corresponding to 0.4% by weight or approximately 0.6 mol/dm³. This is somewhat higher than the values given by Thoma et al. [174] on the basis of radiotracer measurements. Whilst it is clear that a large proportion of this water must be present as separate water molecules, some dimers and trimers must also be present. Harrison et al. [175,176] have recently thoroughly investigated the water uptake of plasticized PVC ISE membranes. Their final estimate for the extent of water uptake (about 20 wt% water in a 40-micron thick surface layer) appears to be exaggerated. Even so, the qualitative observations about water distribution in the membrane as a function of soaking time remain very interesting. The question of the formation of colloidal droplets of water in the membranes due to the presence of ionic impurities is dealt with in the next section. It is important to realise that most measurements on PVC membranes reported in the literature relate to membranes that are saturated with water at 0.6 mol/dm³. Membranes that are made up are not contacted with water will generally contain a smaller concentration of water as a result of contact with moist air. By combining permeation data with the estimated equilibrium concentration of water, it should be possible to evaluate the diffusion coefficient for water molecules in a PVC membrane (see Section 3.6). For PVC (33%)–DOS we have recently [173] estimated D as equal to 3×10^{-7} cm²/s.

The influence of the presence of water on the properties of PVC-based membranes is unclear. It appears to have little influence on the dielectric constant; however, it is probable that small ions such as Li⁺ are present in wet membranes as the hydrated ions (unless, of course, they are complexed by an ionophore).

Another area of uncertainty in respect to plasticized PVC is the value of complexation constants within it. For example, for the reaction

$$K^+ + val \rightleftharpoons K^+ val \tag{3.66}$$

there is, at present, no reliable data, though a recent analysis of Warburg impedances [177] suggested a value of the equilibrium constant lying between 10^5 and 10^7 dm³/mol for PVC (33%)–DOS. For this sort of value, if we consider a 10^{-2} M valinomycin + 10^{-3} M K⁺ membrane, we would expect virtually of all of the K⁺ to be in the form of the valinomycin complex with free K⁺ present at no more than 10^{-6} M.

3.7.2.4 BULK PROPERTIES OF MEMBRANES USED AS ION-SELECTIVE ELECTRODES

A typical membrane used to sense K⁺ ions consists of PVC (33%)–DOS and 10^{-2} M valinomycin + 10^{-3} M KBPh₄. The actual species present in the membrane can be evaluated from the previous discussion as

- 9×10^{-3} M valinomycin
- 7×10^{-4} M K + val BPh₄ (ion pair)
- 3×10^{-4} M K⁺ val
- 3×10^{-4} M BPh₄
- $< 10^{-6}$ M K⁺ (free)

The free valinomycin will have a diffusion coefficient of 1.5×10^{-8} cm²/s. The diffusion coefficient of the K⁺ val–BPh₄ ion pair is not known, but is likely to be close to 1×10^{-8} cm²/s. The ionic conductivity of K⁺ val is 0.07 cm²/molΩ, while that of BPh₄ is 0.01 cm²/molΩ. For Nerstian behaviour, we require that only the K⁺ ion crosses the water/membrane interface, i.e., that ions such as Na⁺ and Cl⁻ in a contacting water phase remain in that phase while BPh₄ remains in the membrane phase. The composition of the membrane can be varied somewhat from that described here, provided:

(1) That the solubility of valinomycin is not exceeded (about 2×10^{-2} M)

(2) That the KBPh₄ concentration is less than that of the valinomycin, since, if it is greater, it is spontaneously lost to the water

(3) That the KBPh₄ concentration comfortably exceeds that of ionic impurities in the membrane

In the past, K^+ sensing membranes have been made up using only valinomycin and no added salt. These membranes have depended on the presence of impurities in the PVC or the plasticizer. In general, it is necessary to condition such electrodes by contacting them with aqueous K^+ solutions so as to allow a degree of ion exchange to occur. In 1977, Thoma et al. [174] showed that PVC plasticized with an apolar plasticizer and containing valinomycin as electro-active material absorbed potassium ions from a 10^{-3} M KCl solution while the chloride absorption from the same solution was almost negligible. It was concluded that the membrane behaved as a low capacity (of the order of 10^{-4} M) ion exchanger. There was, however, no obvious reason for this observation, and therefore, an exchange reaction between membrane-solved water and aqueous phase potassium ions was postulated:

$$L(M) + H_2O(W) + K^+(W) = KL^+(M) + OH^-(M) + H^+(W) \qquad (3.67)$$

where L is the ligand valinomycin and W and M denote the aqueous and membrane phase, respectively. Recent studies [137–141] suggest a different mechanism. It has been found that inorganic salts like KCl or CsCl, when deliberately incorporated into the valinomycin/PVC membrane, increase the cation exchange capacity without otherwise affecting ISE performance. This effect is due to the somewhat surprising immobilisation of salts within the membrane. Water from the aqueous bathing solution diffuses easily into the membrane and solvates the impurities where they are. Colloidal inclusions of aqueous salt solution are apparently formed, which are virtually immobile. Ion exchange occures between these inclusions and the bulk aqueous phase. Based on this work, the authors have suggested that the intrinsic ion-exchange capacity of the membranes may also be due to ionic impurities. The selectivity of the ion exchange by netural carrier/PVC membranes has been assumed for a long time and attributed to selective complexation by the electroactive material (ligand or carrier). This assumption was underpined by complexation studies in different solvents but not in the membrane matrix itself. Recently, radiotracer studies [138] have provided direct proof. It has been shown [139–141] that the ion-exchange selectivity constants are in fair agreement with the respective potentiometric selectivity constants.

3.7.2.5 PARTITION OF SALTS AND IONOPHORES BETWEEN MEMBRANE AND WATER

Any substance is expected to reach, after some time, an equilibrium distribution between the membrane and aqueous phase. It appears, however, that this expectation is sometimes fulfilled in unusual ways. It has been mentioned before that hydrophilic salts are not leached out of the membrane into the aqueous bathing solution. Rather, it is water that penetrates the membrane and solvates the salts in situ. Ionophores and plasticizers may be very slowly leached out, too, and this process has been found to limit electrode lifetime [145]. The leaching out of netural substances is an exudation process, and it was noted previously that the progress of exudation cannot always be rationalized. A very interesting example for this is the exudation of surfactants [139]. The process was followed by impedance measurements. With a freshly cleaned surface, the formation of a surface layer occurred in the matter of a few hours. After this time, the process was apparently halted. The surface layer could be removed, however, by contacting the membrane surface with inert gas bubbles, apparently because of *trans*-adsorption of the layer onto the bubble surface. Subsequently, formation of a fresh layer on the membrane surface was observed. It is concluded that the exudation rate may depend on the availability of *trans*-adsorption sites.

Notwithstanding these complications, there may be substances that distribute into or from the membrane as if it were a simple homogeneous liquid, and adsorption may also be virtually absent or not influencing transfer rates. The distribution of tetraphenylborate additives is apparently such a case.

3.7.3 The Polymer/Water Interface

Since the polymers we are generally dealing with are essentially homogeneous, single-phase, viscous liquids, we would expect the water/polymer interface to exhibit the features associated with other liquid/liquid interfaces such as the nitrobenzene/water interface [178]. Thus, so far as charged species are concerned, ions present on the aqueous side of the interface must be balanced by equal and opposite charges on the polymer side of the interface. Because of thermal motion, the two charged layers will generally be diffuse in nature. In addition, there may be an ion-free inner layer, which separates the two space charge regions. In most circumstances, the concentrations of ions in the aqueous phase are significantly higher than those in the polymer phase so that the spread of the space charge into the polymer phase will be greater than the corresponding situation in the aqueous phase. Thus, the Debye length for a 10^{-3} M 1:1 electrolyte in PVC (33%)–DOS is 3 nm, whereas for a 10^{-1} M 1:1 electrolyte in water, it is 1 nm. This means that in a typical plasticized PVC membrane of a thickness of 0.1 mm, the space charge region spreads only a very small distance into the bulk of the membrane.

There is, however, likely to be a significant difference between the liquid/liquid interface and the polymer/water interface in that, with the polymer case, either polymer or plasticizer may be accumulated at the interface so that the interfacial composition is different from the bulk composition. In addition, impurities present in the polymer may accumulate at the interface. Recent measurements [179] on the $CaCl_2$(water)/$AsPh_4BPh_4$–PVC (33%)–NPOE interface have shown that it is completely polarizable over a potential range of about 0.4 V.

This fact enabled the double layer capacity (C_{dl})/potential relation to be explored for this potential region. The value of C_{dl} shows a distinct minimum within the polarizable region, which is a function of the salt concentration within the polymer phase. This observation strongly suggests the existence of a diffuse layer of charge within the polymer phase as outlined previously.

ATR-IR spectroscopic studies [180–182] of the membranes near the aqueous interface have also revealed acculumation of plasticizer and certain ions under specific conditions.

3.7.3.1 CHARGE TRANSFER ACROSS THE POLYMER/WATER INTERFACE

In a situation where an ion is present in both the water phase and the polymer phase and is at equilibrium across the interface, there will be an exchange of ions across the interface, which can be characterized by an exchange current (i_0). When the potential across the interface is changed from its equilibrium value a net current, (i), will flow, which is proportional to the deviation of the interfacial potential (η) from its equilibrium value, provided that the deviation is sufficiently small. This enables a charge transfer resistance, R_{ct}, to be defined as $R_{ct} = \eta/i$. It is R_{ct} that is directly measurable using impedance measurements. There is not necessarily any simple relationship between i_0 and R_{ct}, given the back to back diffuse double layer structure of the interface. However, it is unlikely that the relationship

$$i_0 = RT/zFR_{ct} \tag{3.68}$$

is incorrect by more than an order of magnitude. It would, however, perhaps be safer to refer to R_{ct} values rather than i_0 values.

Measured R_{ct} values must be strongly dependent on the concentrations of the exchanging ion in both the water phase and the membrane phase. Therefore, for the purposes of fundamental investigations, it is best if R_{ct} values are measured under conditions where the concentrations are well defined [183] rather than where at least one of the concentrations has been obtained by an unknown extent of ion exchange [184]. At the present time, there are few reliable reports in the literature that meet this criterion. However, the situation over the transfer of K^+ between water and PVC (33%)–DOS in which valinomycin is dissolved is reasonably clear [185]. For aqueous KCl solutions of between 10^{-1} M and 10^{-3} M and with a membrane containing 1 mM $KBPh_4$ and 10^{-2} M valinomycin, the exchange rate for K^+ is sufficiently rapid as to make R_{ct} too samll to measure against the bulk resistance of the membrane, indicating an exchange current in excess of 10^{-5} A/cm². When valinomycin is substituted [177] by di-benzo-18-crown-6 (DB18C6), a similar situation is found; however, when impedance measurements are made on membranes in which the ionophore:ion ratio is 1:1, a Warburg impedance is found for valinomycin/K^+, DB18C6/K^+, DB18C6/Na^+. The simplest explanation for this Warburg impedance is that we are seeing a two-step reaction with

$$M^+(W) \rightleftharpoons M^+(M) \tag{3.69a}$$

followed by

$$M^+(M) + L(M) \rightleftharpoons ML^+(M) \tag{3.69b}$$

and that it is the diffusion of low concentration species $M^+(M)$ that gives rise to the Warburg impedance.

Ion transfer measurements have recently been extended [186] to the transfer of Ca^{2+} across the water/PVC (33%)–NPOE interface with either ETH129 or ETH1001 dissolved in the polymer. Here again there is evidence for rapid Ca^{2+} transfer across the interface, provided that the ion/ionophore ratio is at least 1:4 for ETH1001 and 1:6 for ETH129. This difference from the K^+/valinomycin case is presumably related to the fact that in the Ca^{2+} case, more than one ionophore molecule is complexed with the Ca^{2+} ion. It is important to note that slow ion exchange at the polymer/water interface may be due to the presence of impurities in the PVC, which diffuse to the interface [139].

It is interesting for the analytical applications that a very low concentration of the ion to which the ISE is selective, amperometric determinations can be made [187,188]. The experimental set-up for such measurements is virutally the same as the widely available amperometric detectors for liquid chromatography. The working electrode of the detector is simply replaced with low resistance ISE of conventional composition. Detection limits in the amperometric mode appear to be substantially lower than in the potentiometric mode.

3.7.3.2 RESPONSE OF THE WATER/POLYMER INTERFACE TO CONCENTRATION CHANGES

The ionic charges on either side of the water/polymer interface generate a potential across the interface. For example, on the aqueous side the charge may arise from an excess of cations (K^+) over anions (Cl^-), which is balanced in the polymer phase by an excess of anions (BPh_4^-) over cations (K^+val). If we imagine that the aqueous phase is rapidly changed (except for the space charge layer) to a new aqueous phase with a different concentration of the primary ion (K^+), then at $t = 0$ the interfacial potential is just that which was appropriate to the previous aqueous phase. The adjustment of the charges at the interface will occur via the net crossing of the interface by the primary ion (K^+). The relaxation time for this adjustment process will be given by $C_{dl}R_{ct}$, which for $C_{dl} = 1$ $\mu F/cm^2$ and $R_{ct} = 10^4$ Ωcm^2 gives $\tau = 10^{-2}$ sec. It is doubtful whether this relaxation process can be measured experimentally, since in general, the slow diffusion of ions through the stagnant aqueous layer is likely to be much slower than any other process. However, recent measurements [189] using an activity step of 10^{-2} M KCl to 10^{-3} M KCl show a relaxation time of about 10 msec for a PVC/valinomycin membrane, which could possibly involve the $C_{dl}R_{ct}$ time constant.

3.7.3.3 THE RELATIONSHIP BETWEEN INTERFACIAL RATE CONSTANTS AND HOMOGENEOUS EQUILIBRIUM CONSTANTS

When we consider the ion transfer reaction

$$M^+(W) + L(M) \leftrightarrow ML^+(M) \qquad (3.70)$$

which has an exchange current i_0, it is interesting to speculate as to how this reaction varies with the nature of M^+. For example, in the case of valinomycin, how would we expect i_0 to vary between $M^+ = K^+$ and $M^+ = Na^+$? It is important to realise that, when the ion is changed, the equilibrium interfacial potential difference is also changed. Thus, the conditions under which the two i_0 values are measured are different. A theoretical prediction of the variation of i_0 with the complexation constant must depend on the details of the model of the water/polymer interface. However, provided that the mechanism of transfer is the same in the two cases, the change of equilibrium interfacial potential on going from one ion to the other must be expected to more or less offset the change in complexation constant between the two ions, so that the exchange currents should be approximately the same. However, the unidirectional currents, i, measured at the same interfacial potential difference might be expected to follow a relationship of the type

$$i_1/i_2 = (K_1/K_2)^a \qquad \text{where } 0 < a < 1 \qquad (3.71)$$

In this section we have been able to relate the behaviour of the water/polymer interface to the much better understood liquid/liquid interface. However, the properties of the bulk membrane, although related to the properties of viscous liquid solvents, differ from them in a number of important ways as discussed previously.

3.8 SENSING EFFECTS BASED ON OPTICAL BEHAVIOUR

Polymer materials (for instance, PMMA) are widely used in fibre-optic sensors as optical waveguides or claddings. However, polymer materials alone are not optoactive in the sense that their optical parameters cannot be influenced by the environment. Therefore, fibre-optic sensors generally use only composite polymers, in which an optoactive material is entrapped into the polymer matrix. A short summary was given about the most important fibre-optic sensor structures in Section 2.6. The sensors use an optrode structure in reflection, and a porous fibre or a fibre with sensitive cladding in the transmission mode of operation. The physical-chemical processes that are the bases of the sensor operation take place in the optrode "tip" or in the micropores or in the sensitive coating films.

A very simple process was developed for the fabrication of porous polymer waveguides by Zhou and Siegel [191]. The basic principle is the polymerization of a mixture of monomers that can be cross-linked in the presence of an inert and soluble component solvent. After polymerization, the inert solvent should not be chemically bound into the polymer network and should, therefore, be easily removable from the polymer to leave an interconnected porous structure. For fibre materials, methyl methacrylate and triethylene glycol dimethacrylate can be used as monomers. Octane was used as the inert solvent. After the heat annealing necessary for polymerization in glass capillaries, the fibres were put in a solution to remove any remaining inert solvent.

In optrode-type structures, a lot of different polymers are used. PVC and PTFE can be used as membranes, for example. Typical optrode tip polymers are epoxies, polyacrylamide, and Nafion®. The latter is an interesting example because it is a polymer electrolyte, and its ion content can be changed by ion-exchange processes as a function of different parameters of the environment (see Sections 3.3 and 3.7).

Copolymerization of polymers is often used to bind indicator dye materials into the polymer matrix. This approach converts a water-soluble indicator dye into a water-insoluble form, preventing the leakage of the dye from the matrix. Covalent binding of the dye into matrices seems to be the most efficient process for immobilization. Indicator dyes (also including the conventional pH indicator materials) are often used to produce an optoactive matrix. These materials are known to change their colour and may be other optical properties under different environmental conditions. In a few cases, fluorescence quenching, chemi-, and bioluminescence effect also can be achieved. Organopolysiloxanes can change their refractive index and/or thickness due to the swelling when they adsorb and/or absorb molecules from the surrounding medium.

3.8.1 Colourimetric Effects

The colourimetric changes can be followed by absorption or reflexion spectra changes, or more simply using a monochromatic light source, by absorbance or reflectance value changes.

A *carbon monoxide* fibre-optic was developed by Zhou and Siegel [191] with a porous polymer optical fibre and palladium chloride reagent, which was dissolved in the monomer solution before forming the porous polymer fibre. During the measurement, the carbon monoxide first diffuses into the porous polymer fibre, and then reacts with the indicator to create the metallic palladium black particles, which cause the light-scattering loss and a decrease in observed light intensity. Although this probe is nonreversible, the approach can be utilized for alarm systems that detect carbon monoxide leaks. In fact, there is no equilibrium between the carbon monoxide and the indicator; the palladium chloride will keep reacting with the carbon monoxide until all amount is consumed. This mechanism results in a totally different dynamic response from the reversible sensors.

A lot of chemical fibre-optic sensors have been built up using conventional acid-base or pH-indicators. Their operation is based on the colour changes of the indicators when they are protonated or deprotonated. In the framework of this section, it is not possible to give a detailed description about the wide variety of different indicators and their reaction mechanisms. Only a few polymer-indicator dye composite examples will be mentioned, which have already been used as fibre-optic sensors.

pH-sensors can be fabricated directly from the acid-base indicators or from their combination. Recently, agarose, a typical polysaccharide material, was used by Hao et al. [192] instead of polyacrylamide to immobilize phenol red indicator. Phenol red in the agarose gel matrix shows almost identical optical and chemical properties to those in water. The agrose gel matrix appears to allow relatively fast proton transfer. Phenol red has a colour-change range of pH 6–8.

Figure 3.50 shows the typical relationship between absorbance measured at the absorption peak and pH values. It is well demonstrated that pH values can be measured only in a limited region, depending of the type of indicator.

The solution to overcome this problem is the co-immobilization of two or several type indicator dyes into the same polymer matrix. Boisde et al. [193] described pH sensors with poly(N-vinylimide-azole) (PVI) grafted onto an optical fibre with immobilized dyes, such as chlorophenol red, bromophenol blue, bromothymol blue, etc. It was possible to get sensors in the range of pH 1–10.

Different type *ion sensors* can be produced when an ion-selective ionophore and a proton-selective (chromo)-ionophore (i.e., a pH indicator) as a functional additive is entrapped in a polymer matrix. The potential that arises from the complexation of reactive ions (I) to be detected by the ionophore (Io) in the membrane is compensated by the coextraction of protons by the indicator ionophore (Ind).

The process can be illustrated by the reaction scheme [194]

$$I_{aq}^- + H_{aq}^+ + Io_{mem}^+ + Ind_{mem}^- \longleftrightarrow Io^+ I_{mem}^- + Ind^- H_{mem}^+ \qquad (3.72)$$

The ion concentration in the solution is therefore correlated with the

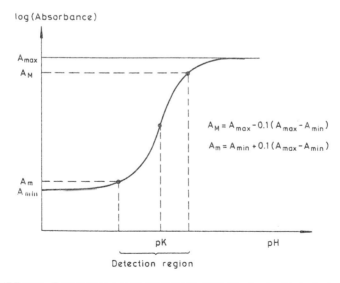

FIGURE 3.50. Typical absorbance-pH characteristics of colourimetric indicators [193].

amount of protons in the membrane and can be detected spectroscopically by the protonation of the indicator. Lumpp et al. [194] applied the mentioned mechanism in a nitrate ion sensor using a bromocresol-purple/nickel(II)-phenanthroline mixture in PVC.

The counter reaction, the deprotonation of a pH indicator entrapped in a polymer matrix, can also be used for the detection of ions. If a neutral charge carrier (C) is the ionophore in the membrane, a conventional ion-exchange mechanism can be supposed:

$$I^+_{aq} + C_{mem} + Ind^-H^+_{mem} \longleftrightarrow CI^+_{mem} + Ind^-_{mem} + H^+_{aq} \qquad (3.73)$$

This mechanism was used for the detection of ammonium, NH^+_4, ions by Reichert et al. [195]. They developed such sensor optrodes with Teflon® membrane and bromophenol blue indicator. Zhou et al. [196] used the porous fibre structure with bromocresol purple indicator for ammonia-vapour detection. The latter sensor type is also based on the detection of ammonium ions originating from ammonia-vapour. These ions are present in the water adsorbed in the porous polymer structure.

Enzyme optrodes were also developed for the detection of urea [197]. The reaction of urease enzyme entrapped in the polymer matrix (for example, polyacrylamide) with urea in the analyte causes a pH change in the membrane (similarly to the electric sensitive structures, see Section 4.8.2), which is transduced by the indicator into a change in the absorption spectrum.

Direct operation indicators (chromo-ionophores) can also be applied, which show absorbance spectrum changes through a complexation process of ions from the analyte. Czolk et al. [198] developed an *optochemical sensor for* the detection of *heavy metal ions* such as Cd(II), Pb(II), or Hg(II). The sensitive layer consists of 5,10,15,20-tetra(p-sulphonalophenyl)porphyrin covalently immobilized into a polymer matrix. The complexion of the mentioned ions directly causes an absorbance spectrum change without any protonization or deprotonization process.

Synthetic ion carriers such as crown ethers are known to bind alkali and alkaline-earth cations in a rather specific and reversible manner and, thus, have been used in sensing layers for alkali ions. These host molecules may also incorporate a chromogenic group in their structure, thus, allowing a straightforward optical transduction of the chemical recognition process for those ions. Such chromo-ionophores can be immobilized in plasticized PVC membranes or on polystyrene [199,200].

A *relative humidity optrode* sensor was suggested by Sadaoka et al. [201]. They used hydrolyzed Nafion®-dyes composites. Nafion® is an acidic type polyelectrolyte (see Section 3.3.2). It can be assumed that the acid strength

of the solid acid is weakened by the sorption of water molecules, and this change induces colour change in the indicator dye [200] (crystal violet in that case).

Another fibre-optic humidity sensor utilizes the colour change of Co^{2+}, Cu^{2+}, or V^{5+} salts when they are hydrated or dehydrated by ambient air. The salts were embedded in gelatine or poly(vinyl pyrrolidone) and deposited onto the surface of the waveguide to get a cladding-based evanescent-wave reflectometric sensor (see Section 2.6.3) [202].

3.8.2 Fluoroescence and Luminescence Effects

Other types of optrodes utilize not only the colourimetric changes, but the *fluoroescence quenching* of the indicator dyes as well [203]. Several sensor types were developed for the measurement of the *partial pressure of* molecular *oxygen* in liquids, for instance, in blood. The fluorescent dye, such as perylene dibutyrate, is absorbed to organic beds contained within a hydrophobic gas permeable membrane, such as porous polyethylene tubing. The dye is excited with blue light (468 nm), and it emits radiation at 514 nm (green). The oxygen partial pressure according to the Stern-Volmer equation is [203]

$$p_{o_2} = K[(\text{blue/green}) - 1]^m \qquad (3.74)$$

A *fluorescence-based* p_{co_2} optrode with a nanolitre-sized droplet of concentrated dichlorofluorescein (which fluoresces as a function of the solvent pH and can also be used for pH sensors) in a buffered solution was also described [204].

A fibre-optic *fluorosensor for* the determination of *relative humidity* was presented by Posch and Wolfbeis [205]. It is based on the quenching of fluorescence of silicagel-adsorbed perylene dyes by water vapour.

Most investigations on *fluorescence pH optrodes* were performed with 1-hydroxy-pyrene-3,6,8-trisulphonate (HPTS) [206] as a fluorescent indicator. The deprotonated form of HPTS bound to cellulose can be excited at 475 nm to give a fluorescence emission at 530 nm. This emission increases with increasing pH.

An optrode using a positively charged fluorescent dye attached on a PVC membrane has been constructed by Kawabata et al. [207] for potassium ion detection. Its response is ascribed to the movement of the dye caused by ion exchange with the cation.

Biosensitive optrodes can also be fabricated by the immobilization of biological compounds (enzymes or substrates) at the tip of fibres, associated with colourimetric (see above) or fluorometric detection.

Another approach is based on the measurement of an enzymatically cata-lysed light emission resulting in a *luminescence effect*. A fibre-optic probe for hydrogen peroxide based on the luminol-mediated chemiluminescence reaction with horseradish peroxidase immobilized in a polyacrylamide gel on the tip of an optical fibre was described by Freeman and Seitz [208] a few years ago. Also, a biosensor type for ATP and NADH was proposed by Gautier et al [209]. Preactivated polyamide membranes were selected for the immobilization of enzymes. A familiar example of chemiluminescence comes from the firefly, which uses the enzyme luciferase. In the presence of Mg^+ and the substrate luciferin, ATP is utilized in the following reaction:

$$ATP + luciferin + O_2 \xrightarrow{\text{luciferase}} AMP + oxyluciferin$$

$$+ PPi + CO_2 + light\ (560\ nm) \tag{3.75}$$

Luminescent porphyrin probes, in particular, phosphorescent Pt-octa-ethylporphine-polystyrene optrodes, were shown to be effective for use in quenched-luminescence oxygen sensors and enzyme biosensors [210].

3.8.3 Sensing Effects Based on Light Refraction

It is evident to build up sensor elements that are based on the light refrac-tion (see Sections 2.6.3 and 2.6.4) and/or interferometry. The optical pathlength in a given medium is determined by the refractive index and the geometric pathlength. The changes of both parameters results in a change of the phase shift, which can be detected by interferometry.

On the other hand, waveguide effective refractive index or transmission effectiveness variation also results from the refractive index change of the cladding. This means that film materials that could not be applied before may be used for fibre-optic or optical integrated chemical sensors. For example, the addition of indicators such as dyes or fluorescence materials to the film is not necessary. The physical-chemical changes take place with the direct participation of the sensing polymer film material.

Despite these advantages, it is difficult to find appropriate film materials because of the strict requirements. The polymeric cladding, or film materials, must exhibit some physical and optical properties. They must allow diffusion of vapours to and from sites of selective interaction, their refractive index should be slightly lower than the index of the core, and they should be able to vary as the vapour to be detected is absorbed in the polymer. Moreover, they must be transparent for the wavelength used and homogeneous to prevent light scattering.

Polysiloxane polymers seem to present these properties and can be used successfully in chemical-sensor applications. Polysiloxanes have been investigated for several years as sorbent dielectric films in gas sensors. Their structure and behaviour was described in Sections 3.1.4.2 and 3.5.2.5 in connection with their ability to change the permittivity or the film thickness when absorbing molecules from the surrounding gases. The close connection between the refractive index (n) and permittivity (ϵ_r) is well known and can be expressed as

$$n^2 = \epsilon_r \tag{3.76}$$

Therefore, the refractive index of the polymeric film, and/or cladding, varies with the permittivity as the vapour to be detected is absorbed in it. The Clausius-Mosotti equation

$$\frac{n^2 - 1}{n^2 + 2} = \frac{4 \cdot \pi}{3} \cdot \frac{\varrho}{M} \cdot p_m \tag{3.77}$$

can be used as an approximation. It can be seen qualitatively that a variation of the molecular polarizability, p_m, of the density ϱ, and/or of the molecular weight M, leads to a variation of the refractive index of the polymer film. Moreover, polysiloxanes may show a considerable swelling when they absorb gas molecules; therefore, the films can alter their thickness according to the absorbed amount of gases (see Section 3.5.2.5).

One of the often used methods to detect small thickness changes of transparent layer is the optical reflection mode interferometry, which can also be used in fibre-optic sensors. By measuring the shift of the intensity maxima formed by constructive interference of both the incident and the reflected light beams in the polymer film, the thickness changes can be determined sensitively.

Several materials in various arrangements have been investigated for a lot of hydrocarbons and chlorinated hydrocarbons [116,211,212]. Poly-(dimethylsiloxane) (PDMS) was examined in several studies and seems to be a good candidate for applications in sensors that operate on the mentioned bases. It shows both swelling and refractive index shift when exposed to organic solvent vapours (see also Section 3.5.2.5). Figure 3.51 shows the relative change of the film thickness as a function of n-heptane concentration [211]. A very good linearity has been found in the case of heptane between a few ppm and approximately 1000 ppm.

Measurements using other organic solvents have shown that a very low detection limit in the 1 ppm range in the gas phase can be achieved for certain chlorinated hydrocarbons, such as C_2Cl_4, which results in a six times bigger change than heptane [21].

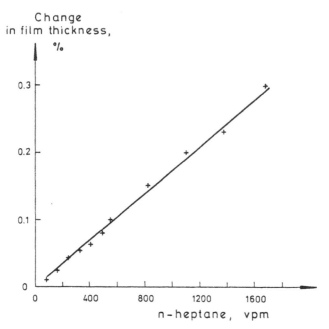

FIGURE 3.51. Calibration function for interferometric sensors in the case of *n*-heptane. (Redrawn from the figure of Gauglitz et al. [211], with the kind permission of Elsevier Sequoia S. A., Lausanne, Switzerland, publisher of *Sensors and Actuators A* and *Sensors and Actuators B.*)

Table 3.16 gives a summary about the refractive-index changes for a few organic compounds. The values were calculated from the phase shift values measured with an integrated optical Mach-Zehnder interferometer (see Section 2.6.4) at identical organic solvent and hydrocarbon concentrations. The measurements were carried out in a wavelength range between 750 and 820 mm [212].

3.9 SENSING EFFECTS IN ELECTROCONDUCTING CONJUGATED POLYMERS[5]

The electroconducting polymers concerned here are not composite plastics filled with conductive particles (see Section 3.2), but organic materials

[5]This section has been contributed by Gérard Bidan, *CEA(Centre d'Etudes Nucléaires de Grenoble DRFMC/SESAM, Laboratoire d'Electrochimie Moléculaire, BP 85 X, F-38041 Grenoble Cedex, France.* He received his Ph.D. in 1980. Since 1985, he has been the head of the Laboratoire. His research interests lie in the field how to apply electrochemistry for the synthesis of functional molecular materials, their characterization and application possibilities, including also the development of functionalized electroconducting conjugated polymers.

that exhibit intrinsic electronic conductivity, reflecting their delocalized electronic structure along the conjugated backbone. The common characteristic of electroconducting conjugated polymers (ECPs) is a one-dimensional organic backbone based on the alternation of single and double bonds, which enables a superorbital to be formed for electronic conduction. Polyacetylene, polyaromatic, and polyheterocyclic chains can provide such a structure (see Figure 3.52).

Macroscopic conduction through these polymers takes place by charge hopping both along the polymer chains and also between the macromolecules that make up individual fibres and between the fibres themselves. However, in the neutral (undoped) state these materials are only semiconducting. The electronic conductivity appears when the material is doped, i.e., when electrons or holes are injected into the superorbital. For reasons of electroneutrality, counter-ions called dopants are simultaneously inserted into the polymer matrix. [*Here an interesting analogy can be revealed between more conventional inorganic semiconductors and ECPs — that both of them can be doped to get material with lower resistivity*].

One of the increasingly studied subjects in connection with electroconducting polymers (ECPs) is their application in the field of sensors [213], mainly in electrochemical sensors. This is due to three key properties of ECPs:

- Film deposition possibly using electrosynthesis (see Section 1.4.4): the electrochemical oxidation of a monomer solution (pyrrole,

TABLE 3.16. *Refractive Index Changes of a Polymer Coating for Different Organic Compounds (after Gauglitz and Ingenhoff [212], with the Kind Permission of Elsevier Sequoia, Lausanne, Switzerland, Publisher of Sensors and Actuators A and Sensors and Actuators B).*

Substance	Refractive Index Change
n-Pentane	0.0002
n-Hexane	0.00032
n-Heptane	0.00042
n-Octane	0.00116
Cyclohexane	0.00334
Iso-octane	0.00076
Dichloromethane	0.00038
Trichloromethane	0.0025
Tetrachloromethane	0.006
Perchloroethylene	0.0053
Trichloroethylene	0.00436
Toluene	0.00562
Meta-xylene	0.0038
Ethanol	0.0009

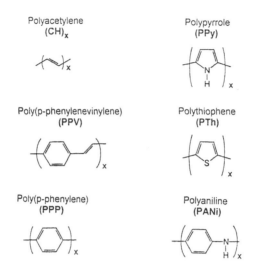

FIGURE 3.52. Bond-alternant backbone of the most studied electroconducting conjugated polymers (ECPs).

thiophene, or aniline derivatives; see Figure 3.52) under appropriate conditions gives a doped conductive polymer film deposited at the surface of the electrode. The control of the growth rate and film thickness is simply achieved by counting the electrical charge used for the synthesis (see Figures 1.32 and 3.53).

- Behaviour as a straightforward transduction matrix: this layer is a remarkable transduction matrix sensitive to gases, vapours, ions, and biomolecular systems, resulting in a straightforward conductance, impedance, or redox potential change "via" the modulation of their doping level (see Figure 3.54). Another advantage of this doping/dedoping mechanism is the possibility to associate ionic movements stoichiometrically with electrical currents (see Figures 1.32 and 3.55). This has been exploited for the detection of electroinactive ions or for the electrocontrolled delivering of drugs (see Sections 3.94 and 3.95).
- Functionalization possibility for specific reactions: for specific interactions, the intrinsic properties of ECPs are not sufficient; it is necessary to introduce additional properties. The functionalized ECP layer acts as the sensitive component of the modified electrode-based sensor.

The functionalization can be realized in several ways, as shown in Figure 3.56:

- The covalent grafting of a specific molecule (SM) to the ECP

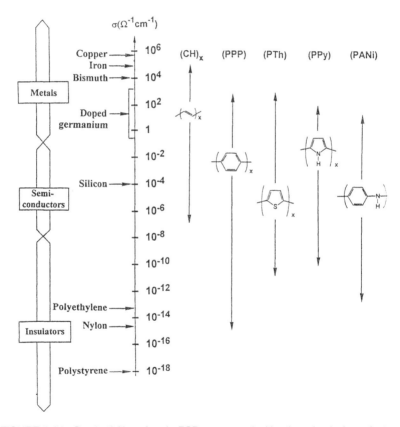

FIGURE 3.53. The "electropolymerization" of thiophene, pyrrole, aniline, and their derivatives is, in fact, a condensing oxidation, which follows an electronic stoichiometry allowing Coulombic control of the film thickness to be achieved. This is illustrated here by a poly(SM-functionalized pyrrole) electrosynthesis (δ = doping level, SM = specific molecule with the desired property, A = dopant).

FIGURE 3.54. Conductivities of main ECPs compared with other classical conductors, semiconductors, and insulators. The arrows indicate the ranges of conductivity from the dedoped state (lower value) to the doped state (upper value).

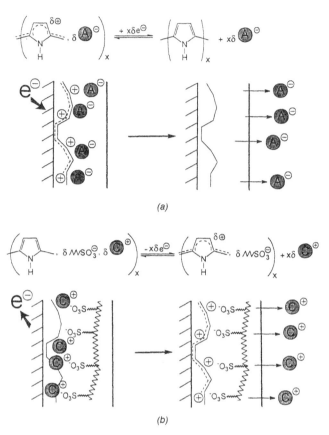

(a)

(b)

FIGURE 3.55. Reversible redox reaction corresponding to the electrical switching of poly-pyrrole (PPy) between the doped and the dedoped state: (a) *p*-doping by mobile anions: the dedoping results mainly in the release of doping ions into the solution; (b) doping by immo-bilized anions, exemplified here by SO_3^-: a covalent binding to PPy skeleton or to ionomers such as poly(styrene sulphonate) or to bulky groups such as persulphonated macrocycles may be formed; consequently, the doping results mainly in the release of counter-ions (C^+).

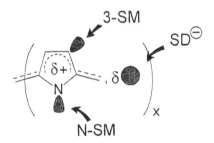

FIGURE 3.56. The various ways to functionalize polypyrrole: grafting of a specific mole-cule, SM, at the *N*-position or at the 3-position; or introduction of a specific dopant ion, SD⁻.

backbone [214–216]; in order to accomplish this, a suitable monomer is synthesized and subsequently electropolymerized (see Figure 3.52). The *N*-functionalization of polypyrrole (PPy) has been the most widely exploited method [217,218], but more recently, the grafting at the 3-position of polythiophene (PTh) [219] and PPy has also been used [220]. In this way, redox mediators (ferrocene [221], quinone [222], and viologene derivatives [217,222], transition metals, coordination complexes (ruthenium polypyridyl [223], porphyrins [224], catenates [225], etc.), and enzymes [226] have been immobilized at the surface of an electrode.

- The inclusion of suitable dopants: [226,227] the key feature of the electropolymerization mechanism is the simultaneous growth and doping. This enables negatively charged species (acting as dopants) to be irreversibly captured providing their size is roughly superior to 1 nm. This has been demonstrated as an efficient way to incorporate sulphonated metalloporphyrins or metallophthalocyanines [229], heteropolyanions [230], and enzymes [231,232]. The entanglement of the polymer network around dopant particles prevents further expulsion by dedoping.

3.9.1 Gas and Vapour Sensing

The first evaluations of ECPs as sensitive components in chemical sensors are based on their redox interaction with some gases, inducing a variation of the doping level, resulting in a quite straightforward conductance monitoring of gas sensor response over several orders of magnitude [233]. Interdigitated electrodes covered by a PPy film have been tested by Miasik et al. for the detection of NH_3, NO_2, and H_2S gases [234]. Since PPy is *p*-doped, electron-donating gases like ammonia reduce the carrier density; hence, there is a resistance increase on exposure to ammonia. In contrast, NO_2 withdraws electron density and increases conductivity. As expected, the sensor shows an increasing response as the concentration is raised from 0.01% through 0.1 to 1% in air, but the response curve is nonlinear.

Yoneyama et al. have shown that electropolymerized PPy films exhibit noticeable gas sensitivities to electron acceptor gases such as PCl_3, SO_2, and NO_2 at room temperature, especially when PPy is reduced electrochemically before exposure [235–237]. On the base of FT-IR spectra investigations, the NO_2-sensing mechanism was found mainly due to chemical doping by NO_2 to give the dopant anion NO_2^-:

$$PPy^\circ + NO_2 \text{ (gas)} \leftrightarrow PPy^+ + NO_2^- \text{ (ion)} \tag{3.78}$$

where PPy° symbolizes the dedoped polypyrrole while PPy⁺ the p-doped material.

These authors also investigated the gas-sensing properties of PTh film [238], but they found a more irreversible behaviour in the conductivity change after exposure to NH_3 and H_2S. The chemical events are not clearly elucidated. For instance, the increase in conductivity of poly (p-phenylene vinylene) [239], upon exposure to ammonia gas, contrasts with the compensation effect (dedoping) behaviour observed with PPy and PTh.

Thin layers of poly(paraphenylene azomethine), $(-\phi-CH=N-\phi-N=CH-)_n$ (PPCN), prepared by vapour phase polymerization, exhibit a good reversible conductivity response, which is sensitive to the partial pressure of doping gas such as I_2 [240]. When this PPCN is modified by introduction of side branches bearing specific functional groups (aromatic, nitro, etc.), the selectivity and sensitivity can be strongly influenced (according to the various gases tested: O_2, CO, CO_2, NO, SO_2, H_2, CH_4, and NH_3) [241]. Good reversibility and response time shortened to a few minutes have been obtained by means of 5- to 30-μm thick layer arrangements on interdigitated electrodes.

PPy doped with amphiphilic anions exhibits an electrochemical behaviour very sensitive to the water contant of the cycling electrolyte. This has been shown by Bidan et al. for self-doped PPy [242] (i.e., grafted by an alkylsulphonate chain) or by Hwang et al. in the case of dodecylsulphonate (DS^-) doped PPy (PPy^+, DS^-) [243]. These latter authors attempt to build a sensitive humidity sensor (1 mm in size) by using a reduced ($PPy°$, DS^-Na^+) layer deposited on an electrode array. The sensor is reported with a stability in air for six months, with a linear change of the resistance on the relative humidity (from 20 Ω to 7 MΩ for RH from 0 to 100%, respectively); however, the reproducibility is poor.

Geniés et al. have shown that a four-probe electrode device coated with polyaniline (PANi) exhibits a gas-sensing response to methanol, ethanol, acetone, and acetonitrile vapours in nitrogen carrier gas (see Figure 3.57) [244].

The presence of basic amino and imino sites along the PANi chains induces important pH and moisture effects on conductivity [245]. This has been exploited by Toppare et al. [246] to build up an NH_3 gas sensor based on PANi and a composite PANi-poly(bisphenol-A carbonate) using a two-probe conductivity measurement technique. The changes in resistance are more well defined and reproducible for the PANi/polycarbonate composite compared to the pure PANi. Relative resistance changes of 9.2, 4.7, and 3.1 were measured at the concentration levels of 0.1%, 0.05%, and 0.025% vol. ammonia, respectively, with good repeatability (applying 60-sec concentration pulses).

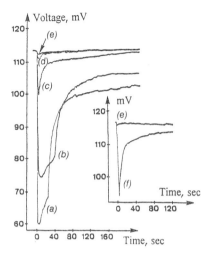

FIGURE 3.57. Film response for different solvents (1000 ppm) in nitrogen carrier gas: (a) MeOH, (b) EtOH, (c) acetone, (d) MeCN, (e) benzene, (f) 100 ppm MeOH (from Geniés et al. [244]).

Another promising assembly for ECP-based chemical sensors is their arrangement in a Field Effect Transistor (FET) device [247] (see Section 2.2.1).

Such electronic devices (SGFETs), operating in the solid state have been developed by Janata et al. for gas sensor applications (see also Section 4.6.2, Figure 4.48). They showed that a poly(N-methylpyrrole)–based FET responds to lower aliphatic alcohols at room temperature and with a time response of seconds [248]. They also demonstrated that, under suitable electropolymerization conditions, where aromatic solvents such as nitrotoluenes are incorporated into the PPy matrix, SGFETs based on PPy/3-nitrotoluene exhibit selective sensitivity to vapour of aromatic compounds [249]. The sensitivity to alcohols and aromatics is explained in terms of N-H or N-π interactions.

Yoshino et al. [250] have investigated gas-sensitive and temperature-dependent Schottky gated field effect transistors utilizing poly(3-butylthiophene). Air, water vapour, ethanol, and chloroform gases show enhancement of source-drain current compared with that in vacuum.

Microelectrode arrays were modified by Wrighton et al. to build up microelectrochemical "transistors" that operate in solution according to redox reactions electrochemically monitored (compared to the "electrostatic" monitoring of FET). The conducting polymers act merely as electrical conductors. By this way, using platinized poly(3-methylthiophene), H_2 and O_2 can be detected [251].

Polyaniline-based microlectrochemical transistor can operate in the solid state by using a poly(vinyl alcohol)/phosphoric acid solid-state electrolyte; in this way water vapours can be detected [252] (see Figure 3.58).

Volume changes in conjugated polymers, due to doping/undoping (i.e., by ion and solvent absorption/desorption) have been observed and exploited in artificial muscle devices [253]. Inganäs et al. have built NH_3 sensors based on a bipolar strip assembly consisting of a conjugated polymer layer [PPy, PTh, or poly(alkylthiophene)] of 8–40 μm of thickness on a polymer substrate (e.g., polyethylene or polyimide) [255]. When the strip is submitted to NH_3 vapour, the small volume change followed by the swelling of the conjugated polymer induces a significant bending of the strip.

Bartlett et al. presented the modelling of the rapid response of gas sensor–chemiresistors [256]. This first approach deals with the gas-polymer interaction limited to the diffusion reaction.

Charlesworth et al. showed that fractional change in the resistance varies lineraly with fractional mass uptake up to about 5% mass change, and the number and type of adsorption sites are the same for three investigated solvents (i.e., water, methanol, and dichloromethane in dry nitrogen at 22°C) [257]. The positive correlation between the specific resistance change (i.e., resistance divided by mass) and the dielectric constant of the solvent can be explained by the variation of the hopping process responsible for the conductivity in ECPs.

In conclusion, interactions between gases, vapours, and ECPs are not fully understood. Redox processes, solvation, acid-base reactions, and aromatic charge-transfer complexes may occur simultaneously. Consequently, specificity of these sensors is low, and the response curves are nonlinear. However, they present the advantage of an easy elaboration with a fast and straightforward response. Some directions to improve selectivity can be suggested, for instance, the functionalization of the ECP by specific interacting groups. For example, bromo-substituted thiophene has been reported specific to bromine vapours [258,259].

3.9.2 Membranes for the Separation of Gases

More recently, the selective gas separation (H_2/N_2, O_2/N_2, CO_2/CH_4, etc.) using ECP-based membranes appeared to be very promising. Integrated devices combining gas separation and detection seem to be feasible.

The techniques used for the separation of gases (see also Section 3.6) have significantly progressed in recent years. The replacement of the conventional method of cryogenic separation by the method of membrane separation results in significant savings in energy costs and in considerable industrial use. One of the most important commercial businesses is the sep-

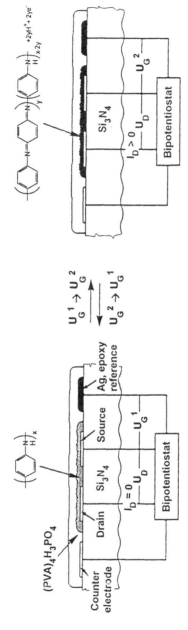

FIGURE 3.58. Solid-state polyaniline-based transistor employing a poly(vinyl alcohol)/phosphoric acid [(PVA)₄/H₃PO₄] solid-state electrolyte. For step variation in U_G between -0.3 V (off state, $I_D = 0$) and $+0.4$ (on state, $I_D > 0$) versus Ag, the turn on/turn off times are about 50 msec. At fixed $U_G = +0.15$ V versus Ag and $U_D = 50$ mV, this H₂O-sensor device turns on and off reversibly as the atmosphere is varied from H₂O saturated to H₂O free, respectively. (From Wrighton et al. [252], reproduced with permission from *J. Am. Chem. Soc.*, 109, p. 6628, ©1987 American Chemical Society.)

215

aration of N_2 from air to be used for protection purposes in the production and containment of inflammable materials. The by-product enriched in O_2 can be utilized as a catalyst for combustion or in medical applications. Kaner et al. have reported that manageable films of self-supported PANi can obtain remarkable selectivity for pairs of commerically interesting industrial gases [260–264].

The highest reported selectivity values for nonconjugated polymers are: 313 for poly(chloro-trifluoroethylene), 16 for cellulose nitrate, and 60 for polyamidefluoride, while these values were 3590 for H_2/N_2, 30 for O_2/N_2, and 336 for CO_2/CH_4, respectively, in the case of PANi. According to the authors, the selectivity was due to the control of porosity at the atomic level. The procedure consisted of the following steps: at first, permeable emeraldine-based films were produced by doping, and then the dedoping of the films was performed from a solution of PANi in 1-methyl-2-pyrrolidone. The size of the imprinted cavities, which were partially filled by redoping (see Table 3.17), were shown to depend on the doping ion that

TABLE 3.17. Effect of Redoping on Oxygen and Nitrogen Permeability through a Polyaniline Membrane (from Kaner et al. [264]).

Level of Membrane Doping		Permeability, Rel. Value		Separation Factor α_{O_2/N_2}
		O_2	N_2	
As cast		0.18	0.02	9
Doped		0.018	0.003	6
Dedoped		0.5	0.06	8.3
R e d o p e d	0.015 M	0.16	0.07	2.3
	0.0175 M	0.04	0.0015	27
	0.02 M	0.0125	0.00125	10

was used (see Section 3.9.4, ion-sieving effect). The resulting permeabilities varied in a controlled manner [260] with the kinetic diameter of the gas (He $<$ H_2 $<$ CO_2 $<$ O_2 $<$ Ar $<$ N_2 $<$ CH_4). This explanation of selectivity based on the open volume in the membrane [263,264] does not exclude other possible interactions, such as those between O_2 and polarons of the PANi membrane, as suggested by Geniés et al. [265], according to the results of electron spin resonance spectra.

To obtain the best compromise between permeability and selectivity, it is necessary to associate the selective membrane to a thin porous material at high pressure (above 150 bars). For this purpose, asymmetric membranes have been developed by depositing thin films of PPy or poly(N-methylpyrrole)$-$PMePy$-$(several μm thick) on membranes of alumina Anopore (Martin et al. [266,267]) or ceramic Membralox (Iwasaki et al. [268]). The coefficients of selectivity observed by these authors with PMePy for the O_2/N_2 couple were eight and three, respectively [266,268], which were significantly less than that obtained for PANi membranes.

3.9.3 Electrochemical Memory Effect

The microelectrochemical transistors with ECPs developed by Wrighton et al. [251,269] are miniaturized sensors for gases (see Section 3.9.1) or ions (see Section 3.9.4). Kaneto et al. [270] have described a similar structure that operates as an electrochemical memory using poly(3-methyl-thiophene) sensing film and a polymeric electrolyte based on P(EO/PO) copolymer. The application of a voltage difference of -5 V between the electrode and drain results in a drain current for a source-drain voltage of 0.2 V. It is important that the cumulative effect of the reading pulses and erasing pulses allows four stable states of this memory, which correspond to four levels of doping.

3.9.4 Ion Detectors

The detection of electroinactive ions by ECP-based electrochemical devices originates from the so-called "ion-sieving" effect, or memory effect [271,272]. This means that it is possible to vary the permeability of electropolymerized films via their controlled anodic growth [273,274] (see also Section 3.9.2)

A PPy film can prevent anions of larger size from penetrating the matrix. The cut-off size of anions is determined by the size of anion that was used for the electropolymerization of pyrrole.

Shirakawa et al. presented a relationship between anodic peak current of differential pulse voltammograms and the diameter of electrolyte anions

[271]. Dong et al. presented a potentiometric chloride sensor based on Cl⁻ doped PPy. The device shows a Nerstian response for 10^{-1}–10^{-4} M Cl⁻ with a slope of 58–60 mV/decade [275]. However, NO_3^-, HCOO⁻, Br⁻, and I⁻ ions strongly interfere. These authors have presented a potentiometric sensor built from a chlorine-doped PPy-coated carbon-fibre microdisc electrode, which may be used to detect chloride in vivo [276].

Heineman et al. introduced the concept of an electrochemical detector based on the repetitive doping-dedoping of PPy for the electrochemical determination of electroinactive anions by flow-injection analysis. At the doping potential of the conductive polymer, the presence of electroniactive ions in solution allows the oxidation of the polymer and causes electron flow (see Figure 3.59). The amperometric response at 0.9 V (versue Ag/AgCl) of the PPy-coated electrode was linear over the concentration range 10 μm to 1 mM [277].

Bulky ions such as glycine, proteins, and polyelectrolytes could be used for mobile buffers since these substances are too big to enter as dopants. The electrode was stable for over two weeks in an anaerobic atmosphere. Ye and Baldwin tested a PANi-modified electrode for flow-injection analysis of electroinactive anions [278]. The stability and sampling seem better than in the case of the PPy-based detector. Regeneration between each ion injection is not necessary, providing that the potential is held to $+0.2$ V (versus Ag/AgCl) where PANi is not fully oxidized or reduced. In favourable cases (i.e., NO_3^- or ClO_4^- injected into a glycine mobile phase), the detection limit (signal/noise = 3) is 0.1 ppm or lower with a linear response (correlation coefficient > 0.99) extended to concentrations at least three orders of magnitude higher.

Wallace et al. showed that the selectivity of this ionic sensor operating in

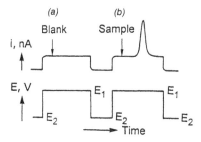

FIGURE 3.59. Schematic representation of polypyrrole electrode detector response to (a) a blank and (b) a sample containing anions. The electrode is held at E_2 potential level between two measurements for undoping and then stepped to E_1 before sample injection to detect anions by doping. (From Heineman et al. [277], redrawn with permission from *Anal. Chem.*, 58, p. 1979, ©1986 American Chemical Society.)

flow-injection mode can be controlled by various factors, i.e., the nature of the counter-ion incorporated during synthesis, the nature of the eluent, and the frequency of the applied potential pulse [279,280]. Martinez et al. developed a microsensor for electroinactive cations in flow-injection analysis based on the same principle [281]. They inverted the sign of the mobile detected ions (i.e., cations) by using a polypyrrole-dodecyl sulphate as sensitive layer. As a matter of fact, the dodecyl sulphate acts as an immobilized doping anion (see Figure 3.55).

In an approach similar to those developed for gas sensors, the specificity to ions may be obtained by functionalization of the ECP by species offering a specific interaction with the ions to be analysed.

An attempt in this direction was made by Wallace et al. who have derivatized a PPy-coated electrode by the carbodithioate ligand. They showed the ability of this poly(pyrrole-N-carbodithioate) electrode to uptake Cu^{2+} ions from solution, allowing a detection limit of 1 ppm [282]. These authors also tested the analytical ability of this electrode toward Hg^{2+} detection [283], but the device suffers from very low stability towards cycling.

Bidan et al. have shown that, during the electropolymerization of PPy, it is possible to form cavities of the catenand type [225]. In solution, these cavities are formed by embedding coordinated groups (L) around a metal center (M^{n+}) (see Figure 3.60), according to the strategy developed by Sauvage [284]. These complexes, which are immobilized in the polymer matrix, can be reversibly demetallized and remetallized by chemical methods. Recently, Blohorn et al. have shown that these ECP with "pseudocatenate" cavities present a selectivity for the ion Cu^I: the film previously demetallized in 90% and then remetallized in a 1:4 mixture of $Cu(CH_3CN)_4BF_4$ and $Co(BF_4)_2$ [265] exhibits an electrochemical response characteristic of a Cu^I complex, whereas that associated with Co is very weak.

Bidan et al. have immobilized heteropolyanions of the Keggin structure $(XMe_{12}O_{40})^{n-}$ with X = P, Si and Me = W, Mo and n = 3, 4 in the ECP, which give electrocatalytic properties to the ECP [230,286–290]. Recently, they have established that the iron-substituted heteropolyanion $[Fe^{III}PW_{11}O_{39}(H_2O)]^{4-}$ (specific to the reduction of nitrite to ammonium ions) retains its electrocatalytic properties when immobilized and associated with PMePy. [289,290]. An amperometric sensor made of a carbon electrode modified by a PMePy/$[Fe^{III}PW_{11}O_{39}(H_2O)]^{4-}$ film and maintained at -1.2 V (versus $Ag/10^{-2}$ M $[Ag^+]$) exhibits a catalytic current that varies linearly with the NO_2^- concentration in the range 5×10^{-6} to 3×10^{-2} M (see Figure 3.61). Note that the interaction of the ion-substituted heteropolyanion occurs by inner-sphere with the nitrite ion by

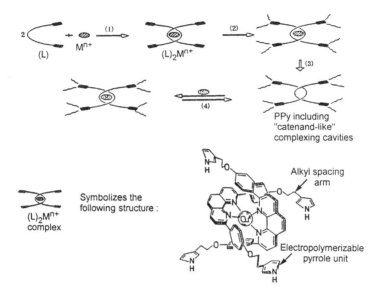

FIGURE 3.60. Synthesis strategy towards polypyrrole film containing complexes of entwined ligands based on a three-dimensional template effect induced by a transition metal (dashed circle); reaction (1): preparation of the $(L)_2M^{n+}$ complex, the ligand L consists of a 2,9-diphenyl-1,10-phenanthroline subunit bearing two electropolymerizable pyrrole nuclei; reaction (2): electro-polymerization of $(L)_2M^{n+}$; reaction (3): after demetallization the preformed cavity is maintained in the polymer matrix; reaction (4): remetallization may be performed using another metal (dotted circle) ($M^{n+} = Cu^+$ in the detailed structure).

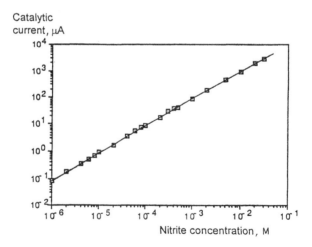

FIGURE 3.61. Curve of the catalytic current, I_{cat} as a function of nitrite concentration. I_{cat} was measured at -1.20 V from cyclic voltammograms obtained for a glassy carbon electrode modified by a poly(N-methylpyrrole) film doped with $[Fe^{III}PW_{11}O_{39}(H_2O)]^{4-}$. The film is prepared by electrolysis, passing 90 mCcm^{-2} at 0.60 V (versus Ag/10^{-2} M [Ag$^+$]) in a CH$_3$CN solution containing 5×10^{-2} M tetrabutylammonium salt of $[Fe^{III}PW_{11}O_{39}(H_2O)]^{4-}$ and 10^{-2} M N-methylpyrrole, as previously described [289,290].

220

formation of a nitrosyl $[Fe^{II}(NO)PW_{11}O_{39}]^{5-}$ complex (see Figure 3.62) in the PMePy bulk, which ensures an excellent selectivity with no interference from the nitrate ions.

The sensitivity to cations has been accomplished by grafting polyether chains on PPy [291–293] or PTh [294–296]. The selective recognition of Li$^+$, Na$^+$, or K$^+$ was improved by grafting crown-ethers onto PPy [297,298] or PTh [299,300] or by the inclusion of 18-dibenzo-6-crown-tetra-sulphonate as doping ion [302]. In particular, Bäuerle et al. [301] have shown that the combination of a conducting polymer and crown-ether units leads to very sensitive and selective cation-responsive materials, which may be used in amperometric metal cation sensors. The transduction phenomenon responsible for the resistance or electroactivity (i.e., cyclic voltammetric response) change seems to be the twisting of the conducting polymer backbone induced by the cation recognition. Recently, Marsella and Swager have been designing conducting polymers (i.e., functionalized PTh) that undergo stimulus-induced conformational changes [300] (see Figure 3.63), and an ionochromic response is reported.

Another approach to improve the ionic selectivity of ECPs is to split the transduction phenomenon between the ionic recognition and the redox events (i.e., to produce a coupling between ionic and electronic fluxes). Cadogan et al. built an all-solid state sodium-selective electrode based on calixarene ionophore in a poly(vinyl chloride) membrane (i.e., the selective component) with a polypyrrole sublayer acting as the driving force by coupling electrical signal with ionic flow [301].

Rault-Berthelot et al. [303,304] have shown that poly(18-dibenzo-6-crown-ether) (an example of electroactive polymer, conjugated partially) presents remarkable properties for extracting Ag$^+$ and Au^{3+}.

Lemaire et al. have grafted chiral groups on polythiophenes in order to recognize chiral ions stereoselectively [such as $(+)$- or $(-)$-10-camphor-sulphonate], which are used as doping agents [305,306]. These structural

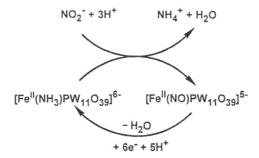

FIGURE 3.62. Reduction pathway of nitrite ions into ammonium ions on the reducing sites of $[Fe^{II}(NO)PW_{11}O_{39}]^{5-}$ immobilized in the poly(*N*-methylpyrrole).

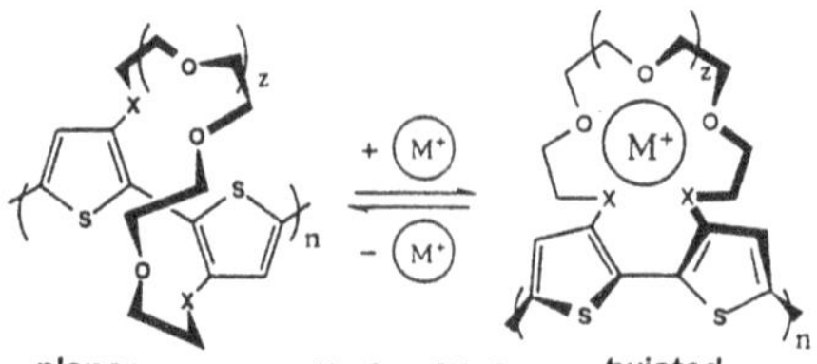

planar
bithiophene
unit

X = O or -CH$_2$-O-
z = 1,2

twisted
bithiophene
unit

FIGURE 3.63. Twisting of the polythiophene backbone induced by the conformation change of the crown ether macrocycle from the uncomplexed to the cation-complexed state. (From Swager et al. [300], reproduced with permission from *J. Am. Chem. Soc.*, 115, p. 12214, © 1993 American Chemical Society.)

interactions are revealed by a significant variation of the doping level and of the voltammogram's shape with the ion chirality.

An ECP-based pH electrode has also been evaluated. Pickup has electrosynthesized poly(3-methylpyrrole-4-carboxylic acid), which includes a carboxylate group directly grafted onto the polymer backbone. The pH sensitivity of this group is transmitted, via donating-withdrawing effects, in terms of variation of the formal doping potential of the polymer [307]. The Wrighton microelectrochemical "transistor" based on platinized poly(3-methylthiophene)(PMeTh) [251] is also a pH-sensitive device. The same device based on PPy grafted by the pH-sensitive bipyridinium redox centre demonstrates a pH-sensitive amperometric response [308].

3.9.5 Electrocontrolled Delivering of Biological Substances of Drugs and Ionic Compounds

The advantage of the reversibe doping/dedoping mechanism of ECPs is to associate ionic movement with electrical currents stoichiometrically. This coupling between ion movements and electrical current can be used in a manner inverse to that used for ion detection. The polymeric conductor in the doped state is considered as a reservoir of anions that can be adjusted in a quantitative fashion by pulses of current for dedoping by known quantities of electricity [see Figure 3.55(a)]. Miller et al. have validated this concept by studying the release of ferricyanide ion [309,310] and observing biological substances such as glutamate ions released in aqueous solution during electrochemical reduction of a PPy film in which the doping agent was previously exchanged by electrochemical cycling [311]. Even dopamine [312] and dimethyldopamine [313] (positively charged under acidic conditions) are released by oxidation of a composite PMePy/poly(styrene sulphonate) [see Figure 3.55(b)].

Miller et al, have also used poly(3-methoxythiophene) to trap and release glutamate [314] and salicylate anions [315]. More recently, Dong et al. have used PPy film that was electrosynthesized in the presence of *p*-toluene sulphonate anion and then exchanged by ATP (5′-adenosine triphosphate) to release the neurotransmitter in an electrically controlled manner [316]. The *N*-methylphenothiazine (NMP), a model compound for neuroleptics of the phenothiazine family, was incorporated after processing in the anionic compound *N*-butylsulphonate phenothiazine [317].

In order to overcome the limitation due to the obligation for using charged drugs, Bidan et al. have recently presented an approach to trap/deliver neutral drugs based on host-guest interaction. The neutral guest NMP is encapsulated in the host heptasulphonated-cyclodextrin ($-CDSO_3^-$) tailor-made to dope PPy [318,319]. Figure 3.64(b) shows the entrapment of NMP in $(PPy^+, -CDSO_3^-)$ compared to the quasi no-incorporation [Figure 3.64(a)] in (PPy^+, ClO_4^-).

3.9.6 Membranes with Electrocontrollable Permeability

The movements of ions across membranes containing fixed ionic sites depend on the nature and number of the charged sites. As early as 1976 Yu and Buvet et al. [320] showed that the capacity of ionic exchange of an ECP could be controlled by application of an electrical potential [321]. Murray et al. have demonstrated that the PPy could be used as an ionic conducting membrane that exhibits different permeabilities to cations and anions according to redox state [322]. Since PPy is associated with a polyanionic polyelectrolyte [323], one can observe that the polarity of mobile charges is reversed from a negative to a positive sign due to the oxidation and reduction of PPy (see Figure 3.55) [325]. The association of a polypyrrole doped by Cl^- (PPy^+, Cl^-) with a polypyrrole incorporating poly(vinyl sulphate) (PPy^{7+}, PVS^-) permits de-ionization of aqueous KCl solutions ranging from 10^{-3} M or 10^{-4} M to 10^{-4} M or 10^{-5} M, respectively [324]. The global reaction can be written as

$$PPy^o + (PPy^+, PVS^-) + KCl \longrightarrow (PPy^+, Cl^-) + (PPy^o, PVS^-, K^+)$$
$$(3.79)$$

The oxidation of neutral PPy (PPy^o) and the reduction of PPy doped by the polyelectrolyte (PPy^+, PVS^-) under galvanostatic conditions (I = 0.05 mA/cm²) causes incorporation of Cl^- and K^+ ions as doping and pseudo-doping ions, respectively (see Figure 3.55).

The ion-exchange capacity varies according to the hydrophilic or hydrophobic modifications of the membrane. The permeation properties of meth-

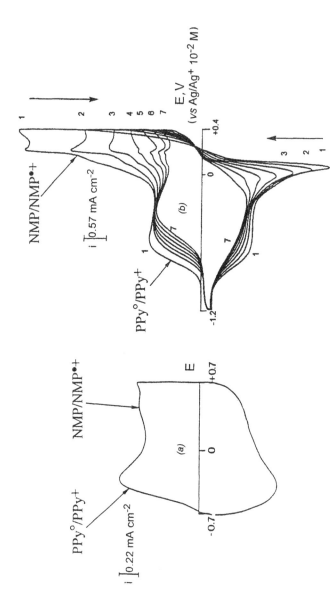

FIGURE 3.64. (a) Cyclic voltammetric curve at 50 mV/s in CH₃CN 0.5 M LiClO₄ of a PPy film synthesized using 0.163 C/cm² at +0.8 V versus Ag/10⁻² M [Ag⁺] from CH₃CN 0.5 M LiClO₄, containing 4 × 10⁻³ M pyrrole and 2 × 10⁻³ M NMP; (b) Successive cyclic voltammetric curves at 10 mV/s in H₂O 0.5 M LiClO₄ of a (PPy⁺, −CDSO₃⁻ [NMP]) modified electrode prepared by dipping a (PPy⁺, −CDSO₃) film (0.3 C/cm²) in CH₃CN 0.1 M NMP for 1.5 hours, then rinsing in CH₃CN and H₂O 0.5 M LiClO₄ for forty-five minutes.

anol in vacuum of an assembly of porous PTFE film on a grid of Au/PPy or PANi/electrolyte containing methanol have been analysed as a function of the inserted gold electrode potential [326]. The conversion from the reduced state to the oxidized state resulted in a decrease in permeability by a factor of 2.5 for PPy, as opposed to an increase by a factor of 1.2 for PANi. This technology (pervaporation) can be applied to the separation of liquids or gases, particularly the separation of water/alcohol mixtures (see Section 3.6.2).

The controlled growth of PANi, PPy, and polyphenol films has been exploited by Wang et al. for changing permeability type in solution. Thus, the electrochemical detection of catechol, H_2O_2, acetaminophene, or hydrazine, during in-line chromatography, is possible in the presence of electrical substances that typically interfere. The deposit of ECP on the electrode used for detection plays the role of selective membrane by size exclusion [327]. Developing the concept of Yu and Buvet, Nagaoka et al. recently reported the electrochemical chromatographic separation of ionic compounds using a conductive stationary phase that consists of glassy carbon particles coated with polyaniline of polypyrrole film [328]. Halide ions (i.e., Cl^-, Br^-, and I^-) with SCN^- were separated successfully on a PANi-coated column at a stationary phase potential of 0.3 V (versus Ag/AgCl).

The discrimination between dopamine and interfering anions in biological solutions (e.g., ascorbic acid) has been accomplished by depositing a 0.5-μm film of overoxidized PPy^+, DDS^- (DDS^- = dodecylbenzene sulphonate) on a glassy electrode. Gao et al. have attributed the high sensitivity and selectivity to the preconcentration of analyte and discrimination of charge [329]. In a similar fashion, Brajter-Toth et al. have shown that a glassy carbon electrode covered by overoxidized PPy exhibited selective permeability to substances such as dopamine due to hydrophobic or electrostatic interactions [330].

3.9.7 Radiation Detectors

Yoshino et al. [331,332] have shown that conducting polymers such as polyacetylene or polythiophene exposed to various gases like SF_6, which are usually not effective as dopants, undergo an insulator-metal transition under irradiation by electron beam, γ-rays, or X rays. The radiation doses can easily be monitored by measuring the changes of electrical conductivity or the absorption spectra. Much higher doses of radiation (> 150 Mrad) can be detected than with conventional CTA (cellulose triacetate) detectors (< 10 Mrad).

3.9.8 Electrochemical Biosensors

Amperometric biosensors transduce the diffusion-controlled flux of a substrate to an enzyme electrode into an electrical current. Biosensors based on ECP/enzyme structures have been a topic of considerable interest in very recent years [333–338]. A remarkable review of Bartlett et al. on the immobilization of enzymes in electropolymerized films (among them ECPs) has been recently presented [334]. As a matter of fact, ECPs bring improvements in two respects:

(1) *Immobilization of the enzyme* – Various techniques for immobilizing an enzyme at the surface of the electrode have been developed: encapsulation, covalent binding, adsorption, and entrapment [339]. The proteins of glucose oxidase (GOD) (isoelectric point $= 4.3$), as well as those of other enzymes, have a negative surface charge at physiological pH because of an excess of glutamate and aspartate over lysine or arginine [340]. Consequently, by suitable choice of the electropolymerization conditions of the ECP, enzymes acting as negative dopants are entrapped through Coulombic interactions. This one-step technique is easy to perform and well reproducible.

(2) *Electrical wiring* – The redox centres of a redox enzyme are located far enough from the outermost surface to be electrically inaccessible. Consequently, most enzymes do not exchange electrons with electrodes on which they are immobilized. The concept of "electrical wiring" of redox enzymes, as introduced by Heller [340,341], covers two pathways: the electron transfer from the surface of the electrode to the electron relay (or mediator) located near the surface of the enzyme, followed by the electron transfer from this relay to the redox centres of the enzyme in its bulk (see Figure 3.65 and Section 4.8.2 for further explanation).

It is important to emphasize at this point that the redox mediator (O_2, ferrocene, quinone) is usually a low molecular mass molecule able to diffuse inside the reduced enzyme (in this case, inside GOD) and to closely approach the active site to oxidize.

The immobilization of the mediator (and the enzyme), a technique by which its addition into the reacting medium can be avoided, must allow movements free enough in order to establish an electrical connection.

To comply with such requirements, Heller et al. [342] have developed three-dimensional wired structures consisting of positively charged redox polymers such as a poly(vinyl pyridine) complex of Os(bpy)$_2$Cl. Following the same "electrical wiring" idea, Skotheim et al. have reported the use of electron-transfer relays based on highly flexible polysiloxanes grafted by ferrocene mediators [343]. Ferrocene mediator has also been grafted via a flexible arm at the outer surface of the glucose oxidase [344]. Conducting polymers such as PPy provide a three-dimensional electroconducting structure (at operating potentials where PPy is doped), as well as an easy functionalization by the redox mediator, either by covalent grafting or by inclusion of specific dopants (see the introduction of Section 3.9). This is in accordance with both the functions required by the "electrical wiring."

3.9.8.1 GLUCOSE SENSORS

In their pioneering works, Umana et al., [345], Bartlett et al. [336,337], and Foulds et al. [346] have reported the electrochemical entrapment of glucose oxidase (GOD) into PPy [345–348] and PMePy films [336,337] to build up amperometric biosensors. The determination of the apparent Michaelis-Menten constant shows that the immobilized enzyme activity is comparable to those of the enzyme in solution [346].

The redox relay (Mox/Mred, see Figure 3.65) may be the natural electronic mediator O_2. However, the use of other oxidizing mediators to regenerate GOD has been strongly investigated. Among them, ferrocene derivatives [349,350], quinone derivatives [351–353], and organic conducting salts such as TTF-TCNQ [354–356] have been the most employed in various biosensor devices.

When O_2 mediator is involved, the regeneration of the enzyme is the following reaction:

$$GODH_2 + O_2 \longrightarrow GOD + H_2O_2$$

$$H_2O_2 \longrightarrow O_2 + 2H^+ + 2e^- \text{ (at the electrode)}$$

$$(3.80)$$

where $GODH_2$ refers to the reduced form of GOD.

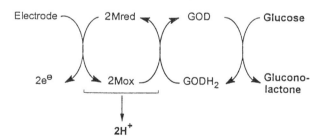

FIGURE 3.65. Electrical wiring of a redox enzyme; flow of electrons from the electrode through a monoelectronic relay (Mox/Mred) to the glucose oxidase enzyme GOD; $GODH_2$ is the reduced form.

The amperometric detection of hydrogenperoxide has been exploited in various ways:

(1) The direct detection of the oxidation current of the produced H_2O_2: It seems that H_2O_2 has to diffuse to the electrode/ECP interface in order to be oxidized, since there is no redox involvement of the PMePy [335] or PPy [357–359] matrices. The simultaneous incorporation of flavin coenzyme with GOD into PPy film results in an increase of catalytic activity for glucose oxidation (H_2O_2 oxidation at 0.6 V versus Ag/AgCl). However, the immobilized flavin coenzyme does not act as an electron transfer mediator between GOD and PPy film [360].

(2) An alternative is the determination of H_2O_2 by Mo^{VI} catalysed reduction with iodide ions, the produced iodine being amperometrically detected at 0.2 V (versus Ag/AgCl) [345].

(3) In order to decrease the potential of detection of H_2O_2 (and consequently to decrease the PPy matrix degradation) horseradish peroxidase (HRP) has been introduced simultaneously with GOD. This bienzyme electrode has been developed by Tatsuma et al. [347] and functions as a glucose sensor on the basis of H_2O_2 reduction catalysed by HRP at +150 mV versus Ag/AgCl. Two configurations of the bienzyme electrode have been examined: a homogeneous GOD/HRP/PPy and a bilayer electrode GOD/PPy//HRP/PPy assembly; this latter configuration gives a higher efficiency.

(4) A potentiometric detection of H_2O_2 has also been reported, with the advantage of reducing the PPy matrix degradation [348].

(5) An assembly of two platinum electrodes interdigitated with a PANi/GOD film responds to different concentrations of glucose according to variations in electrical resistance (measured at 1.33 kHz for a current of 5 nA). Rather than an oxidation (doping) by the produced H_2O_2, this resistance change may result from the formation of gluconic acid, which modifies the local pH [361]. The dependence of the PANi conductivity with pH is well known [362] but does not exhibit a linear response for glucose concentrations more than 10 mM.

(6) A PANi matrix appears less sensitive to degradation by O_2 or H_2O_2 than PPy or even PMePy. Shinohara et al. detect the O_2 consumption (turned into H_2O_2) amperometrically, the electrode being held at the reduction potential of O_2 (-0.5 versus Ag/AgCl). But at this potential, PANi is not electroconductive; this implies that the PANi/glucose oxidase layer is permeable to oxygen that is deposited on the platinum electrode [363].

A recent improvement in PANi/glucose oxidase electrodes was reported by

Shaolin et al. [364] by introducing glucose oxidase at a controlled pH of 5.5 after the electrosynthesis of PANi. The linear response is extended from 0.1 mM to 50 mM. Surprisingly, it seems that the enzyme, in spite of its size, is effectively incorporated within the film, and the lifetime during storage in phosphate buffer was reported longer than sixty days. This may be due to a different porous structure of PANi compared to PPy. The recent covalent binding of glucose oxidase at the surface of amino-functionalized polypyr-role films [365] indicates that the enzyme does not penetrate the PPy bulk. On the other hand, Sauvage et al. reported the immobilization of GOD by dipping a Ni-cyclam–modified polypyrrole film in a phosphate buffer solu-tion containing 1 mg/ml GOD [366]. They reported stability in the sensor responses for at least one week. This discrepancy can be explained on the basis of the results of Tamiya et al. [367]. They showed that adsorption of GOD onto PPy film strongly depends on the electropolymerization condi-tions: PPy electrosynthesized from aqueous solution readily incorporates GOD in contrast to inactive PPy film synthesized from CH_3CN medium. Yaniv et al. also showed that electropolymerization potential has a profound effect upon the microdistribution of GOD within PPy film on the base of scanning tunneling microscopy experiments [368].

At this point, two key questions may arise:

(1) Does the enzyme incorporation in ECPs result in a really more efficient assembly than the use of other polymeric and inert matrices; what is the interest in terms of surface concentration of the active enzyme?

(2) How do you improve the electrical wiring (in consequence of the answer to the upper question): either by using new mediators or by inducing direct electronic transfer?

Bartlett and Whitaker have modelled the function of a PMePy/GOD elec-trode with oxygen mediator [337]. Marchesiello et al. have modified this model, also taking into consideration the conducting behaviour of the PPy matrix [369,370], i.e., the possibility of direct electron exchange with the mediator benzoquinone (see Figure 3.66).

It was demonstrated that the ECP sensor is more sensitive by a factor of approximately 2.5 and exhibits a rapid amperometric response, which is in-dependent of the film thickness compared to sensors using inert matrices (i.e., without including electroactive species). The thickness of the reaction layer is limited by the diffusion coefficient of glucose at the PPy interface $(D = 4.3 \times 10^{-10} \text{ cm}^2\text{sec}^{-1})$, which is significantly less than that in solu-tion $(D = 7 \times 10^{-6} \text{ cm}^2\text{sec}^{-1})$.

Table 3.18 shows that sensors based on polypyrrole possess relatively weak sensitivities for the immobilized volume, since the catalytic current corresponds to the third of an enzymatic monolayer. Different attempts have

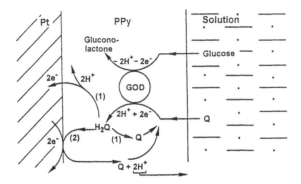

FIGURE 3.66. Schematic representation of the electrocatalytic oxidation of glucose by a PPy-GOD electrode in the presence of benzoquinone (Q) mediator: pathway (1): oxidation of the reduced form of the mediator (H_2Q) "via" the PPy matrix; pathway (2): oxidation of H_2Q after diffusion to the electrode/PPy interface (from Geniés et al. [369]).

been made to increase the surface concentration (i.e., incorporate the active enzyme to the ECP bulk).

For example, Lowe et al. have grafted polymerizable pyrrole groups onto the lysyl groups located at the outer part of the enzyme GOD. The modified enzyme exhibits the same properties as the unmodified enzyme. In addition, the stability of this PPy-GOD is six times better than that of the PPy/GOD. Moreover, this new procedure immobilizes an amount of the enzyme that is approximately 50% more [371,372].

TABLE 3.18. Sensitivity (in Equivalent Monolayers) for Various Glucose Sensors from Literature Data. The GOD Enzyme is Immobilized in Different Polymers.

Source	Polymer and Mediator	Equivalent in Monolayers
Marchesiello and Geniés [353,369]	PPy Benzoquinone	0.3
Foulds and Lowe [346]	PPy O_2	0.28
Palmisano et al. [375]	PPD* O_2	1.25
Heller et al. [341]	Redox and polycationic polymer**	4.7

* PPD = poly(o-phenylenediamine).
** $[Os(bpy)_2Cl]^{1+/2+}$ completed by poly(vinyl pyridine).

Another approach, reported by Schuhmann et al. [365], consisted of grafting the enzyme to polypyrrole films bearing free amino groups. However, the PPy film is substituted in an inhomogeneous manner, as it was previously observed by Skotheim for PPyFc [373]. Recently, Schumann et al. have also explored the possibility of grafting GOD to the surface of an electrodeposited poly(*p*-aminophenyl-2-azulene) film [374]. The small effect of volume (an increase of 40% in the current for an increase in the thickness by a factor of ten) was interpreted to be due to an increase in the surface ruggedness. The electrode of polyazulene-GOD exhibited a rapid response to the enzyme, but the direct transfer was not detected [374].

Cosnier et al. have developed polypyrroles that are formed on a precursor gel of monomeric pyrrole amphiphile, i.e., bound by long chains (C_{12}) of alkylammonium [376]. In addition to the preimmobilization of the enzyme under nondenaturing conditions in a monomer gel, the presence of extra positive charges added to doping charges and the hydrophobic character of the gel favour the retention of the enzyme in the resulting polymer. For polyphenoloxidase, a threshold of 2×10^{-9} and 5×10^{-8} M has been obtained for catechol and dopamine detection, respectively [377]. Moreover, Marchesiello et al. have shown that the absorption of glucose oxidase to the surface of PPy and PMeTh films results in amperometric sensors that are more sensitive (especially in the case of PMeTh) and significantly more stable over time than sensors based on immobilization of the enzyme in the PPy bulk during polymerization of the pyrrole [378]. This method of adsorption permits the use of polymers that cannot be synthesized in water such as PMeTh.

In order to overcome the degradation of the PPy matrix in the presence of H_2O_2 at the potential value of 0.7 V (versus Ag/Ag^+) and also in order to operate independently of oxygen tension fluctuations, the use of oxidizing mediators to regenerate the immobilized GOD has been investigated.

Thus, *p*-benzoquinone has been added to the GOD solution analysed by PPy-GOD–based sensors [232,353,379]. However, in order to be compatible with in vivo bioapplications, the electron mediator must be immobilized. A one-step enzyme entrapment in a PPy grafted by a ferrocene electron mediator has been realized by Foulds et al. [380,381], but the co-immobilized GOD-ferrocene assembly exhibited a reduced catalytic efficiency, reflecting the steric hindrance and low mobility of the mediator, preventing the electrical wiring from being accomplished [340].

Recently, Schuhmann et al. succeeded in opening the field of miniaturized reagentless biosensors [382]. They realized the covalent binding of enzyme (GOD) and redox mediator (osmium complex) on the sensor surface, using adequately grafted pyrrole monomers. The electrochemical terpolymerization of *N*-methylpyrrole (as a spacer unit), pyrrole-modified glucose oxidase, and [Os(2,2′-bipyridine)$_2$(3-{Pyrrol-1-ylmethyl}pyridine)Cl]$^-$ leads to ternary copolymers simultaneously integrating redox relays and enzyme activity. This device involves a direct electrical communication of the enzyme with the electrode via the polymer-bound redox relays as it could be deduced from the encouraging results obtained previously by Heller et al. using other osmium complexes [342]. Moreover, the electrochemical building method based on functionalized ECPs allows the miniaturization of this amperometric enzyme sensor into individual spots deposited onto a gold substrate with a diameter of about 150 μm [382].

In order to gain in mobility while keeping immobilization within the film, GOD and anionic mediators were simultaneously incorporated as dopants during the PPy electrosynthesis. Yoneyama et al. have recently entrapped GOD and ferrocene carboxylate anions in PPy [383], but there was one serious drawback in that case: ferrocene carboxylate anions did not show high reversibility in PPy films.

Better results were obtained using hydroqinone-sulphonate ions (HQS) as mediator [384]. They suggested, from spectroscopic measurements, that HQS adsorbes on the protein shell of GOD. The incorporation of this HQS-adsorbed GOD complex performs the wiring of the enzyme to PPy; however, the quantity of the enzyme that is immobilized remains small, and at the end of one week the amperometric response falls by 70%. Treatment of the GOD-HQS complex by urea (a mild denaturing agent) improves the sensitivity of the PPy/GOD-HQS film [385].

Tarasevitch et al. have introduced another redox quinonic component, tiron in PMePy simultaneously with GOD and a fluorinated sulphonated polyanionic electrolyte (altion). Its addition to the polymerization solution causes the formation of elastic smooth film of a regular structure and good adhesion. The tiron incorporated does not seem to play the role of redox mediator to recycle GOD but has a catalytical activity in H$_2$O$_2$ electrooxidation. The comparable idea of co-immobilization of the GOD and ferrocyanide does not result in any significant improvement [386].

The best way to eliminate problems with electronic mediators should be based on direct oxidation of the GODH$_2$ without interference of mediators. Recent results are very promising. Aizawa et al. showed, by differential voltamperometry with a pulsed current, that direct electron transfer is possible between the PPy and the GOD [387]. Kimura et al. used this idea for the measurement of glucose in the absence of mediators, but in the process the denaturation of the enzyme causes the sensor to lose its selectivity (i.e., other sugars will also be detected) [388]. Recently, Nolte et al. have built a biosensor in which the nondenatured GOD exchanges electrons directly with the polymer conductor [PPy, PMeTh or poly(3-hexylthiophene)], which is electrochemically incorporated into the pores of a polycarbonate filtration membrane [389–392]. The pores are partially filled, which results in the formation of microtubules of ECP, 1 μm in diameter, containing GOD introduced by simple soaking. The sensor can operate continuously for two weeks without alteration of the response. The sensor is selective for glucose concentrations ranging from 0.1 to 250 mM (linear response up to 25 mM). The surface interaction between PPy and GOD is electrostatic in nature; the concave form and internal ruggedness of the PPy microtubule should fit to the enzyme dimension to allow the electron transfer. An alternative method is to recoat the internal walls of the pores with PPy (the interspheric space created from stacking 220-nm thick latex membranes). These pores are then filled with GOD by simple soaking. This behavior also suggests direct electron transfer between the enzyme and the PPy [393]. Lu et al. proposed that a direct electron transfer from the active redox center of GOD is possible in PANi, on the basis of a similar redox behaviour exhibited by 7-acetyl-10-methyl-isoalloxazine (a model compound of GOD active redox center) immobilized into PANi [394]. Bannister et al. presented a similar approach to mediatorless sensors based on ECPs [395]. Voltammetric studies of a poly(indole-5-carboxylic-acid) modified glassy carbon electrode showed a catalytic effect to ascorbate and NADH, without the use of electron-transfer mediators.

3.9.8.2 BIOSENSORS BASED ON OTHER ENZYME SYSTEMS

Glucose oxidase served as the test enzyme for use of ECPs in the fabrication of biosensors. Other enzymes were subsequently immobilized in polypyrrole (e.g., peroxidase for the reduction of H_2O_2 [347,396,397], alcohol dehydrogenase for the measurement of ethanol [398], pyruvate oxidase [399], tyrosinase or polyphenoloxidase for the measurement of phenol derivatives [400], NADH dehydrogenase for oxidation of NADH [401], glucose dehydrogenase for the measurement of glucose [402], glutamate dehydrogenase [403,404], and urease for detection of urea [405]). Very low concentrations of uric acid can be detected by the PANi/uricase electrode [406–408]. The co-immobilization of two or three enzymes in PPy has also been published (e.g., cholesterol oxidase and cholesterol esterase for the measurement of cholesterol in blood [409,410] and invertase, mutarotase, and glucose oxidase for the measurement of saccharose [411]).

These studies illustrate the simplicity of using PPy or other ECPs as immobilization matrices, since all of the partners needed in the preparation of biosensors (enzymes, cofactors, and mediators) can be incorporated in the ECP films simultaneously [338,382,398,402,410].

Another significant example is that of nitrate reductase immobilized by electrostatic capture or covalent grafting to the surface of the PTh film activated by the mediator viologen. Such an assembly has applications in the quantitative analysis of nitrate [412]. In addition, other "ECP-enzyme collaborations" have been exploited.

The enzyme function can be electrochemically tuned by controlling the amount of cosubstrate or coenzyme inserted in the ECP film. Variation of the potential applied to the ECP-enzyme electrode modifies the doping level, and, consequently, the anion amount that penetrates the ECP membrane will also be changed [413]. Thus, glutamate dehydrogenase activity was controlled by the amount of L-glutamate and NADP$^+$ oxireductase in PPy [404]. Similarly, pyruvate oxidase, which requires the presence of phosphate anion, has its activity modulated by the loss (dedoping) or gain (doping) of phosphate anions in the PPy film [399].

Uchida et al. have developed a penicillin sensor based on a resistive assembly of a Wrighton-type electrode [252] interdigitated by a copoly(pyrrole/N-methylpyrrole) film that is covered by a membrane containing some penicillinase [414,415]. The protons produced by the enzyme modify the conductivity of the copolymer connecting the electrode bands.

Bartlett et al. developed a similar enzyme "switch" made of microband carbon electrodes interconnected by a layer of PANi, which was covered with an electrochemically deposited layer of poly(1,2-diaminobenzene) including GOD [416]. The mediator TTF$^+$/TTF interacts by redox exchange to the PANi layer, thus modulating its conductivity.

3.9.8.3 BIOELECTRODES

Many biosubstances that interfere in biocatalysis have been immobilized in ECP film deposited on an electrode for the realization of a biomimetic electrode. Monica et al. have shown that immobilized cytochrome c exhibits quasi-reversible electron transfer in overoxidized PPy films [417,418]. Moreover, Aizawa et al. reported reversible electron transfer for pyroquinoline quinone (a cofactor for certain oxireductases) at the PPy surface [419], a reaction that is irreversible at the surface of conventional electrodes. An important electrocatalytic effect was reported for the oxidation of NADH on the PMeTh surface [420]. Cosnier et al. realized "regiospecificity" for the electrochemical reduction of NAD$^+$ to a biochemically active cofactor (1,4-NADH) in a PPy film that was grafted to the complex [Rh^{III}(C$_5$Me$_5$)(bby)Cl]$^+$(PPy-Rh^{III}).[421,422]. These authors have exploited this electrochemical regeneration of NADH for the enzymatic reduction of pyruvate to L-lactate using the electrochemical chain: electrode//PPy-Rh^{III}/NADH/pyruvate reductase//pyruvate. Calvo et al. have included the anionic natural coenzyme flavin mononucleotide (FMN) in the PPy; however, anionic retention decreases at the potential where

FMN is electroactive since PPy is in its reduced state [423]. On the other hand, Cosnier and Innocent have obtained efficient incorporation of FMN in PPy (cationic and amphiphilic) by ionic exchange [424]. The redox properties of FMN, particularly the catalytic reduction of O_2, were conserved.

3.9.8.4 THE ECP/BIOLOGY INTERFACE: MOLECULAR RECOGNITION

The advantages discussed in the beginning of Section 3.9.8, in addition to the possibility for application to molecular interaction of recognition [425], prevail over the problems encountered to express the enzyme activity in volume. Consequently, there is a tendency to go beyond construction of enzymatic sensors and integrate assemblies of biological substances to the ECP bulk to create biomimetic electrodes (see Section 3.9.8.3). Wallace et al. have exploited the coupling between molecular interactions of biological substances in PPy [279] and the modulation of permselectivity (see Section 3.9.6) that is controlled electrochemically [426]. Accordingly, application of a potential to the stationary phase of ECP was used in chromatography systems for the separation of caffeine and theophylline [427] or to improve the efficiency of separation based on antibody/antigen affinity, e.g., an antibody against human serum albumin (HSA) immobilized in the PPy phase [428]. The activity of the protein HSA that was incorporated was measured by the ELISA test. Shimidzu has used ionic exchange after electrosynthesis to introduce single-stranded DNA sequence into a PPy film [429]. This method of introduction suffers from a lack of irreversibility. A potential response in the mV range was observed in the presence of a solution containing the target of complementary sequence.

Recently, Bidan, Téoule et al. have irreversibly immobilized an oligonucleotide on a polypyrrole matrix by chemical binding, which resulted in excellent stability and reversible hybridization of the DNA fragment into the PPy bulk [430].

Tripathy et al. [431] developed a biotinylated poly(3-hexylthiophene-*co*-3-methanolthiophene) in a Langmuir-Blodgett configuration, which is to immobilize a photoactive antenna protein via a streptavedin bridging protein. Such a molecular architecture may serve a signal transduction role for potential biosensor.

In conclusion, the topic of interface between ECPs and biology is rich in recent developments and potential applications, including the electrocontrolled delivering of medically important substances, chromatography by molecular recognition, enzymatic sensors, biomimetic electrodes, and immunosensors.

The physico-chemical modulation of the electrical conductivity of ECPs, their straightforward response, and easy electrodeposition as film and functionalization, either by entrapment of anions or by covalent grafting, make these materials attractive candidates to build up chemical (gas) sensors and electrochemical sensors (ionic detectors and biosensors). However, the structure of ECPs, as well as the mechanisms and parameters that influence it has not clearly been understood so far. Consequently, real progress cannot be accomplished to overcome their lack in selectivity and their slow degradation.

In the case of biosensors, the biological component is also a limiting factor for the lifetime of the device. The lifetime of the ECP matrix may be considered of secondary importance, in comparison to the innovating contributions of ECPs [338]:

- simultaneous incorporation of all components needed in the preparation of biosensors (enzymes and mediators)
- possibilities for the direct electrical wiring between enzyme and ECP
- an immobilized ECP matrix as a potentiometric or amperometric transducer with possibilities for miniaturization in electrochemical transistor assemblies.

3.10 REFERENCES AND SUPPLEMENTARY READING

1. Kittel, Ch., *Introduction to Solid State Physics,* John Wiley and Sons, Inc., New York (1981).
2. Cady, W. G., *Piezoelectricity,* McGraw-Hill Book Co., New York, (1946).
3. Kawai, H., "The Piezoelectricity of Polyvinylidene Fluoride," *Japan J. Appl. Phys.,* 8 (1969), pp. 975–976.
4. Sessler, G. M. and Berraissoul, A., "Tensile and Bending Piezoelectricity of Single-Film PVDF Monomorphs and Bimorphs," *IEEE Transactions on Electrical Insulation,* Vol. 24, No. 2 (1989), pp. 249–254.
5. Gross, B., Gerhard-Multhaur, R., Berrassioul, A. and Sessler, G. M., "Electron-Beam Poling of Piezoelectric Polymer Electrets," *J. Appl. Phys.,* 62 (4) (1987), pp. 1429–1432.
6. Chatigny, J. V., "Piezofilm – Ein Sensorik-Basismaterial," *Elektroniker* Nr. 7 (1988).
7. Sessler, G. M., "Acoustic Sensors," *Sensors and Actuators A.* 25–27 (1991), pp. 323–330.
8. Bauer, F., "Properties Ferroelectriques et response sous choc du Polymere PVF_2 et des Copolymeres VF_2/C_2F_3H Polareises," *Conference Capteurs 86 Technologie et Applications,* Paris (1986), pp. 242–251.
9. Abdullar, M. I. and Das Gupta, D. K., "Electrical Properties of Ceramic/Polymer Composites," *IEEE Trans. on Electrical Insulation,* Vol. 25, No. 3 (1990).
10. Möckl, T., Magosi, V. and Eccardt, C., "Sandwich-Layer Transducer–A Versatile

Design for Ultrasonic Transducers Operating in Air," *Sensors and Actuators, 21–23* (1990), pp. 687–692.

11. Baker, W. O., "Polymers in the World of Tomorrow," *Proc. of the Symp. at the 2. Chemical Congress of the North American Continent,* Las Vegas (1980), pp. 165–202.

12. Whatmore, R. W., "Pyroelectric Devices and Materials," *Rep. Prog. Phys., 49* (1986), pp. 1335–1386.

13. Ruppel, W., "Pyroelectric Sensor Arrays on Silicon," *Sensors and Actuators A, 31* (1992), pp. 225–228.

14. Ploss, B. and Bauer, S., "Characterization of Marterials for Integrated Pyroelectric Sensors," *Sensors and Actuators A, 25–27* (1991), pp. 407–411.

15. Okuyama, M., Togami, Y., Taniguichi, H., Hamakawa, Y., Kimita, M. and Denda, M., "Basic Characteristics of an Infrared CCD with a Pyroelectric Gate," *Sensors and Actuators A, 21–A23* (1990), pp. 465–468.

16. Biddle, M., Rickert, S. E., Lando, I. B. and Laschewsky. "The Use of the Langmuir-Blodgett Technique to Obtain Ultra-Thin Polar Films," *Sensors and Actuators, 20* (1989), pp. 307–313.

17. Norton, H. N., *Sensor and Analyzer Handbook,* Prentice Hall, Inc., Englewood Cliffs, NJ (1982).

18. Morioka, Y. and Kobayashy, J., "Sorption Hysteresis and Network Structure of Pores in Porous Substances. Part 1.—Simulation of Capillary Condensation in Adsorption Process," *Nippon Kagaku Kaishi* (1979), pp. 157–163.

19. Dubinin, M. M. and Astakhov, V. A., "Description of Absorption Equilibria of Vapours on Zeolites over Wide Ranges of Temperature and Pressure," *Molecular Sieve Zeolites, Vol. II,* Gould, R. F., ed., in *Adv. in Chemistry Series,* No. 102, Amer. Chem. Society, New York (1971).

20. Boltshauser, T., Chandran, L., Baltes, H., Base, F. and Steiner, D., "Humidity Sensing Properties and Electrical Permittivity of new Photosensitive Polyimides," *Sensors and Actuators B, 5* (1991), pp. 161–164.

21. Nakaasa Instrument Co. Ltd., Catalogue, c 911-84001-15K-2′D (1984), p. 78.

22. Boltshauser, T. and Baltes, H., "Capacitive Humidity Sensors in SACMOS Technology with Moisture Absorbing Photosensitive Polyimide," *Sensors and Actuators A, 25–27* (1991), pp. 509–512.

23. Denton, D. D., Senturia, S. D., Anolik, E. S. and Scheider, D., "Fundamental Issues in the Design of Polymeric Capacitive Moisture Sensors," Digest of Techn. Papers, *3rd Int. Conf. on Solid-State Sensors and Actuators (Transducers '85),* Philadelphia, PA (1985), pp. 202–205.

24. Ralston, A. R. K., Buncick, M. C., Denton, D. D., Boltshauser, T. E., Funk, J. M. and Baltes, H. P., "Effects of Aging on Polyimide: A Model for Dielectric Behaviour," Digest of Tech. Papers, *1991 Int. Conf. on Solid State Sensors and Actuators (Transducers '91),* San Francisco, CA (1991), pp. 759–763.

25. Matsuguchi, M., Sadaoka, Y., Nosaka, K., Ishibashi, M., Sakai, Y., Kuroiwa, T. and Ito, A., "Effect of Sorbed Water on the Dielectric Properties of Acetylene-Terminated Polyimide Resins and Their Applications to a Humidity Sensor," *J. Electrochem. Soc., 140/3* (1993), pp. 825–829.

26. Hamann, C., Kampfrath, G. and Müller, M., "Gas and Humidity Sensors Based on Organic Active Thin Films," *Sensors and Actuators B, 1* (1990), pp. 142–147.

27. Lin, J., Möller, S. and Obermeier, E., "Thin-Film Gas Sensors with Organically

Modified Silicates for the Measurement of SO_2," *Sensors and Actuators B*, 5 (1991), pp. 219–221.

28. Endres, H. E. and Drost, S., "Optimization of the Geometry of Gas-Sensitive Interdigital Capacitors," *Sensors and Actuators, B* 4 (1991), pp. 95–98.

29. Haug, M., Schierbaum, K. D., Endres, H. E., Drost, S. and Göpel, W., "Controlled Selectivity of Polysiloxane Coatings: Their Use in Capacitance Sensors," *Sensors and Actuators A*, 32 (1992), pp. 326–332.

30. Ruschau, G. R., Newnham, R. E., Runt, J. and Smith, B. E., "0-3 Ceramic/Polymer Composite Chemical Sensors," *Sensors and Actuators*, 20 (1989), pp. 269–275.

31. Forlani, F. and Prudenziati, M., "Electrical Conduction by Percolation in Thick Film Resistors," *Electrocomp. Sci. and Techn.*, 3 (1976), p. 77.

32. Lundberg, B. and Sundqvist, B., "Resistivity of a Composite Conducting Polymer as a Function of Temperature, Pressure, and Environment: Applications as a Pressure and Gas Concentration Transducer," *J. Appl. Phys.*, 60 (3) (1986), pp. 1074–1079.

33. Harsányi, G., "Polymer Thick-Film Technology: A Possibility to Obtain Very Low Cost Pressure Sensors," *Sensors and Actuators A*, 25–27 (1991), pp. 853–857.

34. Harsányi, G. and Hahn, E., "Thick-Film Pressure Sensors," *Mechatronics*, Vol. 3 No. 2 (1993), pp. 167–171.

35. Lachinov, A. N., "Polymer Films as a Material in Sensors," *Sensors and Actuators A*, 39 (1993), pp. 1–6.

36. Hu, K. A., Moffatt, D., Runt, J., Safari, A. and Newnham, R., "V_2O_3-Polymer Composite Thermistors," *J. Am. Ceram. Soc.*, 70 (8) (1987), pp. 583–585.

37. Bruschi, P. and Nannini, A., "Low Temperature Behaviour of Ion-Beam-Grown Polymer Metal Composite Thin Films," *Thin Solid Films*, 196 (1991), pp. 201–213.

38. Bruschi, P., Cacialli, F. and Nannini, A., "Sensing Properties of Polypyrrole-Polytetrafluoroethylene Composite Thin Films from Granular Metal-Polymer Precursors," *Sensors and Actuators A*, 32 (1992), pp. 313–317.

39. Neuburger, G. G. and Warren, P. C., "Chemically Actuated Electronic Switch," *Sensors and Actuators B*, 1 (1990), pp. 326–332.

40. Yamazone, N. and Shimizu, N., "Humidity Sensors: Principles and Application," *Sensors and Actuators*, 10 (1986), pp. 379–398.

41. Watanabe, M., Sanui, K., Ogata, N., Kobayashi, T. and Ohtaki, Z., "Ionic Conductivity and Mobility in Network Polymers from Poly(propylene oxide) Containing Lithium Perchlorate," *J. Appl. Phys.*, 57 (1) (1985), pp. 123–128.

42. Sorensen, P. R. and Jacobsen, T., "Conductivity, Charge Transfer and Transport Number—An AC-Investigation of the Polymer Electrolyte LiSCN-Poly(ethylene oxide)," *Electrochimica Acta*, Vol. 27. No. 12 (1982), pp. 1671–1675.

43. Sakai, Y., Sadaoka, Y., Matsuguchi, M. and Hirayama, K., "Water Resistive Humidity Sensor Composed of Interpenetrating Polymer Networks of Hydrophilic and Hydrophobic Methacrylate," *Transducers '91, Int. Conf. on Solid State Sensors and Actuators*, San Francisco, CA (1991), pp. 562–565.

44. Huang, P. H., "Halogenated Polymeric Humidity Sensors," *Sensors and Actuators*, 13 (1988), pp. 329–337.

45. Yan, H. and Lu, J., "Solid Polymer Electrolyte Based Electrochemical Oxygen Sensor," *Sensors and Actuators*, 19 (1989), pp. 33–40.

46. Tsuchitani, S., Sugawara, T., Kinjo, N., Ohara, S. and Tsunoda, T., "A Humidity Sen-

sor Using Ionic Copolymer and Its Application to a Humidity Temperature Sensor Module," *Sensors and Actuators,* 15 (1988), pp. 375–386.

47. Sakai, Y., Sadaoka, Y. and Matsuguchi, M., "A Humidity Sensor Using Cross-Linked Quaternized Polyvinylpyridine," *J. Electrochem. Soc.,* Vol. 136, No. 1 (1989), pp. 171–174.

48. Vacik, J. and Kopecek, J., "Specific Resistances of Hydrophilic Membranes Containing Ionogenic Groups" *Journal of Appl. Polymer Sci.,* Vol. 19 (1975), pp. 3029–3044.

49. Sheppard Jr., N. F., Tucker, R. C. and Salehi-Had, S., "Design of a Conductimetric pH Microsensor Based on Reversibly Swelling Hydrogels," *Sensors and Actuators B,* 10 (1993), pp. 73–77.

50. Christensen, W. H., Sinha, D. N. and Agnew, S. F., "Conductivity of Polystyrene Film upon Exposure to Nitrogen Dioxide, a Novel NO_2 Sensor," *Sensors and Actuators B,* 10 (1993), pp. 149–153.

51. Maseeh, F., Tierney, M. A., Chu, W. S., Joseph, J., Kim, H. L. and Otagawa, T., "A Novel Silicon Micro Amperometric Gas Sensor," *Transducers '91, Int. Conf. on Solid State Sensors and Actuators,* San Francisco, CA (1991), pp. 359–362.

52. Otagawa, T., Madou, M., Wing, S. A., Rich-Alexander, J., Kusanagi, Sh., Fujioka, T. and Yasuda, A., "Planar Microelectrochemical Carbon Monoxide Sensors," *Sensors and Actuators B,* 1 (1990), pp. 319–325.

53. Miura, N., Harada, T., Shimizu, Y. and Yamazone, N., "Cordless Solid-State Hydrogen Sensor Using Proton-Conductor Thick Film," *Sensors and Actuators B,* 1 (1990), pp. 125–129.

54. Dunmoro, F. W., "An Electrometer and Its Application to Radio Meteorography," *J. Res. Nat. Bur. Std.,* 20 (1938), pp. 723–744.

55. Takaoka, Y., Maebashi, Y., Kobayashi, S. and Usui, T., "Kanshitu Soshi (Humidity Sensitive Device), Jpn. Patent, 58-16467 (1983).

56. Kinjo, N., Ohara, S., Sugawara, T. and Tsuchitani, S., "Changes in Electrical Resistance of Ionic Copolymers Caused by Moisture Sorption and Desorption," *Polymer J.,* 15 (8) (1983), pp. 621–623.

57. Sakai, Y., Sadaoka, Y. and Ikeuchi, K., "Humidity Sensors Composed of Grafted Copolymers," *Sensors and Actuators,* 9 (2) (1986), pp. 125–131.

58. Sakai, Y., Sadaoka, Y. and Fukumoto, H., "Humidity-Sensitive and Water-Resistive Polymeric Materials," *Sensors and Actuators,* 13 (3) (1988), pp. 243–250.

59. Sakai, Y., Rao, V. L., Sadaoka, Y. and Matsuguchi, M., "Humidity Sensor Composed of a Microporous Film of Polyethylene-graft-poly-(2-acrylamido-2-methylpropane sulfonate)," *Polym. Bull.,* 18 (12) (1987), pp. 501–506.

60. Sakai, Y., Sadaoka, Y., Matsuguchi, M. and Rao, V. L., "Humidity Sensor Using Microporous Film of Polyethylene-graft-poly-(2-hydroxy-3-methacryloxypropyl trimethyl-ammonium chloride)," *J. Mater. Sci.,* 24 (1) (1989), pp. 101–104.

61. Sakai, Y., Sadaoka, Y., Matsuguchi, M., Rao, V. L. and Kamigaki, M., "A Humidity Sensor Using Graft Copolymer with Polyelectrolyte Branches," *Polymer,* 30 (6) (1989), pp. 1068–1071.

62. Sakai, Y., Matsuguchi, M., Sadaoka, Y. and Hirayama, K., "A Humidity Sensor Composed of Interpenetrating Polymer Networks of Hydrophobic Mathacrylate Polymers," *J. Electrochem. Soc.,* 140 (2) (1993), pp. 432–436.

63. Sadaoka, T., Sakai, Y., Akiyama, H., "A Humidity Sensor Using Alkali Salt-Poly(ethylene oxide) Hybrid Films," *J. Mater. Sci.,* 21 (1) (1986), pp. 235–240.

64. Sakai, Y., Sadaoka, Y., Matsuguchi, M., Moriga, N. and Shimada, M., "Humidity Sensors Based on Organopolysiloxanes Having Hydrophilic Groups," *Sensors and Actuators*, 16 (1989), pp. 359–367.

65. Grate, J. W., Martin, S. J. and White, R. M., "Acoustic Wave Microsensors, Part I," *Anal. Chem.*, 65 (1993), pp. 940A–948A.

66. Grate, J. W., Martin, S. J. and White, R. M., "Acoustic Wave Microsensors, Part II," *Anal. Chem.*, 65 (1993), pp. 987A–996A.

67. Grate, J. W. and Abraham, M. H., "Solubility Interactions and the Selection of Sorbent Coating Materials for Chemical Sensors and Sensor Arrays," *Sensors and Actuators B*, 3 (1991), pp. 85–111.

68. Jencks, W. P., *Catalysis in Chemistry and Enzymology,* Dover Publications, Inc., New York (1987).

69. Israelachvilli, J. N., *Intermolecular and Surface Forces,* Academic Press, New York (1992).

70. Hunter, C. A. and Sanders, J. K. M., "The Nature of the pi-pi Interactions," *J. Am. Chem. Soc.*, 112 (1990), pp. 5525–5534.

71. Abraham, M. H., "Scales of Hydrogen-Bonding: Their Construction and Application to Physicochemical and Biochemical Processes," *Chem. Soc. Rev.*, 22 (1993), pp. 73–83.

72. Abraham, M. H., Whiting, G. S., Doherty, R. M. and Shuely, W. J., "Hydrogen Bonding. Part 13. A New Method for the Characterisation of GLC Stationary Phases – The Laffort Data Set," *J. Chem. Soc., Perkin Trans.*, 2 (1990), pp. 1451–1460.

73. Abraham, M. H., Whiting, G. S., Doherty, R. M. and Shuely, W. J., "Hydrogen Bonding. XVI. A New Solute Solvation Parameter, $\pi H/2'$ from Gas Chromatographic Data," *J. Chromatogr.*, 587 (1991), pp. 213–228.

74. Abraham, M. H., Greller, P. L., Prior, D. V., Duce, P. P., Morris, J. J. and Taylor, P. J., "Hydrogen Bonding. Part 7. A Scale of Solute Hydrogen-Bond Acidity Based on log K Values for Complexation in Tetrachloromethane," *J. Chem. Soc., Perkin Trans. II* (1989), pp. 699–711.

75. Abraham, M. H., Grellier, P. L., Prior, D. V., Duce, P. P., Morris, J. J. and Taylor, P. J., "Hydrogen Bonding. Part 10. A Scale of Solute Hydrogen-Bond Basicity Using log K Values for Complexation in Tetrachloromethane," *J. Chem. Soc., Perkin Trans.*, 2 (1990), pp. 521–529.

76. Abraham, M. H., Grellier, P. L. and McGill, R. A., "Determination of Olive Oil-Gas and Hexadecane-Gas Partition Coefficients, and Calculation of the Corresponding Olive Oil-Water and Hexadecane-Water Partition Coefficients," *J. Chem. Soc., Perkin Trans. II*, (1987), pp. 797–803.

77. Conder, J. R. and Young, C. L., *Physicochemical Measurements by Gas Chromatography,* John Wiley & Sons, New York (1979).

78. Grate, J. W., Snow, A., Ballantine, D. S., Wohltjen, H., Abraham, M. H., McGill, R. A. and Sasson, P., "Determination of Partition Coefficients from Surface Acoustic Wave Vapour Sensor Responses and Correlation with Gas-Liquid Chromatographic Partition Coefficients," *Anal. Chem.*, 60 (1988), pp. 869–875.

79. Ballantine, D. S., Rose, S. L., Grate, J. W. and Wohltjen, H., "Correlation of Surface Acoustic Wave Device Coating Responses with Solubility Properties and Chemical Structure Using Pattern Recognition," *Anal. Chem.*, 58 (1986), pp. 3058–3066.

80. Grate, J. W., Klusty, M., Barger, W. R. and Snow, A. W., "Role of Selective Sorption

in Chemiresistor Sensors for Organophosphorous Detection," *Anal. Chem.*, 62 (1990), pp. 1927–1924.

81. Rose-Pehrsson, S. L., Grate, J. W., Ballantine, D. S. and Jurs, P. C., "Detection of Hazardous Vapours Including Mixtures Using Pattern Recognition Analysis of Responses From Surface Acoustic Wave Devices," *Anal. Chem.*, 60 (1988), pp. 2801–2811.

82. Snow, A. W., Sprague, L. G., Soulen, R. L., Grate, J. W. and Wohltjen, H., "Synthesis and Evaluation of Hexafluorodimethylcarbinol Functionalized Polymers as Microsensor Coatings," *J. Appl. Poly. Sci.*, 43 (1991), pp. 1659–1671.

83. Grate, J. W., Rose-Pehrsson, S. L., Venezky, D. L., Klusty, M. and Wohltjen, H., "A Smart Sensor System for Trace Organophosphorus and Organosulfur Vapour Detection Employing a Temperature-Controlled Array of Surface Acoustic Wave Sensors, Automated Sample Preconcentration, and Pattern Recognition," *Anal. Chem.*, 65 (1993), pp. 1868–1881.

84. Nieuwenhuizen, M. S. and Nederlof, A. J., "A SAW Gas Sensor for Carbon Dioxide and Water. Preliminary Experiments," *Sensors and Actuators B*, 2 (1990), pp. 97–101.

85. Zhang, G. -Z. and Zellers, E. T., "Coated Surface Acoustic Wave Sensor Employing a Reversible Mass-Amplifying Ligand Substitution Reaction for Real Time Measurement of 1,3-Butadiene at Low and Sub-ppm Concentrations," *Anal. Chem.*, 65 (1993), pp. 1340–1349.

86. Zellers, E. T. and Zhang, G. -Z. "Steric Factors Affecting the Discrimination of Isomeric and Structurally Related Olefin Gases and Vapours with a Reagent-Coated Surface Acoustic Wave Sensor," *Anal. Chem.*, 64 (1992), pp. 1277–1284.

87. Zellers, E. T., White, R. M. and Rappaport, S. M., "Use of a Surface-Acoustic-Wave Sensor to Characterize the Reaction of Styrene Vapour with a Square Planar Organoplatinum Complex," *Anal. Chem.*, 62 (1990), pp. 1222–1227.

88. Zellers, E. T., Hassold, N., White, R. M. and Rapport, S. M., "Selective Real-Time Measurement of Styrene Vapour Using a Surface-Acoustic-Wave Sensor with a Regenerable Organoplatinum Coating," *Anal. Chem.*, 62 (1990), pp. 1227–1232.

89. Grate, J. W., Klusty, M., McGill, R. M., Abraham, M. H., Whiting, G. and Andonian-Haftvan, J., "The Predominant Role of Swelling-Induced Modulus Changes of the Sorbent Phase in Determining the Responses of Polymer-Coated Surface Acoustic Wave Vapour Sensors," *Anal. Chem.*, 64 (1992), pp. 610–624.

90. Abraham, M. H., Hammerton, I., Rose, J. B. and Grate, J. W., "Hydrogen Bonding Part 18. Gas-Liquid Chromotographic Measurements for the Design and Selection of Some Hydrogen Bond Acidic Phases Suitable for Use as Coatings on Piezoelectric Sorption Detectors," *J. Chem. Soc. Perkin Trans.*, 2 (1991), pp. 1417–1423.

91. Barlow, J. W., Cassidy, P. E., Lloyd, D. R., You, C. J., Chang, Y., Wong, P. C. and Noriyan, J., "Polymer Sorbents for Phosphorus Esters: II. Hydrogen Bond Driven Sorption in Fluoro-Carbonol Substituted Polystyrene," *Polym. Eng. Sci.*, 27 (1987), pp. 703–715.

92. Grate, J. W., McGill, R. A. and Abraham, M. H., "Chemically Selective Polymer Coatings for Acoustic Vapour Sensors and Arrays," *Proc. IEEE Ultrason. Symp.* (1992), pp. 275–279.

93. Patrash, S. J. and Zellers, E. T., "Characterization of Polymer Surface Acoustic Wave Sensor Coatings and Semiempirical Models of Sensor Responsed to Organic Vapours," *Anal. Chem.*, 65 (1993), pp. 2055–2066.

94. Amato, I., "Getting a Whiff of Pittcon: Evolving an Electronic Schnozz," *Science,* 251 (1991), pp. 1431–1432.

95. Newman, A. R., "Electronic Noses," *Anal. Chem.,* 63 (1991), pp. 585A–588A.

96. Gardner, J. W. and Bartlett, P. N., *Sensors and Sensory Systems for an Electronic Nose,* Kluwer Academic Publishers, Dordrecht (1992).

97. Göpel, W. and Schierbaum, K. D., "Definitions and Typical Examples," in *Sensors: A Comprehensive Survey, Part I, Vol. II,* eds., Göpel, W., Hesse, J. and Zemel, J. N., VCH Verlag, Weinheim, (Germany) (1991), pp. 1–27.

98. Dickert, F. L., Haunschild, A., Hoffmann, P. and Mages, G., "Molecular Recognition of Organic Solvents and Ammonia: Shape and Donor Properties as Sensor Effects," *Sensors and Actuators B,* 6 (1992), pp. 25–28.

99. Niewenheuzen, M. S. and Venema, A., "Mass-Sensitive Devices," in *Sensors: A Comprehensive Survey, Part 1, Vol. II,* eds., Göpel, W., Hesse, J. and Zemel, J. N., VCH Verlag, Weinheim (Germany), (1991).

100. Schierbaum, K. D., Gerlach, A., Haug, M. and Göpel, W., "Selective Detection of Organic Molecules with Polymers and Supramolecular Compounds: Applications of Capacitance, Quartz Microbalance, and Calorimetric Transducers," *Sensors and Actuators A,* 31 (1992), pp. 130–137.

101. Nahm, W. and Gauglitz, G., "Thin Polymer Films as Sensors for Hydrocarbons," *GIT Fachz. Lab.,* 7 (1990), p. 889.

102. Schierbaum, K. D. and Göpel, W., "Functional Polymers and Supramolecular Compounds for Chemical Sensors," *Proc. of the EMRS 1993 Spring Meeting,* Strasbourg (France), May 4–7, 1993, and *Synth. Metals,* 61 (1993), pp. 37–45.

103. Voronkow, M. G., Mileshkevich, V. P. and Yuzhelevskii, Yu. A., *The Siloxane Bond,* Consultants Bureau, New York (1978), p. 50.

104. Mark, J. E., "Silicon-Containing Polymers," in *Silicone-Based Polymer Science,* eds., Zeigler, J. M. and Gordon Fearon, F. W., ACS, Washington (1990), pp. 47–68.

105. Vögtle, F., *Supramolecular Chemistry,* J. Wiley, Chichester, New York, Brisbane, Toronto, Singapor, 3rd Ed. (1993).

106. Göpel, W. and Lampe, U., "Influence of Defects on the Electronic Structure of Zinc Oxide Surfaces," *Phys. Rev.,* B 22 (1980), p. 6447.

107. Sybyl, Molecular Modeling Software, Vers. 6.0, Tripos Ass. Inc. 1993.

108. Schierbaum, K. D., Weiß, T., Thoden Van Velzen, U. and Reinhoudt, D., "The Interaction of Self-Assembled Monolayers of Hoegberg[4]arene with Organic Molecules: A Quartz Microbalance,"DS, and XPS Study in prep.

109. Schierbaum, K. D., Zhou, R., Hierlemann, A. and Göpel, W., "Modified Polymers for Reliable Detection of Organic Solvents: Thermodynamically Controlled Selectivities and Sensitivities," *Eurosensors VII, Abstracts,* Budapest (Hungary) (1993), p. 101.

110. Riddick, J. A. and Bunger, W. B., *Organic Solvents, Vol. 2, Techniques of Chemistry,* Wiley-Interscience, New York (1970).

111. Zhou, R., Vaihinger, S., Geckeler, K. E. and Göpel, W., "Reliable CO_2 Sensor Based on Silicon-Based Polymers with Quartz Microbalance Transducers," *Eurosensors VII, Abstracts,* Budapest (Hungary) (1993), p. 86.

112. Zhou, R., Patskovsky, S., Noetzel, G., Schierbaum, K. D. and Göpel, W., "Optimierung von Polymersensoren für Kohlenwasserstoffe in Wasser," *1. Dresdener Sensor Symp.,* Dec. 14–16, Dresden (Germany) (1993).

113. Schierbaum, K. D., Gerlach, A., Göpel, W., Müller, W., Vögtle, F., Dominik, A. and Roth, H. J., "Surface and Bulk Interactions of Organic Films with Calixarene Layers," *Fresenius Z. Anal. Chem.*, (submitted for publication).

114. Stewart, J. J. P., MOPAC Manual QCPE #455, United States Air Force Academy, Colorado Springs.

115. Gauglitz, G., Kaspar, S., Krauss, G. and Schurr, O., "Reflektometrisch-interferometrische Sensoren für Gase und Flüssigkeiten," *1. Dresdener Sensor Symp.*, Dec. 14–16, Dresden (Germany) (1993).

116. Haug, M., Schierbaum, K. D., Gauglitz, G. and Göpel, W., "Chemical Sensors Based upon Polysiloxanes: Comparison between Optical, Quartz Microbalance, Calorimetric, and Capacitance Sensors," *Sensors and Actuators B*, 11 (1993), 383–391.

117. Barrer, R. M., *Diffusion in and through Solids*, Cambridge University Press, London (1941).

118. Noszticziusz, Z., Oláh, K., Vajtha, Zs. and Pálmai, Gy., "Carbon-Hydrogen-Monitor Using a Membrane Permeator Combined with a Flame Ionization Detector," *Magy. Kém. Foly.*, 85 (1979), p. 28.

119. Kesting, R. E., *Synthetic Polymeric Membranes*, McGraw-Hill Book Co., New York (1971).

120. Stannett, V. and Yasuda, H. in Raff, R. and Doak, K., eds., *Crystalline Olefin Polymers, Part 2*, Interscience, New York (1965).

121. Small, P., "Factors Affecting the Solubility of Polymers," *J. Appl. Chem.*, 3 (1953), p. 71.

122. Michaels, A., Bixler, H. and Fein, H., "Theoretical Model for Predicting Al_2O_3 Particle-Size Distributions in Rocket," *J. Appl. Phys.*, 35 (1964), p. 3165.

123. Binning, R., Lee, R., Jennings, J. and Martin, E., "Separation of Liquid Mixtures by Permeation," *Ind. Eng. Chem.*, 53 (1961), p. 45.

124. Lang, O. and Stokesberry, D., Natl. Bur. Std. (US) Rept. (1968), p. 9872.

125. Bresler, E. and Wendt, R., "Onsagers Reciprocal Relation. Examination of Its Application to a Simple Membrane Transport Process," *Science*, 163 (1969), p. 264.

126. Ambard, L. and Trautmann, S., *Ultrafiltration*, Charles C. Thomas, Springfield, IL (1960).

127. General Electric Co., *Permselective Membranes*, Schenectady, New York (1969).

128. Myers, A., Tammela, V., Stannett, V. and Szwarc, M., "Permeability of Chlorotrifluoroethylene Polymers," *Mod. Plastics*, 37/6 (1960), p. 139.

129. Hawkins, W. L., "Polymer Chemistry," in *Physical Design of Electronic Systems, Vol. II.*, ed., Everitt, W. L., Prentice-Hall, Inc., Englewood Cliffs, NJ (1970).

130. Yasuda, H., Gochin, M. and Stone, W., "Hydrogels of Poly(hydroxyethyl methacrylate) and Hydroxyethyl Methacrylate – Glycerol Monomethacrylate Copolymers," *J. Polymer Sci.*, A1/4 (1966), p. 2913.

131. Mort, W. E., *The Principles of Ion-Selective Electrodes and of Membrane Transport*, Elsevier, Amsterdam (1981).

132. Cammann, K., *Working with Ion-Selective Electrodes*, Springer, Berlin (1979).

133. Koryta, J. and Stulik, K., *Ion-Selective Electrodes, 2nd ed.*, Cambridge University Press, Cambridge (1983).

134. Bergveld, P. and Sibbald, A., "Analytical and Biomedical Applications of Ion-

Selective Field-Effect Transistors," in *Comprehensive Analytical Chemistry, Vol. XXIII*, ed., Svehla, G., Elsevier, Amsterdam (1988).

135. Janata, J., *Principles of Chemical Sensors*, Plenum Press, New York (1989).

136. *Ionophores for Ion-Selective Electrodes and Optodes*, Bulletin of Fluka Chemie AG, Buchs, Switzerland (1991).

137. Horvai, G., Horváth, V., Farkas, A. and Pungor, E., "Ionorganic Salts Trapped in Neutral Carrier's Ion-Selective Electrode Membranes form Ion-Exchange Sites," *Anal. Lett.*, 21 (1988), p. 2165.

138. Horvai, G., Horváth, V., Farkas, A. and Pungor, E., "Selective Ion-Exchanger Behaviour of Neutral Carrier Ion-Selective Electrode Membranes," *Talanta*, 36 (1989), p. 403.

139. Horvai, G., Horváth, V., Farkas, A. and Pungor, E., "Electrical and Radio-Tracer Investigations of Ion-Selective Membranes," *Proceedings of the 5th Symposium on Ion-Selective Electrodes*, ed., Pungor, E., Akadémia Kiadó, Budapest (1989), pp. 397–405.

140. Horváth, V. and Horvai, G., "Selectivity of Plasticized Poly(vinyl chloride)-Based Ion-Selective Electrodes," *Anal. Chim. Acta*, 282 (1993), p. 259.

141. Horváth, V., Horvai, G. and Pungor, E., "6–9 Ion Exchange at Neutral Carrier Ion-Selective Electrode Membranes," *Fresenius J. Anal. Chem.*, 346 (1993), p. 569.

142. Lindner, E., Tóth, K. and Pungor, E., *Dynamic Characteristics of Ion-Selective Electrodes*, CRC Press, Boca Raton (1988).

143. Eisenman, G., *Ion-Selective Electrodes*, ed., Durst, R. A., National Bureau of Standards Spec. Publ. 314, Washington, DC (1968), p. 1.

144. Oesch, U. and Simon, W., "Kinetic Study of Distribution of Electrically Neutral Ionophores between a Solvent Polymeric Membrane and an Aqueous Phase," *Anal. Chem.*, 52 (1980), p. 602.

145. Dinten, O., Spichiger, U. E., Chaniotakis, N., Gehrig, P., Rusterholz, B., Morf, W. E. and Simon, W., "Lifetime of Neutral-Carrier-Based Liquid Membranes in Aqueous Samples and Blood and the Lipophilicity of Membrane Components," *Anal. Chem.*, 63 (1991), p. 596.

146. Sears, J. K. and Darby, J. R., *The Technology of Plasticizers*, Wiley, New York (1982).

147. Lidner, E., Cosofret, V. V., Kusy, R. P., Buck, R. P., Rosatzin, T., Schaller, U., Simon, W., Jeney, J., Tóth, K. and Pungor, E., "Responses of H⁺ Selective Solvent Polymeric Membrane Electrodes Fabricated from Modified PVC Membranes," *Talanta*, 40 (1993), p. 957.

148. Pick, J., Tóth, K., Pungor, E., Vasak, M. and Simon, W., *Anal. Chim. Acta*, 64 (1973), p. 477.

149. Lindner, E., Niegreisz, Zs., Tóth, K., Pungor, E., Berube, T. R. and Buck, R. P., "Electrical and Dynamic Properties of Non-plasticized Potassium Selective Membranes," *J. Electroanal. Chem.*, 259 (1989), p. 67.

150. Mostert, I. A., Anker, P., Jenny, H. -B., Oesch, U., Morf, W. E., Ammann, D. and Simon, W., "Neutral Carrier Based Silicone Rubber Membranes for Oxonium, Potassium (1⁺), Ammonium, and Calcium (2⁺), Selective Electrodes," *Microchim. Acta*, I (1985), p. 33.

151. Ma, S. C., Chaniotakis, N. A. and Meyerhoff, M. E., "Response Properties of Ion-Selective Polymers Membrane Electrodes Prepared of Aminated and Carboxilated Poly(vinyl chloride)," *Anal. Chem.*, 60 (1988), p. 2293.

152. Rosatzin, T., Holy, P., Seiler, K., Rusterholts, B. land Simon, W., "Immobilization of

Components in Polymer Membrane-Based Calcium-Selective Bulk Optodes," *Anal. Chem.*, 64 (1992), p. 2029.

153. Gehrig, P. M., Rusterholz, B. and Simon, W., "Very Lipophilic Sodium Selective Ionophore for Chemical Sensors of High Lifetime," *Anal. Chim. Acta*, 233 (1990), p. 295.

154. Nishida, H., Takada, N., Yoshimura, M., Sonoda, T. and Kobayashi, H., "Tetrakis [3,5-bis(trifluoromethyl)phenyl] Borate. Highly Lipophilic Stable Anionic Agent for Solvent-Extraction of Cations," *Bull. Chem. Soc. Jpn*, 57 (1984), p. 2600.

155. Lindner, E., Gráf, E., Niegreisz, Zs., Tóth, K., Pungor, E. and Buck, R. P., "Responses of Site-Controlled, Plasticized Membrane Electrodes," *Anal. Chem.*, 60 (1988), p. 295.

156. Moody, G. J., Oke, R. B. and Thomas, J. D. R., "Influence of Light on Silver-Silver Chloride Electrodes," *Analyst*, 92 (1970), p. 910.

157. Moody, G. J., Saad, B. and Thomas, J. D. R., "Glass Transition Temperatures of Poly(vinyl chloride) and Polyacrylate Materials and Calcium Ion-Selective Electrode Properties," *Analyst*, 112 (1987), p. 1143.

158. Cha, S. G., Liu, D., Meyerhoff, M. E., Cantor, A. C., Mdgley, A. R., Goldberg, H. D. and Brow, R. B., "Electrical Performance Biocompatibility, and Adhesion of New Polymer Matrices for Solid State Ion Sensors," *Anal. Chem.*, 63 (1991), p. 1666.

159. Lindner, E., Cosofret, V. V., Ufer, S., Buck, R. P., Kao, J. W., Neuman, M. R. and Anderson, J. M., *J. Biomed. Materials Res.* (accepted paper).

160. Stachwill, T. and Harrison, D. J., "Synthesis and Characterization of New Polyvinyl-Chloride Membranes for Enhanced Adhesion on Electrode Surfaces," *J. Electroanal. Chem.*, 202 (1986), p. 75.

161. Harrison, D. J., Cunningham, L. L., Li, X., Teclemariam, A. and Perman, D., "Enhanced Lifetime and Adhesion of Potassium Ion-Ammonium Ion-, and Calcium Ion-Sensitive Membranes on Solid Surfaces Using Hydroxyl-Modified Polyvinyl-Chloride Matrices," *J. Electrochem. Soc.*, 135 (1988), p. 2473.

162. Lindner, E., Cosofret, V. V., Ufer, S., Buck, R. P., Kusy, R. P., Ash, R. B. and Nagle, H. T., "Design of Ionophore-Free H$^+$-Selective Solvent Polymeric Membranes for Further Biomedical Applications," *J. Chem. Soc. Faraday Trans.*, 89 (1993), p. 361.

163. Dürselen, L. F., Wegmann, D., May, K., Oesch, U. and Simon, W., "Elimination of the Asymmetry in Neutral-Carrier Based Solvent Polymeric Membranes Introduced by Proteins," *Anal Chem.*, 60 (1988), p. 1455.

164. Fiedler, U. and Ruzicka, J., "Selectrode, Universal Ion-Selective Electrode VII Valinomycin-Based Electrode with Nonporous Polymer Membrane and Solid-State Inner Reference Ststems," *Anal. Chim. Acta*, 67 (1973), p. 179.

165. Horvai, G., Graf, E., Tóth, E., Pungor, E. and Buck, R. P., "Donnan Exclusion Failure in Low Anion Site Density Membranes Containing Valinomycin," *Anal Chem.*, 58 (1986), pp. 2735–2741.

166. Armstrong, R. D. and Todd, M., "The Role of PVC in Ion Selective Electrode Membranes," *J. Electroanal. Chem.*, 237 (1987), p. 181.

167. Van den Berg, A., van der Wal, P. D., Skowronska-Ptasinska, M., Sudholter, E. J. R., Reinhoudt, D. N. and Bergveld, P., "Nature of Anionic Sites in Plasticized Poly(vinyl chloride) Membranes," *Anal. Chem.*, 59 (1987), p. 2827.

168. Armstrong, R. D., Covington, A. K. and Proud, W. G., "Solvent Properties of PVC Membranes," *J. Electroanal. Chem.*, 257 (1988), pp. 155–171.

169. Armstrong, R. D. and Todd, M., "Ionic Mobilities in PVC Membranes," *Electrochim. Acta*, 32 (1987), p. 155.

170. Armstrong, R. D. and Asshassi-Sokhabi, H., "Mobility of Na^+ in PVC Membranes Containing Valinomycin and Di-benzo-18-crown-6," *Electrochim. Acta*, 32 (1987), p. 135.

171. Wartman, L. H., "Electrical Resistivity of Plasticized Poly(vinyl chloride) Insulating Materials," *Soc. Plast. Eng.*, 20 (1964), p. 254.

172. Iglehart, M. L., Buck, R. P. and Pungor, E., "Plastiziced Poly(vinyl chloride) Properties and Characteristics of Valinomycin Electrodes, DC Current-Voltage Curves," *Anal. Chem.*, 60 (1988), p. 290.

173. Armstrong, R. D. and Johnson, B., *Corros. Sci.* (accepted paper).

174. Thoma, A. P., Viviani-Nauer, A., Arvanitis, S., Morf, W. E. and Simon, W., "Mechanism of Neutral Carrier Mediated Ion Transport through Ion-Selective Bulk Membranes," *Anal. Chem.*, 49 (1977), p. 1567.

175. Li, X. and Harrison, D. J., "Measurement of Concentration Profiles inside a Nitride Ion Selective Electrode Membrane," *Anal. Chem.*, 63 (1991), p. 2168.

176. Chan, A. D. C., Li, X. and Harrison, D. J., "Evidence for a Water-Rich Surface Region in Poly(vinyl chloride) – Based Ion-Selective Electrode Membranes," *Anal. Chem.*, 64 (1992), p. 2512.

177. Armstrong, R. D. and Todd, M., *J. Electroanal. Chem.*, 245 (1988), p. 131.

178. Samec, Z., Maracek, V. and Homolka, D., "Double Layers at Liquid/Liquid Interfaces," *Farad. Disc.*, 77 (1984), p. 197.

179. Armstrong, R. D., Proud, W. G. and Todd, M., "The Electrical Double Layer at the Polymer-Water Interface," *Electrochim. Acta*, 34 (1989), p. 977.

180. Kellner, R., Fischböck, G., Götzinger, G., Pungor, E., Tóth, K., Pólos, L. and Lindner, E., "FTIR-ATR Spectroscopic Analysis of *bis*-Crown-ether Based PVC-Membrane Surfaces," *Fresenius Z. Anal. Chem.*, 322 (1985), p. 151.

181. Kellner, R., Zippel, E., Pungor, E., Tóth, K. and Lindner, E., "FTIR-ATR Spectroscopic Analysis of *bis*-Crown-ether Based PVC Membrane Surfaces," *Fresenius Z. Anal. Chem.*, 328 (1987), p. 464.

182. Tóth, K., Lindner, E., Pungor, E., Zippel, E. and Kellner, R., "FTIR-ATR Spectroscopic Analysis of *bis*-Crown-ether Based PVC-Membrane Surfaces," *Fresenius Z. Anal. Chem.*, 331 (1988), p. 448.

183. Armstrong, R. D., "The Significance of Exchange Currents for the PVC/Water Interface," *J. Electroanal. Chem.*, 245 (1988), p. 113.

184. Xie, S. and Cammann. K., "Apparent Ion-Exchange Current Densities at Valinomycin-Based Potassium Ion-Selective PVC Membranes Obtained with an AC-Impedance Method," *J. Electroanal. Chem.*, 229 (1987), p. 249.

185. Armstrong, R. D., Lockhart, J. C. and Todd, M., "The Mechanism of Transfer of K^+ between Aqueous Solutions and PVC Membranes Containing Valinomycin," *Electrochim. Acta*, 31 (1986), p. 591.

186. Armstrong, R. D. and Todd, M., "Study of Calcium Ion Selective Electrodes – Interfacial Properties," *J. Electroanal. Chem.*, 266 (1989), pp. 175–177.

187. Horváth, V., Horvai, G. and Pungor, E., "Amperometric Measurements with Ion-Selective Electrode Membranes in a Flow System," *Microchimica Acta*, I (1990), p. 217.

188. Horvai, G. and Pungor, E., "Amperometric Determination of Hydrogen and Hydroxyl Ion Concentrations in Unbuffered Solution in the pH Range 5-9," *Anal. Chim. Acta,* 243 (1991), p. 55.

189. Lindner, E., Toth, K., Pungor, E., Beroube, T. R. and Buck, R. P., "Switched Wall Jet for Dynamic Response Measurements," *Anal. Chem.,* 59 (1987), p. 2213.

190. Armstrong, R. D., "Mechanistic Aspects of the K^+ Ion Selective Electrode Based on Valinomycin-PVC," *Electrochim. Acta,* 32 (1987), p. 1549.

191. Zhou, Q. and Siegel Jr., G. H., "Detection of Carbon Monoxide with a Porous Polymer Optical Fibre," *Int. J. of Optoelectronics,* Vol. 4, No. 5, (1989), pp. 415–423.

192. Hao, T., Xing, X. and Liu, Ch. -Ch., "A pH Sensor Constructed with Two Types of Optical Fibres: The Configuration and Initial Results; *Sensors and Actuators B,* 10 (1993), pp. 155–159.

193. Boisde, G., Blanc, F. and Machuron-Mandard, X., "pH Measurements with Dyes Co-immobilization on Optrodes, Principles and Associated Instrumentation," *Int. J. of Optoelectronics,* Vol. 6, No. 5 (1991), pp. 407–423.

194. Lumpp, R., Reichert, J. and Ache, H. J., "An Optical Sensor for the Detection of Nitrate," *Sensors and Actuators B,* 7 (1992), pp. 473–475.

195. Reichert, J., Sellien, W. and Ache, H. J., "Development of a Fiber-Optic Sensor for the Detection of Ammonium in Environmental Waters," *Sensors and Actuators A,* 25–27 (1991), pp. 481–482.

196. Zhou, Q., Kritz, D., Bonel, L. and Siegel Jr., G. H., "Porous Plastic Optical Fiber Sensor for Ammonia Measurement," *Applied Optics,* Vol. 28 No. 11 (1989), pp. 2022–2025.

197. Gauglitz, G. and Reichert, M., "Spectral Investigation and Optimization of pH and Urea Sensors," *Sensors and Actuators B,* 6 (1992), pp. 83–86.

198. Czolk, R., Reichert, J. and Ache, H. J. "An Optical Sensor for the Detection of Heavy Metal Ions," *Sensors and Actuators B,* 7 (1992), pp. 540–543.

199. Alava-Moreno, F., Pereiro-Garcia, R., Diaz-Garcia, M. E. and Sanz-Mendel, A., "A Comparative Study of Two Different Approaches for Active Optical Sensing of Potassium with a Chromoionophore," *Sensors and Actuators B,* 11 (1993), pp. 413–419.

200. Sandanayake, K. R. A. S. and Sutherland, I. O., "Organic Dyes for Optical Sensors," *Sensors and Actuators B,* 11 (1993), pp. 331–340.

201. Sadaoka, Y., Matsuguchi, M., Sakai, Y. and Murata, Y. "Optical Humidity Sensing Characteristics of Nafion-Dyes Composite Thin Films," *Sensors and Actuators B,* 7 (1992), pp. 443–446.

202. Ballantine, D. S. and Wohltjen, H., "Optical Waveguide Humidity Sensor," *Anal. Chem.,* 58 (1986), pp. 2883–2885.

203. Rolfe, P., "In vivo Chemical Sensors for Intensive-Care Monitoring," *MBEC Biosensors* special feature (May 1990), B34–B46.

204. Wolfbeis, O. S., "Biomedical Applications of Fibre Optic Chemical Sensors," *Int. J. of Optoel.,* Vol. 6, No. 5 (1991), pp. 425–441.

205. Posch, H. E. and Wolfbeis, O. S., "Optical Sensors 13: Fibre-Optic Humidity Sensor Based on Fluorescence Quenching," *Sensors and Actuators,* 15 (1988), pp. 77–83.

206. Leiner, M. J. P. and Hartmann, P., "Theory and Practice in Optical pH Sensing," *Sensors and Actuators B,* 11 (1993), pp. 281–289.

207. Kawabata, Y., Yamamoto, T. and Imasaka, T., "Theoretical Evaluation of Optical

Response to Cations and Cationic Surfactants for Optrode Using Hexadecyl-acridine Orange Attached on Plasticized Poly(vinyl chloride) (Membrane)," *Sensors and Actuators B,* 11 (1993), pp. 341–346.

208. Freeman, T. M. and Seitz, W. R., "Chemiluminescence Fiber Optic Probe for Hydrogen Peroxide Based on the Luminol Reaction," *Anal. Chem.,* 50 (1978), pp. 1242–1246.

209. Gautier, S. M., Loic, J. B. and Coulet, P. R., "Alternate Determination of ATP and NADH with a Single Bioluminescence-Based Fiber-Optic Sensor," *Sensors and Actuators B,* 1 (1990), pp. 580–584.

210. Papkovsky, D. B., "Luminescent Porphyrins as Probes for Optical (Bio)sensors," *Sensors and Actuators B,* 11 (1993), pp. 293–300.

211. Gauglitz, G., Brecht, G., Kraus, G. and Nahm, W., "Chemical and Biochemical Sensors Based on Interferometry at Thin (multi-) Layers," *Sensors and Actuators B,* 11 (1993), pp. 21–27.

212. Gauglitz, G. and Ingenhoff, J., "Integrated Optical Sensors for Halogenated and Non-Halogenated Hydrocarbons," *Sensors and Actuators B,* 11 (1993), pp. 207–212.

213. Bidan, G., "Electroconducting Conjugated Polymers: New Sensitive Matrices to Build up Chemical or Electrochemical Sensors. A Review," *Sensors and Actuators B,* 6 (1992), p. 45.

214. Curan, D., Grimshaw, J. and Perea, S., "Poly(pyrrole) as a Support for Electrocatalytic Materials," *Chem. Soc. Rev.,* 20 (1991), p. 391.

215. Simonet, J. and Rault-Berthelot, J., "Electrochemistry: A Technique to Form, to Modify and to Characterize Organic Conducting Polymers," *Prog. Solid St. Chem.,* 21 (1991), p. 1.

216. Garnier, F., "Functionalized Conducting Polymers. Towards Intelligent Materials," *Adv. Mater.,* 28 (1989), p. 513.

217. Audebert, P., Bidan, G., Lapkowski, M. and Lomosin, D., *Grafting, Ionomer Composites, and Auto-Doping of Conductive Polymers,* Springer Series in *Solid-State Sciences, Vol. 76,* ed., Kuzmany, H., Springer, Berlin (1987), p. 366.

218. Deronzier, A. and Moutet, J. -C., "Functionalized Polypyrroles. New Molecular Materials for Electrocatalysis and Related Applications," *Acc. Chem. Res.,* 22 (1989), p. 249.

219. Lemaire, M., Garreau, R., Roncali, J., Delabouglise, D., Korri Youssoufi, H. and Garnier, F., "Design of Poly(thiophene) Containing Oxyalkyl Substituents, *New J. Chem.,* 13 (1989), p. 863.

220. Andrieux, C., Audebert, P. and Saveant, J. -M., "Studies of Modified Electrodes Starting from Pyrrole Monomers Derivatized by Anthraquinone, Phenothiazine or Anthracene Moieties on the Nitrogen or in the 3-Position, Syntheses, Cyclic Voltammetry and a Study of Charge-Transfer Reactions," *Synth. Met.,* 35 (1990), p. 155.

221. Inagaki, T., Hunter, M., Yang, X., Skotheim, T. and Okamoto, Y., "Electrochemical Synthesis of Ferrocene-Functionalized Polypyrrole Films," *J. Chem. Soc., Chem. Commun.* (1988), p. 126.

222. Audebert, P., Bidan, G. and Lapkowski, M., "Behaviour of Some Poly(pyrrole-anthraquinone) Films in DMSO Electrolytes," *J. Electroanal. Chem.,* 219 (1987), p. 165.

223. Cosnier, S., Deronzier, A. and Roland, J. -F., "Polypyridinyl Complexes of Ruthenium (II) Having 4,4′-Dicarboxy Ester-2,2′-bipyridine Ligands Attached Covalently to Polypyrrole Films. Reinvestigation of the Polypyrrole Electrochemical Response in

Poly[*tris*(*N*-bipyridylethyl)pyrrole] Ruthenium (II) Films," *J. Electroanal. Chem.*, 285 (1990), p. 133.

224. Armengaud, C., Moisy, P., Bedioui, F. and Devynck, J., "Electrochemistry of Conducting Polypyrrole Films Containing Cobalt Porphyrin," *J. Electroanal. Chem.*, 277 (1990), p. 197.

225. Bidan, G., Divisia-Blohorn, B., Lapkowski, M., Kern, J. -M. and Sauvage, J. -P., "Electroactive Film of Polypyrroles Containing Complexing Cavities Preformed by Entwining Ligands on Metallic Centers," *J. Amer. Chem. Soc.*, 114 (1992), p. 5986.

226. Yon-Lin, B., Smolander, M., Cromton, T. and Lowe, C., "Covalent Electropolymerization of Glucose Oxidase in Polypyrrole. Evaluation of Methods of Pyrrole Attachment to Glucose Oxidase on the Performance of Electropolymerized Glucose Sensors," *Anal. Chem.*, 65 (1993), p. 67.

227. Shimidzu, T., "Derivatization of Conducting Polymers with Functional Molecules Directed via Molecular Structural Control towards a Molecular Device," *React. Polym.*, 11 (1989), p. 177.

228. Shimidzu, T., Iyoda, T., Segawa, H. and Dujitsika, M., "Functionalizations of Conducting Polymers by Mesoscopically Structural Control and by Molecular Combination of Reactive Moiety," in *Intrinsically Conducting Polymers: An Emerging Technology* ed., Aldissi, M., Kluwer Academic Publishers (1993), p. 13–24.

229. Velaquez Rosenthal, M., Skotheim, T. and Linkous, C., "Polypyrrole-Phthalocyanine," *Synth. Met.*, 15 (1986), p. 219.

230. Bidan, G., Geniés, E. and Lapkowski, M., "One-Step Electrochemical Immobilization of Keggin-Type Heteropolyanions in Poly(3-methylthiophene) Film at an Electrode Surface: Electrochemical and Electrocatalytic Properties," *Synth. Met.*, 31 (1989), p. 327.

231. Foulds, N. and Lowe, C., "Enzyme Entrapment in Electrically Conducting Polymers: Immobilization of Glucose Oxidase – Polypyrrole and Its Application in Amperiometric Glucose Sensors," *J. Chem. Soc.*, *Faraday Trans.*, *1* 82 (1986), p. 1259.

232. Bartlett, P. and Whitaker, G., "Electrochemical Immobilization of Enzymes. Part II: Glucose Oxidase Immobilised in Poly-*N*-methylpyrrole," *J. Electroanal. Chem.*, 224 (1987), p. 37.

233. Nylander, Armgarth, M. and Lundström, I., "An Ammonia Detector Based on a Conducting Polymer," *Anal. Chem. Symp. Ser.*, 17 (1983), p. 203.

234. Miasik, J., Hooper, A. and Tofield, B., "Conducting Polymer Gas Sensors," *J. Chem. Soc.*, *Faraday Trans*, I, 82 (1986), p. 1117.

235. Hanawa, T., Kuwabata, S. and Yoneyama, H., "Gas Sensitivity of Polypyrrole Films to NO_2," *J. Chem. Soc.*, *Faraday Trans.*, 1, 84 (1988), p. 1587.

236. Hanawa, T. and Yoneyama, H., "Gas Sensitivities of Polypyrrole Films to Electron Acceptor Gases," *Bull. Chem. Soc. Jpn.*, 62 (1989), p. 1710.

237. Hanawa, T. and Yoneyana, H., "Gas Sensitivities of Polypyrrole Films Doped Chemically in the Gas Phase," *Synth. Met.*, 30 (1989), p. 341.

238. Hanawa, T., Huwabata, S., Hashimoto, H. and Yoneyama, H., "Gas Sensitivities of Electropolymerized Polythiophene Films," *Synth. Met.*, 30 (1989), p. 173.

239. Gu, H. B., Takiguchi, T., Hayashi, S., Kaneto, K. and Yoshino, K., "Effects of Ammonia Gas on Properties of Poly(*p*-phenylene vinylene)," *J. Phys. Soc. Jpn.*, 56 (1987), p. 3997.

240. Chague, J., Germain, C., Maleysson, C. and Robert, H., "Kinetics of Iodine Doping and Dedoping Processes in Thin Layers of Poly-*p*-phenylene Azomethine," *Sensors and Actuators,* 7 (1985), p. 199.

241. Weddingen, G., "New Polymer Systems for Gas Sensors," Ext. Abstract, *ICSM '88,* Santa Fe, NM, June 26–July 2 (1988), p. 271.

242. Bidan, G., Ehui, B. and Lapokowski, M., "Conductive Polymers with Immobilized Dopants:Ionomer Composites and Auto-Doped Polymers–A Review and Recent Advances," *J. Phys. D: Appl. Phys.,* 21 (1988), p. 1043.

243. Hwang, L., Ko, J., Rhee, H. and Kin, C., "A Polymer Humidity Sensor," *Synth. Met.,* 55–57 (1993), p. 3671.

244. Boyle, A., Geniés, E. and Lapkowski, M., "Application of Electronic Conducting Polymers as Sensors: Polyaniline in the Solid State for Detection of Solvent Vapours and Polypyrrole for Detection of Biological Ions in Solutions," *Synth. Met.,* 28 (1989), C769.

245. Travers, J. P., Mendaro, C., Nechtschein, M., Manohar, S. K. and Macdiarmid, A. G., "Moisture Effects in the Conducting Polymer, Polyaniline," *Mat. Res. Soc. Symp. Proc.,* 173 (1990), p. 335.

246. Dogan, S., Akbulut, U., Yalcin, T., Suzer, S. and Toparre, L., "Conducting Polymers of Aniline II. A Composite as a Gas Sensor," *Synth. Met.,* 60 (1993), p. 27.

247. Tsumura, A., Koezuka, H., Tsunoda, S. and Ando, T., "Chemically Prepared Poly-*N*-methylpyrrole) Thin Film. Its Application to the the Field-Effect Transistor," *Chem. Lett.* (1986), p. 863.

248. Josowicz, M. and Janata, J., "Suspended Gate Field Effect Transistor Modified with Polypyrrole as Alcohol Sensor," *Anal. Chem.,* 58 (1986), p. 514.

249. Josowicz, M., Janata, J., Ashley, K. and Pons, S., "Electrochemical and Ultraviolet-Visible Spectroelectrochemical Investigation of Selectivity of Potentiometric Gas Sensors Based on Polypyrrole," *Anal. Chem.,* 59 (1987), p. 253.

250. Ohmori, Y., Muro, K. and Yoshino, K., "Gas Sensitive and Temperature-Dependent Schottky Gated Field Effect Transistors Utilizing Poly(3-alkylthiophenes)," *Synth. Met.,* 55–57 (1993), p. 4111.

251. Thackeray, J. and Wrighton, M., "Chemically Responsive Microelectrochemical Devices Based on Platinized Poly(3-methylthiophene): Variation in Conductivity with Variation in Hydrogen, Oxygen or pH in Aqueous Solution," *J. Phys. Chem.,* 90 (1986), p. 6674.

252. Chao, S. and Wrighton, M., "Characterization of a 'Solid-State' Polyaniline-Based Transistor: Water Vapor Dependent Characteristics of a Device Employing a Poly (vinyl alcohol)/Phosphoric Acid Solid-State Electrolyte," *J. Am. Chem.,* 109 (1987), p. 6627.

253. Baughman, R., Shacklette, L., Elsenbaumer, R., Plichta, E. and Becht, C., "Microelectrochemical Actuators Based on Conducting Polymers," *Molecular Electronics* (1991), p. 267.

254. Otero, T., Rodriguez, J., Angulo, E. and Santamaria, C., "Artificial Muscles from Bilayer Structures," *Synth. Met.,* 55–57 (1993), p. 3713.

255. Pei, Q. and Inganas, O., "Conjugated Polymers as Smart Materials, Gas Sensors and Actuators Using Bending Beams," *Synth. Met.,* 55–57 (1993), p. 3730.

256. Gardner, J. and Bartlett, P., "Design of Conducting Polymer Gas Sensors: Modelling and Experiment," *Synth. Met.,* 55–57 (1993), p. 3665.

257. Charlesworth, J., Partridge, A. and Garrard, N., "Mechanistic Studies on the Interactions between Poly(pyrrole) and Organic Vapors," *J. Phys. Chem.*, 97 (1993), p. 5418.

258. Budrowski, C. and Przyluski, J., "Synthesis of Bromine Sensitive Conducting Polymers . . . ," *Synth. Met.*, 41–43 (1991), p. 597.

259. Przyluski, J. and Budrowski, C., "Conducting Polymer Specific Sensor for Bromine Vapours Based on the Poly-bromothiophenes," *Synth. Met.*, 41–43 (1991), p. 1163.

260. Anderson, M., Mattes, B., Reiss, H. and Kaner, R., "Conjugated Polymer Films for Gas Separations," *Science*, 252 (1991), p. 1412.

261. Anderson, M., Mattes, B., Reiss, H. and Kaner, R., "Gas Separation Membranes: A Novel Application for Conducting Polymers," *Synth. Met.*, 41–43 (1991), p. 1151.

262. Kaner, R., "Designing a Slow Leak," *Nature* (London), 352 (1991), p. 23.

263. Mates, B., Anderson, M., Conklun, J., Reiss, H. and Kaner, R., "Morphological Modification of Polyaniline Films for the Separation of Gases," *Synth. Met.*, 55–57 (1993), p. 3655.

264. Mates, B., Anderson, M., Reiss, H. and Kaner, R., "The Separation of Gases Using Conducting Polymer Films," in *Intrinsically Conducting Polymers: An Emerging Technology* ed., Aldissi, M., Kluwer Academic Publishers (1993), pp. 60–74.

265. Rebattet, L., Geniés, E., Allegraud, J. -J., Pineri, M. and Escoubes, M., "Polyaniline: Evidence of Oxygen-Polarons Interaction to Explain the High Selectivity Values in Oxygen/Nitrogen Gas Permeation Experiments," *Polym. Adv. Techn.*, 4 (1993), p. 32.

266. Liang, W. and Martin, C., "Gas Transport in Electronically Conductive Polymers," *Chem. Mater.*, 3 (1991), p. 390.

267. Martin, C., Liang, W., Menon, V., Parthasarathy, R. and Parthasarathy, A., "Electronically Conductive Polymers as Chemically-Selective Layers for Membrane-Based Separations," *Synth. Met.*, 55–57 (1993), p. 3766.

268. Iwasaki, T., Hotonaga, A., Llosokawa, L., Kamo, J., and Kamada, K., "Selective Gas Separation with Conducting Polymer-Porous Media Composite Membranes by Surface Diffusion Mechanism," *Proc. 3rd SPSJ Int. Polymer Conf., Membranes and Interfacial Phenomena of Polymers*, Nagoya, Japan, 26–29 November (1990), 27B19, p. 41.

269. Shu, C. and Wrighton, M., "Synthesis and Charge-Transport Properties of Polymers Derived from the Oxidation of 1-Hydro-1′-(-6(pyrrol-1-yl)hexyl)-4,4′-bipyridinium *bis*-(Hexafluorophosphate) and Demonstration of a pH-Sensitive Miroelectrochemical Transistor Derived from the Redox Properties of a Conventional Redox Center," *J. Phys. Chem.*, 92 (1988), p. 5221.

270. Kaneto, K., Asano, T. and Takashima, W., "Memory Device Using a Conducting Polymer and Solid Polymer Electrolyte," *Jap. J. Appl. Phys.*, 30 (1991), L215.

271. Shinohara, H., Aizawa, M. and Shirakawa, H., "Ion-Sieving of Electrosynthesized Polypyrrole Films," *J. Chem. Soc., Chem. Commun.* (1986), p. 87.

272. Zhang, W. and Dong, S., "Ion-Sieving Effect of Poly(3-methylthiophene) Thin Film Modified Glassy Carbon Electrodes," *J. Electroanal. Chem.*, 284 (1990), p. 517.

273. Wang, J., Chen, S, -P. and Lin, M., "Use of Different Electropolymerization Conditions for Controlling the Size-Exclusion Selectivity at Polyaniline, Polypyrrole and Polyphenol Films, *J. Electroanal. Chem.*, 273 (1989), p. 231.

274. Cosnier, S., Deronzier, A. and Roland, J. -F., "Controlled Permeability of Functionalized Polypyrrole Films by Use of Different Electrolyte Anions Sizes in the Electropolymerization Step," *J. Electroanal. Chem.*, 310 (1991), p. 71.

275. Dong, S., Sun, Z. and Lu, Z., "Chloride Chemical Sensor Based on an Organic Conducting Polypyrrole Polymer," *Analyst,* 113 (1988), p. 1525.

276. Dong, S. and Che, G., "An Electrochemical Microsensor for Chloride," *Talanta,* 38 (1991), p. 111.

277. Ikariyama, Y. and Heineman, W., "Polypyrrole Electrode as a Detector for Electroinactive Anions by Flow Injection Analysis," *Anal. Chem.,* 58 (1986), p. 1803.

278. Ye, J. and Baldwin, R., "Flow Injection Analysis of Electroinactive Anions at a Polyaniline Electrode," *Anal. Chem.,* 60 (1988), p. 1979.

279. Teasdale, P. and Wallace, G., "Molecular Recognition Using Conducting Polymers: Basis of an Electrochemical Sensing Technology," *Analyst,* 118 (1993), p. 329.

280. Sadik, O. and Wallace, G., "Effect of Polymer Composition on the Detection of Electroinactive Species Using Conductive Polymers," *Electroanalysis,* 5 (1993), p. 555.

281. Martinez, R., Dominguez, F., Gonzales, M., Mendes, J. and Orellana, R., "Polypyrrole-Dodecyl Sulphate Electrode as a Microsensor for Electroinactive Cations in Flow-Injection Analysis and Ion Chromatography," *Analytica Chim. Acta,* 279 (1993), p. 299.

282. O'Riordan, D. and Wallace, G., "Poly(pyrrole-*n*-carbodithioate) Electrode for Electroanalysis," *Anal. Chem.,* 58 (1986), p. 128.

283. Imisides, M. and Wallace, G., "Deposition and Electrochemical Stripping of Mercury Ions on Polypyrrole Based Modified Electrodes," *J. Electroanal. Chem.,* 246 (1988), p. 181.

284. Sauvage, J. -P., "Interlacing Molecular Threads on Transition Metals: Catenands, Catenates, and Knots," *Acc. Chem. Res.,* 23 (1990), p. 319.

285. Billon, M., Blohorn, B. and Bidan, G., works in progress.

286. Bidan, G., Geniés, E. and Lapkowski, M., "Modification of Polyaniline with Heteropolyanions: Electrocatalytic Reduction of Oxygen and Protons," *J. Chem. Soc., Chem. Commun.,* 533 (1988).

287. Bidan, G., Geniés, E. and Lapkowski, M., "Polypyrrole and Poly(*N*-methylpyrrole) Films Doped with Keggin-Type Heteropolyanions: Preparation and Properties," *J. Electroanal. Chem.,* 251 (1988), p. 297.

288. Lapkowski, M., Bidan, G. and Fournier, M., "Synthesis of Polypyrroles and Polythiophene in Aqueous Solution of Keggin-Type Structure Heteropolyanions," *Synth. Met.,* 41–43 (1991), p. 407.

289. Fabre, B., Bidan, G. and Lapkowski, M., "Immobilisation d'héteropolyanions mixtes dans des films de polyméres conducteurs électroniques. Utilisation des électrodes modifiées par ces films pour la détection des ions nitrites," Brevet Français CEA du 23 mars 1993, EN 9303588.

290. Fabre, B., Bidan, G. and Lapkowski, M., "Poly(*N*-methylpyrrole) Films Doped with Iron-Substituted Heteropolytungstates: A New Sensitive Layer for the Amperometric Detection of Nitrite Ions," accepted in *J. Chem. Soc. Chem. Commun.*

291. Arbizzani, C., Mastragostino, M., Meneghello, L., Hamaide, T. and Guyot, A., "Poly *N*-(oxyalkyl) Pyrolle Electrodes and (PEO-SEO)$_{20}$LiClO$_4$ Polymer Electrolyte in Lithium Rechargeable Batteries," *Electrochim. Acts,* 37 (1992), p. 1631.

292. Delabouglise, D. and Garnier, F., "Functionalized Conducting Polymers: Enhancement of the Redox Reversibility of Polypyrrole by Substituting Polyether Chains at the 3-Positions," *Adv. Mat.,* 2 (1990), p. 91.

293. Guillerez, S. and Bidan, G., to be published.

294. Roncali, J., Garreau, R. and Lemaire, M., "Electrosynthesis of Conducting Poly-pseudo-crown Ethers from Substituted Thiophenes," *J. Electroanal. Chem.,* 278 (1990), p. 373.

295. Roncali, J., Shi, L., Garreau, R., Garnier, F. and Lemaire, M., "Tuning of the Aqueous Electroactivity of Substituted Poly(thiophene)s by Ether Groups," *Synth. Met.,* 36 (1990), p. 267.

296. Bryce, M., Chissel, A., Smith, N. and Parker, D., "Synthesis and Cyclic Voltammetric Behaviour of Some 3-Substituted Thiophenes and Pyrroles: Precursors for the Preparation of Conducting Polymers," *Synth. Met.,* 26 (1988), p. 153.

297. Bartlett, P., Benniston, A., Chung, L. -Y., Dawson, D. and Moore, P., "Electrochemical Studies of Benzo-15-crown-5 Substituted Poly(pyrrole) films," *Electrochim. Acta,* 36 (1991), p. 1377.

298. Youssoufi, H., Hmyene, M., Garnier, F. and Delabouglise, D., "Cation Recognition Properties of Polypyrrole 3-Substituted by Azacrown Ethers," *J. Chem. Soc., Chem. Commun.* (1993), p. 1550.

299. Bäerle, P. and Scheib, S., "Molecular Recognition of Alkali-Ions by Crown-Ether-Functionalized Poly(alkylthiophenes)," *Adv. mater.,* 5 (1993), p. 848.

300. Marsella and Swager, T., "Designing Conducting Polymer-Based Sensors: Selective Ionochromic Response in Crown Ether Containing Polythiophene," *J. Am. Chem. Soc.,* 115 (1993), p. 12214.

301. Cadogan, A., Gao, Z., Lewenstam, A., Ivaska, A. and Diamon, D., "All Solid-State Sodium-Selective Electrode Based on a Calixarene Ionophore in a Poly(vinyl chloride) Membrane with a Polypyrrole Solid Contact," *Anal. Chem.,* 64 (1992), p. 2496.

302. Walton, D., Hall, C. and Chyla, A., "Functional Dopants in Conducting Polymers," *Synth. Met.,* 45 (1991), p. 363.

303. Angely. L., Questaigne, V. and Rault-Berthelot, J., "The Influence of Poly-dibenzo-crown Ether Treatments on their Complexing Properties: Application to Ag^+ Extraction," *Synth. Met.,* 52 (1992), p. 111.

304. Rault, Berthelot, J. and Angely, L. J., "Complexing Properties of Poly-Dibenzo-crown Ether Resins towards Inorganic Cations. Case of Poly-$DB_{18}C_6$," *Synth. Met.,* 58 (1993), p. 51.

305. Lemaire. M., Delabouglise, D., Garreau, R., Guy, A. and Roncali, J., "Enantioselective Chiral Poly(thiophenes)," *J. Chem. Soc., Chem. Commun.* (1988), p. 658.

306. Lemaire. M., Delabouglise, D., Garreau, R., Guy, A. and Roncali, J., "Synthése et propriétés des deux formes énantioméres de poly(thiophénes chiraux," *J. Chim. Phys.,* 86 (1989), p. 193.

307. Pickup, P., "Poly-(3-methylpyrrole-4-carboxylic acid): An Electronically Conducting Ion-Exchanger Polymer," *J. Electroanal. Chem.,* 225 (1987), p. 273.

308. Shu, C. and Wrighton, M., "Synthesis and Charge-Transport Properties of Polymers Derived from the Oxidation of 1-Hydro-1′-(6-(pyrrol-l-yl)hexyl)-4,4′-bipyridinium *bis*-(Hexafluorophosphate) and Demonstration of a pH-Sensitive Microelectrochemical Transistor Derived from the Redox Properties of a Conventional Redox Center," *J. Phys. Chem.,* 92 (1988), p. 5221.

309. Chen, C., Wei, C. and Rajeshwar, K., "Flow Assay of Ions at Chemically Modified Electrodes: The Polypyrrole/Fe(CN)$_6^{4-}$ Model System," *Anal. Chem.,* 65 (1993), p. 2437.

310. Miller, L., Zinger, B. and Zhou, Q. -X., "Electrically Controlled Release of Fe(CN)$_6^{4-}$ from Polypyrrole," *J. A. Chem. Soc.*, 109 (1987), p. 2267.

311. Zinger, B. and Miller, L., "Timed Release of Chemicals from Polypyrrole Films," *J. Am. Chem. Soc.*, 100 (1984), p. 6861.

312. Miller, L. and Zhou, Q. -X., "Poly (*N*-methylpyrrolylium) poly(styrenesulfonate): A Conductive Electrically Switchable Cation Exchanger That Cathodically Bind and Anodically Releases Dopamine," *Macromol.*, 20 (1987), p. 1594.

313. Zhou, Q. -X., Miller, L. and Valentine, J., "Electrochemically Controlled Binding and Release of Protonated Dimethyldopamine and Other Cations from Poly (*N*-methylpyrrole)/Polyanion Composite Redox Polymers," *J. Electroanal. Chem.*, 261 (1989), p. 147.

314. Blankespoor, R. and Miller, L. "Polymerized 3-Methoxythiophene. A Processable Material for the Controlled Release of Anions," *J. Chem. Soc., Chem. Commun.*, (1985), p. 90.

315. Miller, L., Chang, A. -C. and Zhou, Q. -X., "Conducting Polymers Which Bind and Release Organic Ions in Response to an Electrical Signal," *Stud. Org. Chem.* (Amsterdam), 30 (1987), p. 361.

316. Li, Y. and Dong, S., "Electrochemically Controlled Release of Adenosine 5'-Triphosphate from Polypyrrole Film," *J. Chem. Soc., Chem. Commun.*, (1992), p. 827.

317. Bidan, G., Mendes-Viegas, F. and Vieil, E., "Electrochemical Study of the Binding of *N*-Butylsulfonate Phenothiozine into a Polypyrrole Film, Submitted to *J. Electroanal. Chem.*

318. Bidan, G., Gadelle, A., Lopez, C., Mendes-Viegas, M. f. and Vieil, E., "Incorporation of Sulfonated Cyclodextrins into Polypyrrole: An Approach for the Electrocontrolled Delivering of Neutral Drugs" (in press) *Biosensors and Bioelectr.*

319. Bidan, G., Gadelle, A., Lopez, C. and Mendes-Viegas, M. f., "Polymére conducteur dopé par un sel de cyclodextrine sulfonée et dispositif pour capter et/ou délivrer une substance active comportant ce polymére," CEA Patent EN 9306655, June 3rd, 1993.

320. Messina, R., Sarazin, C., Yu, L. and Buvet, R., "Propriétés de membranes constituées de polyméres semi-conducteurs échangeurs rédox polyaniline et polypyrrole," *L. Chim. Phys.*, 73 (1976), p. 9.

321. Tsai, E., Pajkossy, T., Rajeshwar, K. and Reynolds, J., "Anion-Exchange Behavior of Polypyrrole Membranes," *J. Phys. Chem.*, 92 (1988), p. 3560.

322. Burgmayer, P. and Murray, R., "An Ion Gate Membrane: Electrochemical Control of Ion Permeability Through a Membrane with an Embedded Electrode," *J. Am. Chem. Soc.*, 104 (1982), p. 6139.

323. Shimidzu, T., Ohtani, A., Iyoda, T. and Honda, K., "Charge-Controllable Polypyrrole/Polyelectrolyte Composite Membranes. Part II. Effect of Incorporated Anion Size on the Electrochemical Oxidation-Reduction Process," *J. Electroanal. Chem.*, 224 (1987), p. 123.

324. Shimidzu, T., Ohtani, A. and Honda, K., "Charge Controllable Polypyrrole/polyelectrolyte Composite Membranes, Part III. Electrochemical Deionization System Constructed by Anion-Exchangeable and Cation-Exchangeable Polypyrrole Electrodes," *J. Electroanal. Chem.*, 251 (1988), p. 323.

325. Zhong, C., Doblhofer, K. and Weinberg, G., "The Effect of Incorporated Negative Fixed Charges on the Membrane Properties of Polypyrrole Films," *Farad. Discuss. Chem. Soc.*, 88 (1989), p. 307.

326. Schmidt, V., Tegtmeyer, D. and Heitbaum, J., "Conducting Polymers as Membranes with Variable Permeabilities for Neutral Compounds: Polypyrrole and Polyaniline in Aqueous Electrolyte," *Adv. Mater.*, 4 (1992), p. 428.

327. Wang, J., Chen, S. -P. and Shan Lin, M., "Use of Different Electropolymerization Conditions for Controlling the Size-Exclusion Selectivity at Polyaniline, Polypyrrole and Polyphenol Films," *J. Electroanal. Chem.*, 273 (1989), p. 231.

328. Nagaoka, T., Fujimoto, M., Nakao, H., Kakuno, K., Yano, J. and Ogura, K., "Electrochemical Separation of Ionic Compounds Using a Conductive Stationary Phase Coated with Polyaniline or Polypyrrole Film, and Ion Exchange Properties of Conductive Polymers," *J. Electroanal. Chem.*, 364 (1994), p. 179.

329. Gao, Z., Chen, B. and Zi, M., "Voltammetric Response of Dopamine at an Overoxidised Polypyrrole-Dodecyl Sulfate Film Coated Electrode," *J. Chem. Soc., Chem. Commun.* (1993), p. 675.

330. Witkowski, A. and Brajter-Toth, A., "Overixidized Polypyrrole Films: A Model for the Design of Permselective Electrodes," *Anal. Chem.*, 64 (1992), p. 635.

331. Yoshino, K. and Kaneto, K., "Application of Insulator-Metal Transition of Conducting Polymers," *Mol. Cryst. Liq. Cryst.*, 121 (1985), p. 247.

332. Yoshino, K. and Kaneto. K., "Radiation Effect in Conducting Polymers, *Mol. Cryst. Liq. Cryst.*, 121 (1985), p. 255.

333. Lyons, M., Bartlett, P., Lyons, C., Breen, W. and Cassidy, J., "Conducting Polymer Based Electrochemical Sensors: Theoretical Analysis of Current Response under Steady State Conditions," *J. Electroanal. Chem.*, 304 (1991), p. 1.

334. Bartlett, P. and Cooper, J., "A Review of the Immobilization of Enzymes in Electropolymerized Films," *J. Electroanal. Chem.*, 362 (1993), p. 1.

335. Bartlett, P. and Birkin, P., "The Application of Conducting Polymer in Biosensors," *Synth. Met.*, 61 (1993), p. 15.

336. Bartlett, P. and Whitaker, R., "Electrochemical Immobilization of Enzymes. Part II: Glucose Oxidase Immobilised in Poly-*N*-methylpyrrole," *J. Electroanal. Chem.*, 224 (1987), p. 37.

337. Bartlett, P. and Whitaker, R., "Electrochemical Immobilization of Enzymes. Part I: Theory," *J. Electroanal. Chem.*, 224 (1987), p. 27.

338. Bartlett, P. and Birkin, P., "The Application of Conducting Polymer in Biosensors," *Synth. Met.*, 61 (1993), p. 15.

339. Bullock, C., "Immobilized enzymes," *Ed. in Chem.* (1989), p. 179.

340. Heller, A., "Electrical Wiring of Redox Enzymes," *Acc. Chem. Res.*, 23 (1990), p. 128.

341. Heller, A., "Electrical Connection of Enzyme Redox Centers to Electrodes," *J. Phys. Chem.*, 96 (1992), p. 3579.

342. Gregg, B. and Heller, A. "Redox Polymer Films Containing Enzymes. 1. A Redox-Conducting Epoxy Cement: Synthesis Characterization and Electrocatalytic Oxidation of Hydroquinone," *J. Phys. Chem.*, 95 (1991), p. 5970.

343. Hale, P., Boguslavsky, L., Inagaki, T., Karan, H., Lee, H., Skotheim, T. and Okamoto, Y., "Amperometric Glucose Biosensors Based on Redox Polymer-Mediated Electron Transfer," *Anal. Chem.*, 63 (1991), p. 677.

344. Schumann, W., Ohara, T., Schmidt, M. and Heller, A., "Electron Transfer between Glucose Oxidase and Electrodes via Redox Mediators Bound with Flexible Chains to the Enzyme Surface," *J. Amer. Chem. Soc.*, 113 (1991), p. 1394.

345. Umana, M. and Waller, J., "Protein-Modified Electrodes. The Glucose Oxidase/Poly-pyrrole System," *Anal. Chem.*, 58 (1986), p. 2979.

346. Foulds, N. and Lowe, C., "Enzyme Entrapment in Electrically Conducting Polymers: Immobilization of Glucose Oxidase-Polypyrrole and Its Application in Amperometric Glucose Sensors," *J. Chem. Soc., Faraday Trans.*, 1, 82 (1986), p. 1259.

347. Tatsuma, T., Watanabe. T. and Watanabe, T., "Electrochemical Characterization of Polypyrrole Bienzyme Electrodes with Glucose Oxidase and Peroxidase," *J. Electroanal. Chem.*, 356 (1993), p. 245.

348. Porter, S., "Polypyrrole as a Potentiometric Glucose Sensor," Ext. Abstract, *ICSM '88*, Sante Fe, 26 June–2 July (1988), p. 70.

349. Cass, A., Davis, G., Prancis, G., Hill, O. H., Aston, W., Higgins, I., Plotkin, E., Scott, L. and Turner, A., "Ferrocene-Mediated Enzyme Electrode for Amperometric Determination of Glucose," *Anal. Chem.*, 56 (1984), p. 667.

350. Bourdillon, C. and Majda, M. "Microporous Aluminum Oxide Films at Electrodes. 7. Mediation of the Catalytic Activity of Glucose Oxidase via Lateral Diffusion of a Ferrocene Amphiphile in a Bilayer Assembly," *J. Am. Chem. Soc.*, 112 (1990), p. 1795.

351. Ikeda, T., Shibata, T. and Senda, M., "Amperometric Enzyme Electrode for Maltose Based on an Oligosaccharide Dehydrogenase-Modified Carbon Paste Electrode Containing p-Benzoquinone," *J. Electroanal. Chem.*, 261 (1989), p. 351.

352. Bianco, P., Haladjian, J. and Bourdillon, C., "Immobilization of Glucose Oxidase on Carbon Electrodes," *J. Electroanal. Chem.*, 293 (1990), p. 151.

353. Marchesiello, M. and Geniés, E., "Glucose Sensor: Polypyrrole-Glucose Oxidase Electrode in the Presence of p-Benzoquinone," *Electrochim. Acta*, 37 (1992), p. 1987.

354. Bartlett, P. and Bradford, V., "Modification of Glucose Oxidase by Tetrathiafulvalene," *J. Chem. Soc., Chem. Commun.* (1990), p. 1135.

355. O'Hare, D., Parker, K. and Winlove, C., "The Effect of Flavinmononucleotide on the Sensitivity of Glucose Oxidase Electrodes with the Complex Tetrathiafulvalene-Tetracyanoquinodimethane," *Biolectrochem. and Bioenerg.*, 23 (1990), p. 203.

356. Hale, P. and Skotheim, T., "Cyclic Volummetry at TCNQ and TTF-TCNQ Modified Platinum Electrodes: A Study of the Glucose Oxidase/Glucose and Galactose Oxidase/Galactose Systems," *Synth. Met.*, 28 (1989), p. 853.

357. Yabuki, S. -L., Shinohara, H. and Aizawa, M., "Electro-Conductive Enzyme Membrane," *J. Chem. Soc., Chem. Commun.* (1989), p. 945.

358. Belanger, D., Nadreau, J. and Fortier, G., "Electrochemistry of the Polypyrrole Glucose Oxidase Electrode," *J. Electroanal. Chem.*, 274 (1989), p. 143.

359. Pandey, P., "A New Conducting Polymer-Coated Glucose Sensor," *J. Chem. Soc., Faraday Trans.*, 1, 84 (1988), p. 2259.

360. Katani, A., Kasyu, N. and Sasaki, K., "Effect of Flavin Coenzymes on Current Response for Glucose at Glucose Oxidase/Polypyrrole Modified Electrodes," *Electrochim. Acta*, 39 (1994), p. 7.

361. Hoa, D., Suresh Kumar, T., Punekar, N., Srinivasa, R., Lal, R. and Contractor, A., "Biosensors Based on Conducting Polymers," *Anal. Chem.*, 64 (1992), p. 2645.

362. Jozefowicz, M., Yu, L., Oerichon, J. and Buvet, R., "Propriétés nouvelles des polyméres semiconducteurs," *J. Polym. Sci.*, C22 (1969), p. 1187.

363. .Shinohara, H., Chiba, T. and Aizawa, M., "Enzyme Microsensor for Glucose with an Electrochemically Synthesized Enzyme-Polyaniline Film," *Sensors and Actuators*, 13 (1988), p. 79.

364. Shaolin, M., Huaiguo, X. and Bidong, Q., "Bioelectrochemical Responses of the Polyaniline Glucose Oxidase Electrode," *J. Electroanal. Chem.*, 304 (1991), p. 7.

365. Schujmann, W., "Functionalized Polypyrrole. A New Material for the Construction of Biosensors," *Synth. Met.*, 41–43 (1991), p. 429.

366. Taniguchi, I., Matsushita, K., Okamoto, M., Collin, J. -P. and Sauvage, J. -P., "Catalytic Oxidation of Hydrogen Peroxide at Ni-Cyclam Modified Electrodes and Its Application to the Preparation of an Amperometric Glucose Sensor," *J. Electroanal. Chem.*, 280 (1990), p. 221.

367. Tamiya, E., Karube, I., Hattori, S., Suzuka, M. and Yokoyama, K. "Micro Glucose Sensors Using Electron Mediators Immobilized on a Polypyrrole-Modified Electrode," *Sensors and Actuators*, 18 (1989), p. 297.

368. Yaniv, D. R. and McCormick, L., "Scanning Tunneling Microscopy of Polypyrrole Glucose Oxidase Electrodes," *J. Electroanal. Chem.*, 314 (1991), pp. 353–361.

369. Marchesiello, M. and Geniés, E. A., "A Theoretical Model for an Amperometric Glucose Sensor Using Polypyrrole as Immobilization Matrix," *J. Electroanal. Chem.*, 358 (1993), p. 35.

370. Marchesiello, M., "Utilisation des polyméres conducteurs électroniques pour l'immobilisation de la glucose oxydase á la surface d'électrodes. Application au dosage du glucose par ampérométrie et potentiométrie," Thése de l'Université J. Fourier (1993).

371. Wolowacz, S., Yon Hin, B. and Lowe, C., "Covalent Electropolymerization of Glucose Oxidase in Polypyrrole," *Anal. Chem.*, 64 (1992), p. 1541.

372. Yon-Lin, B., Smolander, M., Cromton, T. and Lowe, C., "Covalent Electropolymerization of Glucose Oxidase in Polypyrrole. Evaluation of Methods of Pyrrole Attachment to Glucose Oxidase on the Performance of Electropolymerized Glucose Sensors," *Anal. Chem.*, 65 (1993), p. 2067.

373. Velazques Rosenthal, M., Skotheim, T. and Warren, J., "Ferrocene-Functionalized Polypyrrole Films," *J. Chem. Soc., Chem. Commun.* (1985), p. 342.

374. Schuhmann, W., Huber, J., Mirlach, J. and Daub, J., "Covalent Binding of Glucose Oxidase to Functionalized Polyazulenes. The First Application of Polyazulenes in Amperometric Biosensors," *Adv. Mat.*, 5 (1993), p. 124.

375. Malitesta, C., Palmisano, F., Torsi, L. and Zambonin, P. -G., "Glucose Fast-Response Amperometric Sensor Based on Glucose Oxidase Immobilized in an Electropolymerized Poly(o-phenylenediamine) Film," *Anal. Chem.*, 62 (1990), p. 2735.

376. Coche-Guerente, L., Deronzier, A., Galland, B., Labbe, P., Moutet, J.-C. and Reverdy, G., "Immobilization of Redox Anions in Poly(amphiphilic pyrrolylalkylammonium) Using a Simple and Monomer Saving One-Step Procedure in Pure Water Electrolyte," *J. Chem. Soc., Chem. Commun.* (1991), p. 386.

377. Cosnier, S. and Innocent, C., "A Novel Biosensor Elaboration by Electropolymerization of an Adsorbed Amphiphilic Pyrrole-Tyrosinase Enzyme Layer," *J. Electroanal. Chem.*, 328 (1992), p. 361.

378. Geniés, E. and Marchesiello, M., "Conducting Polymers for Biosensors, Applications to New Glucose Sensors: GOD Entrapped into Polypyrrole, GOD Adsorbed on Poly(3-methylthiophene)," *Synth. Met.*, 55–57 (1993), p. 3677.

379. Janda, P. and Weber, J., "Quinone-Mediated Glucose Oxidase Electrode with the Enzyme Immobilized in Polypyrrole," *J. Electroanal. Chem.*, 300 (1991), p. 119.

380. Foulds, N. and Lowe, C., "Immobilization of Glucose Oxidase in Ferrocene-Modified Pyrrole Polymers," *Anal. Chem.*, 60 (1988), p. 2473.

381. Lowe, C., Foulds, N., Evans, S. and Yon Hin, B., "Enzyme Entrapment in Conducting Organic Polymers: Application in Amperometric Biosensors, in *Electronic Properties of Conjugated Polymers III*, eds., Kuzmany, H., Mehring, M., Roth, S., Springer series in *Solid State Sciences, Vol. 91*, Springer, Berlin (1989), p. 132.

382. Schuhmann, W., Kranz, C., Huber, J. and Wohlschläger, H., "Conducting Polymer-Based Amperometric Enzyme Electrodes. Towards the Development of Miniaturized Reagentless Biosensors," *Synth. Met.*, 61 (1993), p. 31.

383. Iwakura, C., Kajiya, Y. and Yoneyama, H., "Simultaneous Immobilization of Glucose Oxidase and Mediator in Conducting Polymer Films," *J. Chem. Soc., Chem. Commun.* (1988), p. 1019.

384. Kajiya, H., Sugai, H., Iwakura, C. and Yoneyama, H., "Glucose Sensitivity of Polypyrrole Films Containing Immobilized Glucose Oxidase and Hydroquinonesulfonate Ions," *Anal. Chem.*, 63 (1991), p. 49.

385. Kajiya, Y. and Yoneyama, H., "Enhancement of the Electron Transfer between a Mediator-Adsorbed Glucose Oxidase and an Electrode by Treatment with Urea," *J. Electroanal. Chem.*, 341 M1992), p. 85.

386. Bartlett, P., All, Z. and Eastwick-Field, W., "Electrochemical Immobilization of Enzymes. Part 4. Co-immobilization of Glucose Oxidase and Ferro/Ferricyanide in Poly(N-methylpyrrole) Films," *J. Chem. Soc., Faraday Trans.*, 88 (1992), p. 2677.

387. Yabuki, S., Shinohara, H. and Aizawa, M., "Electro-Conductive Enzyme Membrane," *J. Chem. Soc., Chem. Commun.* (1989), p. 945.

388. De Taxis Du Poet, P., Miyamoto, S., Murakami, T., Kimura, J. and Karube, I., "Direct Electron Transfer with Glucose Oxidase Immobilized in an Electropolymerized Poly (N-methylpyrrole) Film on a Gold Microelectrode," *Anal. Chim. Acta*, 235 (1990), p. 255.

389. Koopal, C., De Ruiter, B. and Nolte, R., "Amperometric Biosensor Based on Direct Communication between Glucose Oxidase and a Conducting Polymer inside the Pores of a Filtration Membrane," *J. Chem. Soc., Chem. Commun.* (1991), p. 1691.

390. Koopal, C., Feiters, M., Nolte, R., De Ruiter, B., Schasfoort, R., Czajka, R. and Van Kempen, H., "Polypyrrole Microtubules and Their Use in the Construction of a Third Generation Biosensor," *Synth Met.*, 51 (1992), p. 397.

391. Czajka, C., Koopal, C., Feiters, M., Gerritsen, J., Nolte, R. and Van Kempen, M., "Scanning Tunneling Microscopy Study of Polypyrrole Films and of Glucose Oxidase as Uses in a Third-Generation Biosensor," *Biochem. Bioenerg.*, 29 (1992), p. 47.

392. Koopal, C., Eijsma, B. and Nolte, R., "Chronoamperometric Detection of Glucose by a Third Generation Biosensor Constructed from Conducting Microtubules of Polypyrrole," *Synth Met.*, 55–57 (1993), p. 3689.

393. Koopal, C., Feiters, M. and Nolte, R., "Third Generation Amperometric Biosensor for Glucose. Polypyrrole Deposited within a Matrix of Uniform Latex Particles as Mediator," *Biochem. Bioenerg.*, 29 (1992), p. 159.

394. Lu, S. -Y., Li, C. -F., Zhang, D. -D., Zhang, Y. and Mo, H. -H., Cai, Q., and Zhu, A. -R., "Electron Transfer on an Electrode of Glucose Oxidase Immobilized in Polyaniline," *J. Electroanal. Chem.*, 364 (1994), p. 31.

395. Somasundrum, G. and Bannister, J., "Mediatorless Electrocatalysis at a Conducting Polymer Electrode: Application to Ascorbate and NADH Measurement," *J. Chem. Soc., Chem. Commun.* (1993), p. 1629.

396. Tarasevich, M. and Bogdanovskaya, V., "Electrochemical Preparation of Polypyrrole Films with Enzymes," *Synth. Met.*, 41–43 (1991), p. 433.

397. Wollenberg, U., Bogdanovskaya, B., Bobrin, S., Scheller, F. and Tarasevich, M., *Anal. Letter.*, 23 (1990), p. 1795.

398. Yabuki, S., Shinohara, H., Ikariyama, Y. and Aizawa, M., "Electrical Activity Controlling System for a Mediator-Coexisting Alcohol Dehydrogenase-NAD Conductive Membrane," *J. Electroanal. Chem.*, 277 (1990), p. 179.

399. Haruyama, T., Shinohara, H., Ikariyama, Y. and Aizava, M., "Modulation of the Function of an Enzyme Immobilized in a Conductive Polymer by Electrochemical Changing of the Substrate Concentration," *J. Electroanal. Chem.*, 347 (1993), p. 293.

400. Cosnier, S. and Innocent, C., "A Novel Biosensor Elaboration by Electropolymerization of an Adsorbed Amphiphilic Pyrrole-Tyrosinase Enzyme Layer," *J. Electroanal. Chem.*, 328 (1992), p. 361.

401. Matsue, T., Kasai, N., Narumi, M., Nishizawa, N., Yamada, H. and Uchida, I., "Electron-Transfer from NADH Dehydrogenase to Polypyrrole and Its Applicability to Electrochemical Oxidation of NADH," *J. Electroanal. Chem.*, 300 (1991), p. 111.

402. Kajiya, Y., Matsumoto, H. and Yoneyama, H. Y., "Glucose Sensitivity of Poly(pyrrole) Films Containing Immobilized Glucose Dehydrogenase, Nicotinamide Adenine Dinucleotide, and Naphthoquinoesulphonate Ions," *J. Electroanal. Chem.*, 319 (1991), p. 185.

403. Yabuki, S., Mizutani, F. and Asai, M., "Preparation and Characterization of an Electroconductive Membrane Containing Glutamate Dehydrogenase, NADP, and Mediator," *Biosensors. Bioelectr.*, 6 (1991), p. 31.

404. Yabuki, S., Mizutani, F., Katsura, T. and Asai, M., "Electrical Control of the Glutamate Dehydrogenase Reaction in Polypyrrole Membranes," *Bioelectrochem. Bioenerg.*, 28 (1991), p. 489.

405. Adelou, S., Shaw, S. and Wallace, G., "Polypyrrole-Based Potentiometric Biosensor for Urea. Part 2. Analytical Optimisation," *Analytica Chim. Acta*, 281 (1993), p. 621.

406. Shaolin, M., Jinquing, K. and Jianbing, Z., "Bioelectrochemical Responses of the Polyaniline Uricase Electrode," *J. Electroanal. Chem.*, 334 (1992), p. 121.

407. Shaolin, M. and Shufan, C., "Effect of Thiourea on the Activity of the Polypyrrole Uricase Electrode and Determination of Thiourea," *J. Electroanal. Chem.*, 356 (1993), p. 59.

408. Shaolin, M., "The Kinetics of Activated Uricase Immobilized on a Polypyrrole Film," *Electrochim. Acta*, 39 (1994), p. 9.

409. Yon, Hin, B. and Lowe, C., "Amperometric Response of Polypyrrole Entrapped Bienzyme Films," *Sensors and Actuators B*, 7 (1992), 339.

410. Kajiya, Y., Tsuda, R. and Yoneyama, H., "Conferment of Cholesterol Sensitivity on Polymeric Films by Immobilization of Cholesterol Oxidase and Ferrocenecarboxylate Ions," *J. Electroanal. Chem.*, 301 (1991), p. 155.

411. Slater, J. and Watt, E., "Use of Conducting Polymer Polypyrrole, as a Sensor," *Anal. Proc.*, 26 (1989), p. 397.

412. Willner, I., Katz, E. and Lapidot, N., "Bioelectrocatalysed Reduction of Nitrate Utilizing Polythiophene Bipyridinium Enzyme Electrodes," *Bioelectrochem. Bioenerg.*, 29 (1992), p. 29.

413. Tanguy, J., Mermillod, N. and Hoclet, M., "Capacitive Charge and Noncapacitive Charge in Conducting Polymer Electrodes," *J. Electrochem. Soc.*, 134 (1987), p. 795.

414. Nishizawa, M., Matsue, T. and Uchida, I., "Penicillin Sensor Based on a Microarray Electrode Coated with pH-Responsive Polypyrrole," *Anal. Chem.*, 64 (1992), p. 2642.

415. Matsue, T., "Electrochemical Sensors Using in Microarray," *Trends in Analytical Chemistry,* 12 (1993), p. 100.

416. Bartlett, P. and Birkin, P., "Enzyme Switch Responsive to Glucose," *Anal. Chem.,* 65 (1993), p. 1118.

417. Caselli, M., Della Monica, M., and Portacci, M., "Electrochemical and Spectroscopic Study of Cytochrome C Immobilized in Polypyrrole Films," *J. Electroanal. Chem.,* 319 (1991), p. 361.

418. Agostiano, A., Caselli, M., Della Monica, M. and Laera, S., "Characterization of Polypyrrole Films Electrodeposited by Water Solutions: Effect of the Supporting Electrolyte and Cytochrome Immobilization," *Electrochim. Acta,* 38 (1993), p. 2581.

419. Shinohara, H., Khan, G., Ikariyama, Y. and Aizawa, M., "Electrochemical Oxidation and Reduction of PQQ Using a Conducting Polypyrrole-Coated Electrode," *J. Electroanal. Chem.,* 304 (1991), p. 75.

420. Atta, N. F., Galal, A., Karagozler, A., Zimmer, H., Rubinson, J. F. and Mark, H., "Voltammetric Studies of the Oxidation of Reduced Nicotinamide Adenine Dinucleotide at a Conducting Polymer Electrode," *J. Chem. Soc., Chem. Commun.* (1990), p. 1347.

421. Cosnier, S. and Gunther, H., "A Polypyrrole $[Rh^{III}(C_5Me_5)(bpy)Cl)]^+$ Modified Electrode for the Reduction of NAD^+ Cofactor. Application to the Enzymatic Reduction of Pyruvate," *J. Electroanal. Chem.,* 315 (1991), p. 307.

422. Cosnier, S., Deronizer, A. and Vlachopoulos, N., "Carbon/Poly{pyrrole-$[Rh^{III}(C_5Me_5)(byp)Cl)]^+$} Modified Electrodes: A Molecularly-Based Material for Hydrogen Evolution (bpy = $2,2'$-Bipyridine)," *J. Chem. Soc., Chem. Commun.* (1989), p. 1259.

423. Battaglini, F., Bonazzola, C. and Clavo, E., "Electrochemistry of Polypyrrole Flavin Electrodes," *J. Electroanal. Chem.,* 309 (1991), p. 347.

424. Cosnier, S. and Innocent, C., "Immobilization of Flavin Coenzyme in Poly(pyrrolealkylammonium) and Characterization of the Resulting Bioelectrode," *J. Electroanal. Chem.,* 338 (1992), p. 339.

425. Bäuerle, P., Götz, G., Hiller, M., Scheib, S., Fischer, T., Segelbacher, U., Bennati, M., Grupp, A., Mehring, M., Stoldt, M., Seidel, C., Geiger, F., Schweizer, H., Umbach, E., Schmelzer, M., Roth, S., Egelhaaf, H. -J., Oelkrug, D., Emele, P. and H. Port, H., "Design, Synthesis and Assembly of New Thiophene-Based Molecular Functional Units with Controlled Properties," *Synth. Met.,* 61 (1993), p. 71.

426. Wallace, G. and Riley, P., "Intelligent Polymers," *Chem. Aust.,* (1991), p. 33.

427. Ge, M. and Wallace, G., "Conducting Polymeric Stationary Phases and Electrochemically Controlled High-Performance Liquid Chromatography," *Anal. Chem.,* 61 (1989), p. 2391.

428. Hodgson, A., Spencer, M. and Wallace, G., "Incorporation of Proteins into Conducting Electroactive Polymers. A Preliminary Study," *Reactive Polymers,* 18 (1992), p. 77.

429. Shimidzu, T., "Functionalized Conducting Polymers for Development of New Polymeric Reagents," *Reactive Polym.,* 6 (1987), p. 221.

430. Teoule, R., Roget, A., Livache, T., Barthet, C. and Bidan, G., "Copolymére nucléotide(s) polymére conducteur électronique, son procédé de préparation et son utilisation," Brevet ORIS/Cis bio international/LSM Patent, EN 93 037 32 (1993).

431. Kamath, M., Lim, J., Chittibabu, K., Sarma, R., Jumar, J., Marx, K. and Tripathy, S., "Biotinylated Poly(3-hexylthiophene-*co*-3-methanolthiophene): A Langmuir Monolayer-Forming Copolymer," *J. M. S. -Pure Appl. Chem.,* A 30 (1993), p. 493.

CHAPTER 4

Sensors Using Polymer Films (Construction, Technology, Materials, and Performance)

In the previous chapters, a summary was given about the technological processes used in the fabrication of inorganic sensor parts and polymer sensor films, about principles of operation of different sensor types that can be realized using sensitive polymer films, and also about the physical and chemical phenomena involved in sensing effects.

In this chapter, sensor structures with polymer films will be described with examples. A short survey will be given about their application possibilities and their performance. The application of the various sensor technologies will also be demonstrated.

The various sensors have been grouped according to quantity to be measured. The different groups are as follows:

- temperature sensors
- mechanical sensors: touch switch devices, deformation sensors, tactile sensors, pressure sensors, accelerometers, vibrometers, etc.
- acoustic sensors: microphones, ultrasonic sensors, hydrophones, etc.
- radiation sensors
- humidity sensors
- gas sensors
- ion-selective sensors
- sensors in medicine and biology
- others (liquid component sensors, material identification, actuators, etc.)

There is still another conventional categorization of sensors according to the nature of interaction, which is the basis of operation, into the groups of physical, chemical, and biosensors. This grouping principle is different from the former one; therefore, it is not used in this chapter. References to the latter categories are made several times, however, according to the definition mentioned above.

4.1 TEMPERATURE SENSORS

There are a lot of conventional and integrated temperature sensors that are widely used in practice and are grouped according to their principle of operation:

- thermocouples
- resistive thermometers (Pt, Ni, Cu)
- NTC and PTC thermistors
- semiconductor based *p-n* junction and spreading resistance thermometers
- capacitive thermometers
- others (IR, electrochemical, etc.)

Without a detailed description, it must be underlined that there is a wide choice in operation principles and properties (sensitivity, operation temperature range, reliability, etc.). The types belonging to the first two groups are already standardized. However, there is a lack of low-cost, reliable PTC thermistors that can be used as overcurrent protecting devices. Composite polymer materials seem to be good candidates for this purpose.

Electrically conductive polymer composites have attracted a great deal of scientific and commercial interest during the last several years (see Sections 1.3.3 and 3.2). In certain compositions, PTC materials have shown a six orders of magnitude resistance change over a 20°C temperature change.

PTC resistors can be used as overcurrent protectors either in temperature or in a current sensing mode. In the former case, the resistor is heated by the element to be measured. In the latter one, it is heated by Joule heat generated by itself, and it acts as an automatic resetting fuse. In that situation, the sensor must be connected in series with the circuit to be protected. Conventionally, PTC overcurrent protecting devices are based on doped semiconducting barium titanate ceramics. The ability to design a PTC device as an overcurrent protector in a power circuit works directly on the ability to produce low-resistivity materials.

A 1-Ωcm resistivity at 20°C is common for conductive polymer composites, while barium titanate–based materials generally exhibit a considerably higher resistivity. Another difference is that conductive polymer composites exhibit mechanical insensitivity to thermal shocks and only slight resistance sensitivity to voltage. Another significant distinction of conductive polymer composites is that they generally persist in the PTC behaviour over the critical temperature (this can also be modified by special fillers; see Section 3.2.3). Barium titanates, on the other hand, become NTC in this region, which leads to thermal runaway in practical circuit applications [1].

A lot of different materials have been investigated for the above-mentioned purposes. The conductive or semiconducting filler materials are carbon black, vanadium, and titanium oxides [2,3]. Examples for polymer matrix materials are as follows: polyethylene (PE), poly(isobutylene), polytetrafluoroethylene (PTFE), and epoxies.

Figure 4.1 shows the temperature resistivity characteristics for a few composite materials. The transition temperature is determined by the melting point of the polymer, by the critical volume fraction of conductive particles, and by the thermal mismatch of the two components. The steepness of PTC effect is influenced by the degree of crystallinity of the polymer. The room temperature and high-temperature resistivity values are determined by the resistivity of the filler and polymer materials, respectively.

Special overcurrent protector devices called PolySwitch™ have been developed by Raychem Co. [1]. PolySwitch™ overcurrent protectors are made from conductive polymer composites. Fabrication of a device includes the creation of the raw material, melt forming it to device shape, attachment of leads, and packaging. The final product looks like a disc capacitor but acts like a solid-state circuit breaker or resettable fuse.

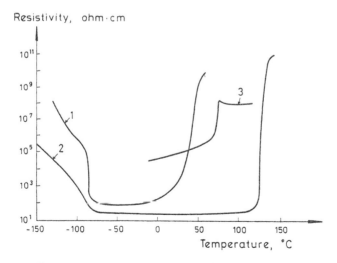

FIGURE 4.1. Typical temperature characteristics of resistivity for different conductive polymer composites [2,3] (1–55 vol% V_2O_3-epoxy; 2–57 vol% V_2O_3-polyethylene; 3–8% carbon black, 69% $C_{36}H_{74}$; poly(isobutylene) composites).

The resistance recovery behaviour is illustrated in Figure 4.2 [1]. When the device temperature is raised near to its melting point and then returned to the ambient temperature, the resistance does not immediately recover to its initial value since the compaction of the polymer is a much slower process than simple cooling alone. The origin of this memory effect is found in the physics of recrystallization of the polymer.

Recently, a temperature-sensitive polymer composite material has been developed using ion implantation for irradiating insulating polyimide films (see Section 1.4.6). Xu et al. [4] have demonstrated that graphite phase appears during ion implantation, and a clean NTC behaviour of the fabricated resistors can be observed according to the hopping conduction mechanism (see Section 3.2.3).

The polyimide was spun on p-type silicon wafer, and after prebaking at 80°C, the samples were subsequently baked at 200°C for 60 min. One hundred seventy keV N$^+$ and 120 keV B$^+$ ion radiation beams were employed to modify the polyimide. Two electrodes on each as-prepared modified polyimide sample were formed in a vacuum evaporation process. The resistivity of the modified layers was about 10^{-2} Ωcm, while their original resistivity value was 10^{17} Ωcm.

The resistivity of the modified layers showed a strong NTC characteristic, with a decrease of resistivity to 50% when the temperature varied from 20°C to 120°C [4]. The fabrication technique is suitable for integrating reliable thermistors onto the surface of semiconductor chips.

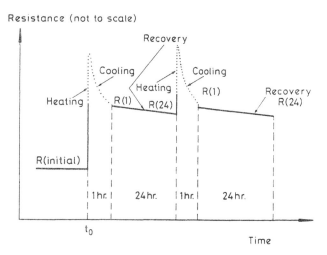

FIGURE 4.2. Resistance recovery of conductive polymer composite switching devices. (Redrawn with permission from the figure of Doljak [1], © 1981 IEEE.)

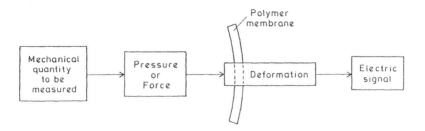

FIGURE 4.3. Operation of polymer-based mechanical sensors.

4.2 MECHANICAL SENSORS

Mechanical sensors are used to measure or detect various mechanical quantities, such as displacement, pressure, force, acceleration, velocity, and vibration. Figure 4.3 demonstrates the general principle of operation of polymer-based mechanical sensors. Mechanical quantities can be easily converted into a concentrated or distributed force, and the flexible polymer membrane will be deformed by the latter. The electrical signal can be generated by the strain of the polymer itself (e.g., piezoelectricity) or by the parameter changes of various elements deposited or mounted on its surface (e.g., piezoresistive effect) (see Sections 3.1 and 3.2). According to the function and operation, the following grouping of mechanical sensors will be used:

- membrane touch switches
- strain gauges
- pressure sensors
- tactile sensors
- accelerometers
- flowmeters
- vibration and pulse sensors

4.2.1 Membrane Touch Switches [5,6]

It is well known that membrane touch switches are widely used in keyboards, but it is less well known that they are primarily based on polymers [5]. Touch switches are, in fact, the most simple force sensors. They give an electrical signal when a given force is applied on their surface.

Manufacturing of membrane switches on a polymer basis was one of the first areas where polymer thick-film (PTF) technology was used in the mid-1970s. Nowadays, PTF in membrane switch panels and keyboards is a mature technology for manufacturing professional products, even for military applications.

Figure 4.4(a) shows the typical structure of a membrane switch panel [6]. It is constructed as a sandwich of two polymeric sheets with a spacer between them. Conductive pads and paths are screen-printed on the inner side of each sheet. All sheets may be made of flexible polyester material, or the bottom substrate may be more rigid, as in the case of a printed circuit board (PCB). When pressure is applied to the top sheet, as shown in Figure 4.4(b), the sheet is flexed through a hole in the spaces to establish electrical contact between the conductive pad of the upper and lower sheets, temporarily closing the circuit on the bottom sheet. Associated circuitry is generally also applied on the bottom layer.

Membrane switch panels are generally custom designed. Thermoplastic PTF pastes are generally used in them. The thermoplastics give the necessary flexibility and are cured at relatively low temperatures (approximately 120°C). Polyester is generally used as the substrate material. Polyethersulphone can be used for temperatures up to 200°C and also with thermosetting pastes [5]. For curing above 200°C, the application of a polyimide sheet is recommended. The conductors are normally made with Ag-based pastes. However, carbon black resistive materials are generally used at the switching contacts. Two crossing conductor layers are

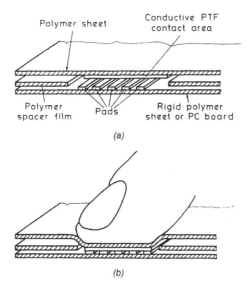

FIGURE 4.4. Typical membrane touch switch configuration: (a) open and (b) closed. (Redrawn with permission from Hicks et al. [6], © 1980 IEEE.)

TABLE 4.1. Typical Specifications of PTF Membrane Touch Switches.

Electrical data:	
Max. switching load (resistive):	1.0 W
Max. voltage:	30 V AC/DC
Max. current:	100 mA
Typical contact resistance of switch:	5 Ω
Typical contact resistance	
including conductors:	max. 100 Ω
Insulation resistance, open circuit:	min. 100 MΩ, 500 V DC
Typical contact bouncing time:	15 msec
Temperature/humidity:	
Temperature range, operation:	$-30 - +70°C$
Temperature range, storage:	$-30 - +70°C$
Accelerated temp./humidity test:	40°C, 95% RH, 150 hours
Mechanical data:	
Switch life cycle:	min. 15 mill. operations
Actuation force:	2.7 N $\pm$ 20%
Switch travel distance:	approx. 0.50 mm
Bending radius, tail:	min. 0.5 mm, max. 3 times
Dimensions:	
Panel thickness, standard:	approx. 0.80 mm
General dimension tolerances:	± 0.3 mm
Size, including tail:	max. 480 $\times$ 320 mm

Reprinted with permission from *Polymer Thick-Film Technology, Design Guidelines and Applications,* ed. Leif Halbo [5], ©Leif Halbo/Lobo Grafisk (1990).

separated by an insulation layer. Several different approaches can be used to process a polymeric dielectric:

- air-dried polymer solution
- two-part epoxy
- peel-apart laminate (dry film laminate)
- UV-curable dielectrics

To illustrate typical specifications for a membrane switch panel, characteristics for a product of this type are given in Table 4.1 [5].

It must be mentioned here that piezoelectric polymer sheets (see Section 3.1.1) with deposited metallization layers can also be used for switches and keyboards. But these devices are not real switching elements (with zero and high resistance values). They are capacitors, and during their operation, they give an active electrical signal when a switching touch is applied; therefore, they are called touch triggers. Their structure is the same as that of the piezoelectric tactile sensors; thus, it will be discussed in Section 4.2.4.

Bell Laboratories was one of the first to utilize the advantages of the piezoelectric polymer film switch. They built keyboards for telephones using piezopolymer films. A 3 × 4 matrix was screen-printed on the film and then impacted for 5 million (+) impacts per key. A constant 8- to 10-Volt output was given from each key, with no button wear or signal degradation. A snap disc actuator was used to provide tactile feel, as well as to give a uniform mechanical input to the keyboard. Their work is being used now by specialised keyboard manufacturers in the design of keyboards for vandalproof systems, remote locations, and harsh environments.

4.2.2 Strain Gauges

Strain gauges are widely used in practice for deformation measurements. Their general structure is shown in Figure 4.5(a). The device is usually stuck onto the surface to be measured. Its substrate is a flexible polymer sheet, which can follow the deformation of the sample surface. The most often used materials are [7,8]:

- polyimides
- epoxies
- glass-reinforced or laminated polyimides and epoxies
- phenolics

The meander-type resistor can be deposited using various metal alloys, such as nickel-chromium, constantan, etc., by thin-film processing techniques (see Section 1.3.1), onto the surface of the substrate. The maximum strains that can be measured using these devices are about 2–3%.

The fractional change of resistance during deformation can be expressed by Equation (3.23). Since the applied materials are not piezoresistive in nature, the resistance change is dominated by a pure geometrical effect. Thus, the value of the gauge factor [see Equation (3.20)] is in the range of 2–3.

For temperature compensation, strain gauges are generally connected into a bridge circuit configuration for practical applications. Figure 4.5(b) shows a few typical layout realization types.

The construction and application of strain gauges mean a special field of sensorics, which cannot be included in the present book. In this section, it is simply the important role of polymers as strain gauge substrates that is emphasized.

4.2.3 Pressure Sensors

Force and pressure sensors applying polymer sheets or films are based on capacitive, piezoresistive, and piezoelectric effect.

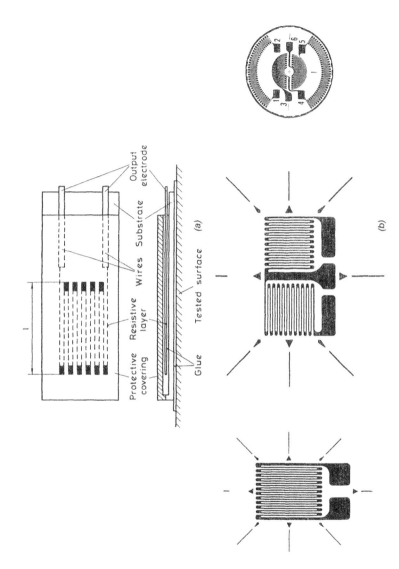

FIGURE 4.5. (a) General structure and (b) typical layouts (single-element, twin-element, and round-membrane bridge configuration) of strain gauges [7].

269

The structure of capacitive membrane-type pressure sensors is very similar to that of membrane touch switches, as shown in Figure 4.6 [9]. Their pressure dependence is obtained by letting the pressure deflect a membrane, thus changing the capacitance to the bottom of the enclosed cavity. The pressure-capacitance relationship thus obtained has an apparent nonlinearity. The nonlinearities can be reduced by a corrugation or weakening of parts of the membrane and by the application of sophisticated electrode structures (see Figure 4.6).

The demonstrated double-comb structure has a capacitance that is approximately directly proportional to the deflection and, thus, the pressure [9]. It is important that the comb teeth do not rotate during deflection since this will change the tooth-to-tooth distance, introducing nonlinearities in the deflection-capacitance conversion. In addition, the membrane must be designed not to have a nonlinear deflection. A stiff membrane centre with a corrugated (or weak) edge is therefore necessary.

Materials and fabrication processes for the mentioned structure are similar to the ones applied for membrane touch switches. Rigid sheets for bottom and spacer loads and a flexible sheet for the membrane can be used. Conductive pads and paths, as well as stiffening layers, can be formulated using polymer thick-film technology.

The greatest drawback of the capacitive pressure sensors is their relatively complicated structure. Much more simple membrane-type pressure sensors can be realized by PTF technology, the operation of which is based on the piezoresistive effect.

The pressure-measurement technique using strain gauges (see Section 4.2.2) attached to deformable substrates has been and still is extensively employed; however, it seems to become inadequate for new user requirements. The deposition of strain gauge or piezoresistive resistors directly on the membrane of the pressure sensor using thin- or thick-film

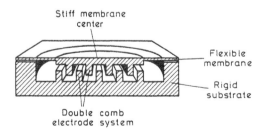

FIGURE 4.6. Schematic view of a capacitive pressure sensor. (Redrawn from Rosengren et al. [9], with the kind permission of Elsevier Sequoia S. A., Lausanne, Switzerland, publisher of *Sensors and Actuators A* and *Sensors and Actuators B*.)

TABLE 4.2. Main Parameters of CERMET and Polymer Thick-Film Resistors and the Calculated Pressure Sensor Quality Factors [9].

	CERMET Thick Films	PTF (typical)	High Reliability PTF
Gauge factor, G	10	10	10
Substrate type	96% alumina	FR4 (glass-epoxy)	FR4 (glass-epoxy)
E(N/mm^2) (Young's modulus)	3.3×10^5	2.1×10^4	2.1×10^4
a (mm) (substrate thickness)	0.6	0.2	0.2
TCR (ppm/°C)	±50	±500	±200
$(\Delta TCR)_{max}$ (ppm/°C)	±5	±50	±20
Stability $\Delta R/R$ (%)	0.3 (1000 h, 150°C)	5 (1000 h, 85°C)	0.5 (1000 h, 85°C)
$\delta(\Delta R/R)$ (%)	0.03	0.5	0.05
Q_T (10^{-3} Km/N)	10	48	120
Q_S (10^{-5} m/N)	16.8	48	480

technology (see Section 1.3), instead of gluing on it strain gauges deposited previously on polymeric foils, can lead to a great improvement in the performance.

It is well known that metal films and semiconductor resistors are characterized by a resistance variation proportional to the applied stress, but more recently it has been revealed that thick-film, even PTF, resistors show a notable sensitivity to deformation (see Section 3.2.2). The good sensitivity of PTF resistors to deformation, together with their cheap production cost, have suggested the development of pressure sensors based on the piezoresistive effect in resistive PTF composites. Although PTF resistors have much worse temperature and long-term stability parameters than the CERMET thick films (see Table 3.4 and 4.2), there is a possibility to obtain pressure sensors with almost the same stability parameters in both cases [10,11].

The sensing element of the PTF pressure sensor consists of a circular edge-clamped epoxy-glass (or flexible polymer) diaphragm on which four PTF resistors connected into a Wheatstone-bridge configuration are screened and cured, as shown in Figure 4.7. The optimum position of the four resistors is also given in the figure. It can be shown [12] that, in order to maximize the loss of the bridge balance and the voltage output, two resistors must be positioned near the centre of the diaphragm and the other two resistors near the edges. The pressure to be measured induces a strain

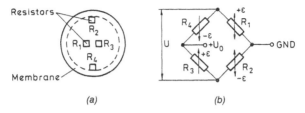

FIGURE 4.7. (a) Structure and (b) circuit connection of the piezoresistive pressure sensor membranes [10].

in the diaphragm, and the resistors change their values: two resistances increase and two decrease:

$$R_i = R_{0i} \cdot (1 \pm S \cdot p) \quad (i = 1,\ldots,4) \tag{4.1}$$

where

p = the pressure
R_{0i} = the nominal resistance values
S = the sensitivity, which depends on the gauge factor of the resistors, on the geometry, and on the elasticity of the diaphragm

Supposing that the nominal values of the resistors are equal, the output voltage of the bridge can be expressed as

$$U = U_0 \cdot S \cdot p = U_0 \cdot \frac{G \cdot K \cdot p}{E \cdot a} \tag{4.2}$$

where

U_0 = the excitation voltage of the bridge
G = the gauge factor
E = the Young's modulus of the substrate
a = the thickness of the substrate
K = constant

The most important parameters of the sensors are the sensitivity, the linearity, and the thermal and long-term zero shifts. The linearity depends on the uniformity of resistances and deformations. Thus, the bridge must be exactly balanced and the resistors well-positioned. To estimate the thermal behaviour of the output voltage, it can be easily shown [10]:

$$\frac{dU}{U_0 \cdot dT} \leq 0.5 \cdot (\Delta TCR)_{max} \tag{4.3}$$

where $(\Delta TCR)_{max}$ indicates the maximum possible difference between the temperature coefficients of resistances, the so-called TCR tracking. The thermal zero shift of the sensor can be characterized by

$$TD = \frac{dU}{U_{FS} \cdot dT} \tag{4.4}$$

where U_{FS} is the maximum, the so-called full-scale output. When comparing different sensors, one can introduce a figure of merit, the quality factor Q_T:

$$Q_T = \frac{G}{E \cdot a \cdot (\Delta TCR)_{max}} \tag{4.5}$$

According to Equations (4.2)–(4.5), Equation (4.4) can also be estimated as

$$TD \leq \frac{0.5}{Q_T \cdot K \cdot p_{max}} \tag{4.6}$$

where p_{max} is the nominal pressure value of the sensor. Following the previous order of ideas, a long-term stability quality factor can also be defined:

$$Q_S = \frac{G}{E \cdot a \cdot \delta(\Delta R/R)_{max}} \tag{4.7}$$

where $\delta(\Delta R/R)_{max}$ indicates the maximum possible difference of long-term drift in resistance values among the resistors.

Table 4.2 shows the main parameters of conventional CERMET thick-film and PTF resistors, as well as the calculated sensor quality factors in the case of alumina and glass-epoxy substrates, respectively. It is well demonstrated that PTF sensors have much higher quality factors than the CERMET ones; thus, they can be used for lower nominal pressure ranges with similar performances [10].

A very simple membrane structure can be used for relative pressure measurements (pressure differences related to the ambient pressure), as shown in Figure 4.8. An absolute pressure sensor needs a closed cavity with a constant pressure or vacuum in it, similar to that shown in Figure 4.6. The most important technological parameters and technical characteristics of a given type of PTF pressure sensor are shown in Table 4.3 in comparison with a CERMET thick-film version. Figure 4.9 shows the typical voltage outputs of the sensors excited with 10 V.

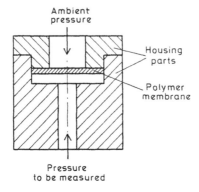

FIGURE 4.8. Structure of the relative pressure sensor [11].

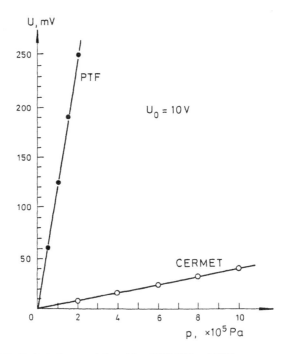

FIGURE 4.9. Typical characteristics of the CERMET and PTF pressure sensors [11].

TABLE 4.3. *Technology Process Parameters and Technical Characteristics of CERMET Thick-Film and PTF Pressure Sensors.*

Parameters	CERMET	PTF
Substrate material	96% alumina	FR4 (epoxy-glass)
Substrate thickness (mm)	0.6	0.2
Substrate radius (mm)	25.4	25.4
Conductor-type	Pd-Ag	Tin-coated copper
Resistor-type		
(sheet Resistivity)	$R_s = 10$ kΩ	$R_s = 10$ kΩ
Drying	150°C/15 min	120°C/15 min
Curing	850°C/10 min	180°C/120 min
Full-scale		
pressure (Pa)	10^6	2×10^5
Supply voltage (V)	10	10
Offset (mV)	±0.1	±0.1
Full-scale output		
(FSO), (mV)	50	250
Response time (ms)	<5	<5
Non-linearity and		
hysteresis	<0.5% FSO	<0.15% FSO
Working		
temperature (°C)	−25−+100	0−+75
Additional		
temperature failure	≤ ±0.05% FSO/°C	≤ ±0.07% FSO/°C
Overpressure	1.5 times FSO	1.5 times FSO
Long-term drift		
(1000 h, 85°C)	≤0.5%	≤0.5%

Piezoelectric force and pressure sensors can also be built up using piezopolymer (see Section 3.1.1) membranes that are metallized to form piezoelectric capacitors [13]. These devices give a voltage signal when the membrane is deformed by the pressure to be measured. Because of the charge neutralization from the environment, the practical application of the devices needs sophisticated circuitry and/or dynamic measurements. The same structures can be used as those used in tactile and vibration sensors (see Sections 4.2.4 and 4.2.5).

As an example, the shock wave sensors can be mentioned. Weapon tests are very expensive, and often, the sensors get destroyed in the course of test. Thus, the reproducibility of sensors is important. By subjecting the PVDF film to a spot biaxially oriented polarization, accurate, predictable piezoactivity results, which is usable up to the plastic yield of the film itself. This corresponds to pressure pulse rise times of nanoseconds, and pressures up to 1 GPa. Piezopolymer films are also finding uses in weapon tests as pressure pulse arrival time sensors.

4.2.4 Tactile Sensors

Tactile sensors have been the subject of research and development for the last fifteen years. To develop robot grippers with properties similar to the human hand is a challenge to design engineers. When an object has to be manipulated by the gripper, during process control, data concerning the orientation and shape of the object can only be deduced from the tactile image. This requires not only appropriate control units, but also efficient tactile sensors measuring the local distribution of forces on the fingertips of the gripper. Small-force sensors are suitable for controlling robotized assembly processes. Either the position of an object can be obtained using binary and scalar force sensors, or the identification of this object can be made using the contact pressure with an array of force sensors in the robot gripper. This can be achieved by integrating many small-force transducers in the gripping area.

The sensor must be as capable as the human hand for sensing physical quantities such as pressure, force, and texture. Information about physical properties, such as hardness and softness in particular, is essential when the gripper is handling delicate, fragile objects so that the robot hand will not apply too much force and crush the object it is handling.

The devices use a solid rigid substrate to support a force-sensitive elastic polymer layer, which is the active film and gives an electrical signal through a kind of transduction mechanism. The signal-detecting circuitry can be hybridized alongside the device on the supporting substrate.

There are a lot of elastic polymer tactile sensor types based on the following effects:

- piezoresistive
- optical
- magnetic
- capacitive
- piezoelectric

Figure 4.10(a) shows the basic structure of a conductive composite polymer rubber-based piezoresistive tactile sensor [14,15]. The elastic rubber layer is a carbon- or silver-loaded polymer composite. Tactile forces imply a localized condensation of conductive particles; thus, the conductivity between two adjacent electrodes increases. Also, the pressure-dependent contact resistance between conductive rubber layer and contact electrodes can be used. Disadvantages of that type are the nonlinear change of resistance and the cross-talk caused by the lateral stiffness of the rubber, which limits the optical resolution to a few millimeters.

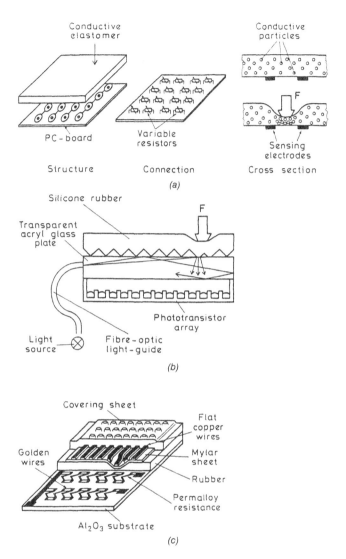

FIGURE 4.10. Rubber- and polymer-based tactile sensors, using different transduction mechanisms: (a) piezoresistive, (b) optical, (c) magnetoresistive, (d) capacitive, (e) piezoelectric, (f) ultrasonic echo [(a,b,c) are redrawn after Wolffenbuttel et al. [14], with the kind permission of Elsevier Sequoia S. A., Lausanne, Switzerland, publisher of *Sensors and Actuators A* and *Sensors and Actuators B;* (d) with permission after Seekircher et al. [18], originally published in the IFAC Proceedings *Robot Control 1988,* by Pergamon Press, Oxford, UK, ©IFAC].

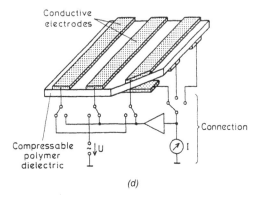

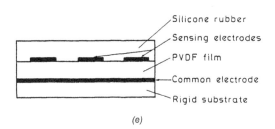

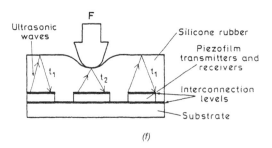

FIGURE 4.10 (continued). Rubber- and polymer-based tactile sensors, using different transduction mechanisms: (a) piezoresistive, (b) optical, (c) magnetoresistive, (d) capacitive, (e) piezoelectric, (f) ultrasonic echo [(a,b,c) are redrawn after Wolffenbuttel et al. [14], with the kind permission of Elsevier Sequoia S. A., Lausanne, Switzerland, publisher of *Sensors and Actuators A* and *Sensors and Actuators B;* (d) with permission after Seekircher et al. [18], originally published in the IFAC Proceedings *Robot Control 1988,* by Pergamon Press, Oxford, UK, ©IFAC].

Figure 4.10(b) shows an optical tactile transducer, which conducts trapped light in a transparent plate under an elastic layer, using the condition of total internal reflection [14,16]. Low-intensity light leaves the plate, and a uniformly black area is seen from the bottom until the elastic polymer film is brought into contact with the transparent layer as a consequence of tactile forces. Diffuse reflection, rather than total internal reflection, takes place, and illuminated areas occur where the object touches the sensor. The large spatial resolution is the major advantage, and the limited dynamics is the main drawback of this type.

Figure 4.10(c) shows the complex structure of a magnetoresistive tactile sensor, the operation of which is based on the megnetic field–dependent resistivity of Permalloy layers [14]. The sensor is composed of a matrix of resistive Permalloy strips on an alumina ceramic substrate covered with an elastic polymer or rubber skin with flat copper wires on top of it. The resistance of each Permalloy strip is sensitive to changes in the strength of the magnetic field, which is induced by the flat copper wires. Tactile forces modulate the localized magnetic field, which is measured with a row-by-row sampling of the network. The required complex mechanical construction limits the minimal size of each element to a few millimeters.

Figure 4.10(d) shows the schematic structure and connection of the capacitive tactile sensor [18]. This sensor applies an elastic polymer mat as a dielectric. (The electrode bars are shaded.) The deformation of the elastic material changes the distance between two parallel capacitor plates, and it can be measured as a change of capacitance. Using the matrix structure shown in Figure 4.10(d), this capacitive method permits the arrangement of many sensors on a small area. The insertion of a dielectric and elastic mat between the electrode bars leads to a tactile sensor. Each intersection of electrodes forms a transducer. The electronic circuitry selects a sensor element and prevents the measurement from being affected by neighbouring elements [17].

The greatest problem of this type of sensor is that the absolute change of capacitance is very small. Assuming a sensor area of 4 mm^2, a thickness of the dielectric of 30 μm, a relative permittivity of five, and a large deformation with a thickness change in the order of 10%, the resulting change of capacitance is 0.6 pF. Accepting an error less than 1%, a resolution of 6 pF is required.

Elastomers seem to be good candidates for dielectrics because of their high sensitivity to the applied force. Their viscoelastic properties, resulting in a high hysteresis, are the main disadvantages.

Seekircher and Hoffmann [18] have optimized the material system of the capacitive tactile sensors. Using peroxide-linked *cis*-polybutadiene as a dielectric, a good compromise between sensitivity and elastic properties can be attained. After degassing, the polybutadiene was vulcanized at 160°C/1.5 min.

The sensor needs flexible electrodes with good adhesion to the dielectrics. The application of electrodes made of conductive rubber (silver-filled *cis*-polybutadiene composite) increases the durability and facilitates the production process. Enhancement of the sensitivity can also be achieved when embedding conductive particles in the dielectric layer. The conductive particles decrease the effective distance between the electrodes and, therefore, effect a higher capacitance and a higher absolute sensitivity of the sensor. It is necessary to coat the conductive particles with a thin insulating film to avoid a short circuit between the electrodes. Good results can be achieved using Al_2O_3-coated aluminium fillings, though the hysteresis is increased. Figure 4.11 shows the typical characteristics of a capacitive tactile sensor element [18].

Piezoelectric tactile sensors are constructed from polyvinylidene fluoride (PVDF) film material, which generates localized charge at the surface when it is subjected to mechanical stresses. Figure 4.10(e) shows the typical structure of piezoelectric tactile sensors using metallized PVDF foil [13].

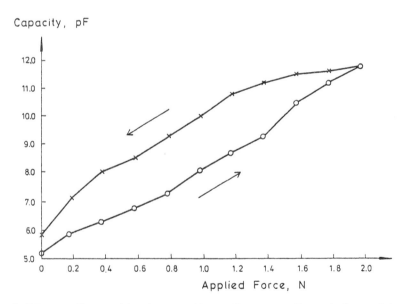

FIGURE 4.11. Characteristic of a capacitive tactile sensor with conductive particles embedded in the dielectric. (Redrawn with permission from the figure of Seekircher et al. [18], originally published in the IFAC Proceedings *Robot Control 1988,* by Pergamon Press, Oxford, UK, ©IFAC.)

The shape of the top electrode can vary according to the application: circular dots, squares, and bars also can be used. PVDF has a large piezoelectric effect, a good linearity, and low hysteresis with respect to elastomers. However, charge leakage through internal resistances prevents the formation of an image of static tactile forces.

Piezoelectric transducers are most often used with charge amplifiers so that the output voltage signal is proportional to the measured stress. According to the finite insulation resistance and amplifier input bias currents, the electrical charge is changed during storage and amplification; thus, the measurement is falsified. By integration, the error increases during the time of measurement. Consequently, a pure static measurement cannot be performed. The application is restricted to situations in which a workpiece has to be quickly gripped, transported, and positioned.

In addition, PVDF also possesses pyroelectric (see Section 3.1) properties; thus, the layers are sensitive to changes and gradients in temperature. Therefore, a temperature difference between the workpiece and the gripper causes a measuring error. In a sensor area of 4 mm², a temperature change of 0.15 K generates a charge equivalent to a force of 1 N. Therefore, the application of heat insulation (using a silicone rubber cover layer) and filtering is necessary [18]. Another drawback of this type of sensor is the narrow temperature range of operation because of the low Curie temperature of PVDF (about 100°C).

Piezopolymer films can also be used in an ultrasonic echo-type tactile sensor configuration [13]. Pieces of piezo film are on the bottom of the transducer with a section of silicone rubber on top of them [see Figure 4.10(f)]. Ultrasonic pulses are transmitted by the piezo films, which bounce off the rubber-air interface and return. The time of flight of the pulses can be determined. Where the rubber is compressed, the time of flight decreases. With a known modulus for the rubber, even the applied force can be determined.

An integrated robotic tactile sensor on a silicon basis was developed by Reston and Kolesar [19]. Twenty-five discrete sensor electrodes, arranged in a 5 × 5 grid, were designed and fabricated from a PVDF film coupled to an integrated circuit. Each of the sensing electrodes were connected to MOSFET amplifiers, which provided sufficient gain to generate usable response signal levels. Two cascaded inverting amplifiers were used, as shown in Figure 4.12(a). To minimize the gate electrode's parasitic line resistance and capacitance, the amplifier should be located as close to the sensor electrode as possible. However, since the gate electrodes must be located beneath the PVDF film, a compromise was attained, and the amplifiers were placed on the periphery of the IC. The complete sensor layout is shown in Figure 4.12(b). The response of the optimum sensor

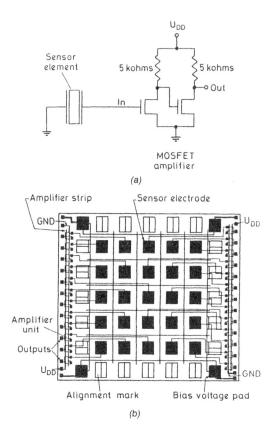

FIGURE 4.12. Integrated tactile sensor array on a silicon chip: (a) circuit diagram of one unit, (b) IC layout. (Redrawn with permission from the figure of Reston et al. [19], ©1990 IEEE.)

configuration (fabricated from a 25-μm thick PVDF film) was linear throughout the load range investigated (0.01–1 N) (see Figure 4.13).

A stress-component-selective tactile sensor array, based on piezoelectric polymer film, and its associated data acquisition system were developed by Domenici and DeRossi [20]. The structure of the sensor is shown in Figure 4.14. The multicomponent sensor is made up of an assembly of seven elemental subarrays sandwiched between two elastic layers. Each subarray consists of six miniaturized individual sensors made of piezoelectric polymers with different characteristics, mounted in a suitable configuration. Miniaturized thermocouples for temperature compensation are vacuum deposited at the centre of each taxel. The individual elements were fabricated from PVDF film with a thickness of 25 μm and 1 mm² metallized

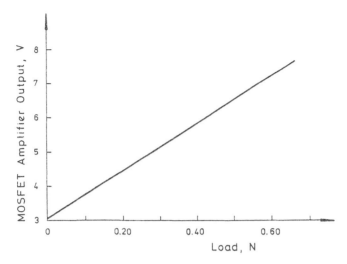

FIGURE 4.13. Amplifier output of the tactile sensor shown in Figure 4.12 with a 25-μm-thick PVDF film and 2 V bias (Redrawn with permission from the figure of Reston et al. [19], ©1990 IEEE.)

surfaces. They were embedded between two slabs of elastomer and fastened by a thin layer of flexible glue. The forty-two miniaturized piezoelectric discs were arranged in groups of six individual elements, each located on the vertices of a hexagon. The seven hexagon subarrays are, in turn, placed to form another hexagon with the seventh subarray in the centre. Contact

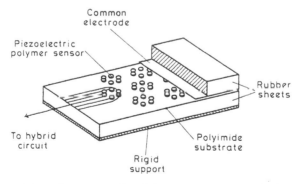

FIGURE 4.14. Schematic representation of the stress-component-selective tactile sensor array. (Redrawn with permission from the figure of Domenici et al. [20], with the kind permission of Elsevier Sequoia S. A., Lausanne, Switzerland, publisher of *Sensors and Actuators A* and *Sensors and Actuators B*.)

electrodes and electrode tracks were obtained on the polyimide substrate by multilayer gold metallization using photolithographic and galvanic growth procedures. Miniaturized thermocouples for temperature compensation were vacuum deposited at the centre of each taxel.

The six independent components of the stress tensor can be calculated from a linear combination of the responses of the six miniaturized elements composing each taxel.

The array was developed with the aim of obtaining a skin-like tactile sensor able to perform fine-form discrimination of objects in contact with it. It is also thought to be instrumental in revealing phenomena related to incipient slippage during object manipulation and grasping.

To validate the individual sensor elements, their sensitivities to the normal and shear stress components were measured. The independent multi-element sensor, connected to dedicated hybrid circuitry comprising multiplexer and charge amplifiers for individual sensors, can be used in reconstructing a map of stress [20].

A tactile stress rate sensor for perception of fine surface features and textures down to the micron was reported by R. D. Howe [21]. Perception of fine surface features is important in robotic manipulation because surface finish affects the frictional properties of objects. Human hands can reliably detect raised ridges less than one micron high by simply sliding their fingertips across them [22]. The sliding motion is essential for this process, since pure normal contact without side-to-side motion does permit such fine discrimination. The developed sensor is also based on a sliding sensing strategy.

The cross section of the stress rate sensor is shown in Figure 4.15. It is roughly cylindrical in shape. The core consists of a rigid structure, around which a thin silicone rubber skin is wrapped. Between the rubber skin and the rigid core, there is a compliant layer of foam rubber of several millimetres' thickness; fluid could also be used in place of the foam. Small strips of PVDF film are molded into the rubber skin beneath the outer contact surface. The outer faces of the film are metallized to provide electrical contact, and the elements are shielded by a grounded conductive layer just under the surface of the rubber skin. In contrast with other piezoelectric sensors, this type is not used with charge amplification, but with FET-input current-to-voltage amplifier. Since the sensor current is the time rate of change of charge, the output voltage, U, is proportional to the time rate of change of the stress or to the stress rate:

$$U = K_1 \cdot \frac{dq}{dt} = K_2 \cdot \frac{d\sigma}{dt} \qquad (4.8)$$

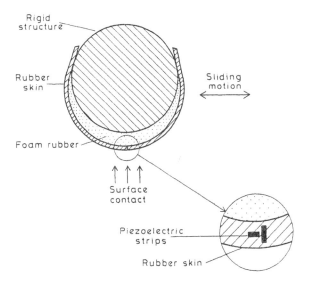

Rigid structure

Rubber skin

Foam rubber

Sliding motion

Surface contact

Piezoelectric strips

Rubber skin

FIGURE 4.15. Schematic structure of the stress rate sensor. (Redrawn with permission from the figure of R. D. Howe [21], ©1991 IEEE.)

where

σ = the stress
q = the electric charge
K_1, K_2 = constants

If the speed of sensor sliding over a test surface, v, is presumed constant, Equation (4.8) can be written as [21]

$$U = K_2 \cdot \frac{d\sigma}{dx} \cdot v \qquad (4.9)$$

Thus, the sensor signal is proportional to the stress gradient in the direction of sliding. The signal can be interpreted with the aid of a solid mechanics model of the contact interaction and a linear deconvolution filter. The ability to detect features as small as a few micrometres was experimentally confirmed [21].

A resonator-type tactile sensor was proposed by Omata and Terunuma [23], which is able to detect, very much like the human hand, the hardness and/or softness of an object.

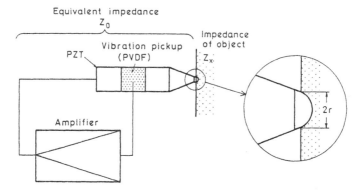

FIGURE 4.16. Structure of the resonant-type hardness/softness tactile sensor. (Redrawn with permission from the figure of Omata et al. [23], ©1991 IEEE.)

Figure 4.16 shows the principle and structure of the sensor [23]. It consists of a piezoelectric ceramic (PZT) transducer and a vibration pick-up made of PVDF film. When an alternating voltage is applied across its electrode, the PZT element is able to vibrate freely in the direction of length. The pick-up detects a vibration, and with a feedback amplifier the system oscillates at the resonance frequency. If the free end of the PZT element is pressed against a surface, as shown in Figure 4.16, the resonance frequency changes according to the acoustic impedance of the object. The change in resonance frequency caused by the stiffness loading effect may be written as

$$\Delta f = \frac{1}{2 \cdot \pi^2} \cdot \left(\frac{k}{Z_0}\right) \tag{4.10}$$

where

Z_0 = the equivalent impedance of the sensor system
k = the stiffness, which can be expressed as

$$k = \frac{2 \cdot r \cdot E}{(1 - v)^2} \tag{4.11}$$

where

v = the Poisson's ratio
E = the Young's modulus
r = the radius of contact area

Figure 4.17 represents the hardness/softness of typical specimens [23]. A foam rubber was employed as the standard hardness when adjusting the feedback gain; therefore, the hardness/softness of the specimen is relative to the foam rubber. The trend shown in Figure 4.17 is in accordance with information provided by the sense of human touch [23].

4.2.5 Other Mechanical Sensors

Peizoelectric polymer films have gained a great interest in the field of various mechanical sensors, which are not involved in the previous types. A survey will be given about the possibilities in this section.

Permanent installation of accelerometers is often required in machine health monitoring, modal analysis, automotive sensors, appliances, and feedback systems. Therefore, a great demand exists for low-cost miniature accelerometers. Micro-accelerometers are based on a piezoresistive or capacitive effect in silicon structures and on piezoelectric effect in ceramics or polymers.

Accelerometers can easily be built up using metallized piezoelectric PVDF films. The typical structure is shown in Figure 4.18. The sensor consists of a PVDF film sandwiched between two metallic electrodes. The lower electrode is fixed to the substrate, while the upper one is provided with a seismic mass, m, which will convert the acceleration, a, into a force, F, according to the relationship $F = ma$. This force is also converted into an electric charge through the piezoelectric effect (see Section 3.1.1) [24].

A special problem of the realization is that the seismic mass may apply a

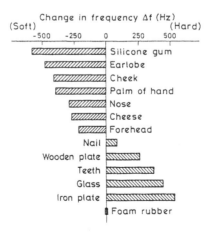

FIGURE 4.17. Change in resonance frequency on different specimens. (Redrawn with permission from the figure of Omata et al. [23], ©1991 IEEE.)

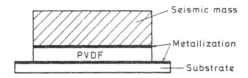

FIGURE 4.18. PVDF accelerometer structure [24].

high stress to the electrodes during measurement. In order to avoid any metal stripping, the nature of the metal and of the glue to stick the mass to the upper electrode must be carefully selected.

The read-out circuitry is generally a charge amplifier with a MOSFET input. The time constant of the sensor, which is determined by the PVDF capacitance, insulation resistance, and the input resistance of the amplifier must be at least one order of magnitude higher than the one of the phenomenon to be followed.

Table 4.4 gives the most important characteristics of a typical PVDF-based accelerometer (AMP, ACH-01) [25]. The sensor consists of two parts connected by a flex cable. The transducer contains not only the piezoelectric sensor itself, but the input FET of the amplifier as well. The interface circuit is a typical charge amplifier (see Figure 4.19).

Flow sensors for the accurate measurement of fluid or gas flowrates are required in many industrial processes. Such measurements usually have to be carried out in two steps. A primary effect converts the flowrate into another physical quantity (differential pressure, frequency, or deformation, for instance), which is then converted into an electrical measurement signal.

TABLE 4.4. *Main Characteristics of a Typical Accelerometer.*

Characteristics ($T = 25°C$)	Units	Typical Values
Sensitivity	mV/g	10
Lower frequency limit	Hz	1
Upper frequency limit	kHz	25
Dynamic range	in g (9.81 m/s²)	0.01–150
Linearity	%	0.1
Transverse sensitivity	%	2
Resonant frequency	kHz	75
Operating temperature	°C	−40– +85
Storage temperature	°C	−60– +110
Maximum shock level	g	500

Reprinted with permission from AMP ACH-01 [25].

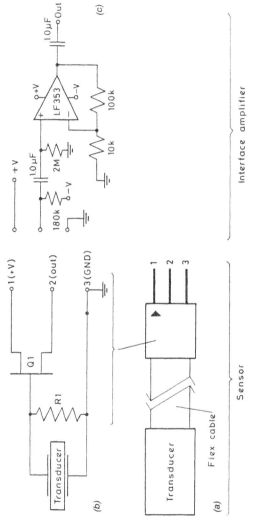

FIGURE 4.19. (a) Outlines, (b) circuit diagram, and (c) operating circuit of a typical PVDF accelerometer [25]. (Reprinted with permission from AMP, ACH-01.)

289

Although many conversion principles and devices have already been utilized, including Venturi pipes, turbine driving, pressure loss on wing-form bodies, ultrasonic echo, vortex shedding frequency, hot-wire anemometry, Coriolis-force mass flow sensors, and electrodynamic measuring methods, there is still a lack of low-cost, reliable flow sensors because the ones based on the former measuring principles need sophisticated electronics. Piezopolymer films seem to overcome this problem.

Based on the relevant piezoelectric properties of PVDF, a low-cost flowmeter was suggested by J. Gutierrez Monreal [26]. The sensor uses a flexural iris orifice made of PVDF film placed as a restrictor on the flow (see Figure 4.20). The structure provides an electrical signal related to the drop of pressure through the diaphragm. At the output of the connected charge amplifier unit, a near linear response was found at different ranges, as shown in the example in Figure 4.21.

A very interesting application of PVDF films is a piezopolymer finger pulse and breathing wave sensor developed by Chen et al. [27]. Pulse sensing is a convenient and efficient way of acquiring important physiological information concerning the cardiovascular system.

Finger pulse pick-ups can be employed in systems that measure blood pressure, heart rate, and blood flow. The sensor can pick up breathing waves simultaneously with pulse waves. It is shown in Figure 4.22 [27]; it has a U-shaped structure, and its open side is closed with Velcro tape. In this way, the sensor can be conveniently and comfortably fixed in the right position on the finger. For the purpose of enhancing the sensor's immunity to electric interference, a hybrid buffer electronics incorporating high-impedance FETs is affixed to it. The pulse-wave signal is sent to the signal processing electronics through the buffer. The PVDF film is in direct contact with the finger; therefore, its metallized surfaces have to be shielded on

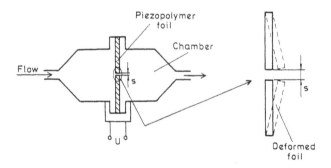

FIGURE 4.20. Typical structure of a PVDF-based flowrate sensor (the idea was suggested by Gutierrez Monreal [26]).

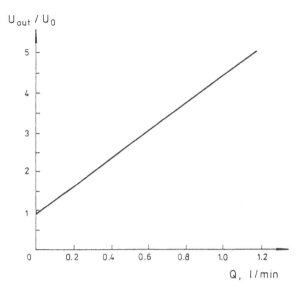

FIGURE 4.21. Typical characteristics of the piezoelectric flowmeter (according to the results of Gutierrez Monreal [26]).

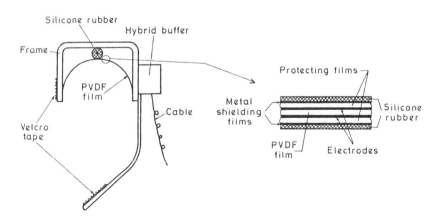

FIGURE 4.22. Structure of the piezopolymer finger pulse and breathing wave sensor. (Redrawn from the figure of Chen et al. [27], with the kind permission of Elsevier Sequoia S. A., Lausanne, Switzerland, publisher of *Sensors and Actuators A* and *Sensors and Actuators B.*)

both sides with thin protecting polymer films and sealed with highly insulating silicone rubber to avoid damage to the surface electrodes through friction and sweat erosion.

Separation of pulse and breathing waves is possible through electronic high- and low-pass filtering, respectively. From recording by PVDF finger pulse sensor, not only can the informative pulse wave and respiration wave be extracited, but also meaningful information concerning the local micro-circulation condition of the measured finger [27].

Piezo films are also used in a variety of medical monitoring systems [28]. One of the most important is the area of apnoe monitors. A sheet of piezo-polymer film is placed under the baby's mattress, and it picks up the respiration and heartbeat of the child by the way of a slight centre of gravity shift during movement. If an apnoe occurs, alarms in the monitor system are activated, waking the child and the parents.

Blood pressure monitors have also been designed using the piezo film as the microphone for listening to the Korotkoff sounds.

Piezoelectric polymer films have introduced new perspectives for vibration sensors and have created an entirely new field of sensorics called the piezofilm measurement technique.

Piezofilm technology, as a whole, is unusual in that the material itself (PVDF) can be the entire sensor, including the physical conductor, required to transmit the signal. This allows customization of transducers to an infinite variety of requirements. Sensors may be simply printed onto film, often hundreds to sheet, and die-cut as needed.

Transducers for use in vibration analysis may be partitioned by type into the domains of displacement, velocity, and acceleration. Piezoelectric polymer film vibration transducers are essentially strain-dependent devices and so fall into the first category, although their practical application bears many similarities with conventional accelerometer techniques [28] (see above). The application of PVDF films is rapidly growing in the field of machinery and health monitoring by vibration analysis. Small film transducers can be mounted in various locations around moving or standing parts of different machines. The sensors can be mounted by a simple sticking process, which is similar to the mounting of strain gauges. The transducers give alternating periodical electrical signal output according to the vibrating deformation of the parts to be examined. Determining whether a shaft is misaligned, a motor is operating smoothly, or a bearing is properly rotating, are examples from that field of application. The great advantage of piezopolymer films is the small mass of sensors that do not modify the frequency response of the object under examination.

Piezopolymer glass break detectors can be used in security systems and car alarms. Since the film gives out an alternating voltage signal, false alarms can be minimized through simple electronics.

The films are widely used in pick-ups for guitars and violins and for lapel and other specialty microphones that call for high fidelity and small size. A lapel microphone typically contains two piezoelectric films, arranged to cancel out clothing noise. But these devices lead us to another field of sensors, to the acoustic sensors, which will be discussed in the next section.

In addition to the flexibility, easy handling, and mounting methods, piezoelectric films have another important property: mechanical vibration can be excited by electrical signals (see Sections 2.3 and 3.1.1). Excitation methods for vibration analysis are conventionally dependent on either electromagnetically driven exciters ("shaker tables" or impact methods such as instrumented impact hammer). Both techniques have their own merits, but neither are wholly appropriate for some experiments. Piezoelectric PVDF film, when driven by a suitable voltage source, can inject continuous low-level mechanical excitation over a very broad frequency range. Displacement levels are very low (normally producing audible, but not directly visible, deformation), but when similar piezofilm transducers are used to monitor the resulting vibration, the dynamic range is adequate to analyse down to several hertz and up to tens of kilohertz [29]. Moreover, the bidirectional transducer capability of piezofilm coupled with the absence of any inherent resonance means that an almost ideal analysis may be performed with the tested object being the sole contributor to any deviation from a flat frequency response.

Resonator-type sensors (see Section 2.3) can also be built up using a pair of piezofilm transducers and a gain stage. This has been achieved with a number of structures: coins, tubes, and spring plates. The action of external influence on such a system (force, temperature, fluid density) may be detected by a change in the self-oscillation frequency. In a more sophisticated arangement, the input signal is generated by a phase-locked loop whose frequency depends on the frequency of the output signal. Here, frequency varies with the load.

Fluid level detection in a container is an example for the measurement technique described above. The sensor consists of two piezopolymer films glued to a metal tube. One film is excited at the natural frequency of the tube. The vibrations are received by the other film. Immersion of the tube in a liquid changes the resonance frequency and forces the output to drop. In a phase-locked loop, the frequency is directly proportional to the height of fluid in the tube.

A special flexible coaxial cable, called piezo cable was also developed as an application of piezoelectric polymers. Its structure is shown in Figure 4.23. This can easily be used as a weigh-in-motion transducer, for instance, for road traffic monitoring [30]. Piezo cables can be put in the road in a sawblade width slot, thus decreasing the installation cost of sensors at toll booths and stoplights. The coax structure automatically solves the shielding

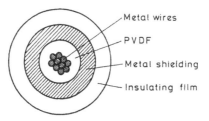

FIGURE 4.23. Structure of the piezo cable.

of the piezofilm. The voltage output is more than sufficient to trigger CMOS circuits, thus allowing low power consumption (battery-operated) counters.

4.3 ACOUSTIC SENSORS

A primary function of these sensors in the information system of the world is transforming the communication modes of human beings into some mode capable of high velocity and quantity. This is achievable by methods such as the analog and digital coding of speech, electromagnetically and/or optically. Techniques of conversion for speech and hearing have been the classic schemes. Here, microphones and vibrating membrane receivers are the crucial elements in the historic evolution of the telephone and, later, of broadcast radio. Acoustic sensors are, in fact, sound-pressure transducers optimized for acoustic measurements.

Polymers have played an important role in these developments. In the condenser microphones, polymer electrets have been used for many years. A second group consists of piezopolymer transducers, which contain a poled polymer, such as polyvinylidenefluoride (PVDF) or one of its co-polymers, as piezoelectric element. But piezopolymer sensors are not being used as low-frequency (audio) microphones; they are common in ultrasonic sensors and hydrophones.

Ultrasonic measurement systems are the other very important and extensively developing area of acoustic sensors. The systems are mainly operating in the pulse-echo mode. They are used in many applications in practice (industrial automation, robotics, nondestructive evaluation, object identification, etc.). Their main tasks are the contactless distance or position measurement in a range of some metres, the presence detection of objects, and object identification by echo profile evaluation.

Hydrophones are a special field of acoustic sensors: they are underwater acoustic receivers. The close acoustic impedance match of piezopolymers to water and their ability to withstand explosive shocks without sensitivity loss result in a good chance to improve the sensitivity and stability of hydrophones by applying polymers instead of conventional piezoceramics. Of recent interest are acoustic devices in which a piezoelectric transducer is directly coupled to an integrated circuit. Such transducers can serve as elements in ultrasonic imaging arrays, particularly when they are miniaturized and shaped by the use of silicon microfabrication methods.

Future fibre-optic telephone systems need a better matching of acoustic sensors to optical transmission systems than that of conventional microphones. Optical-waveguide sensors deliver an optical, rather than an electric, output signal. Another advantage is their insensitivity to electromagnetic interference. Applications of such microphones are expected in the near future. These type sensors are either made of glass fibres or of waveguides on glass or polymer substrates.

Acoustic sensors on a polymer basis will be discussed in the following grouping:

- condenser microphones
- piezoelectric transducers
- hydrophones
- fibre-optic microphones

4.3.1 Condenser Microphones

Condenser microphones are capacitive sound-pressure transducers. A metallic or a metallized polymer diaphragm acts as one electrode, with the backplate as the other electrode of the capacitive sensing element. The basic operation principle of an electret-based condenser microphone is illustrated in Figure 3.8 (see Section 3.1.2). The diaphragm is vibrating under the effect of sound-pressure waves; thus, alternating capacitance with constant electric charge gives an alternating voltage component at the output [31]. The diaphragm is electrically and mechanically bonded to the housing, which is also used as the ground terminal of the housing. The back electrode is retained in a quartz insulator, which forms the rear wall of the sensing cavity and provides an acoustic reference surface.

The possible perforations in the backplate (see Figure 4.24) equalize the pressure in the cavity. A capillary duct provides equalization between sensing cavity and ambient pressure in a controlled manner.

The conventional condenser microphones require a stable DC polarization voltage, applied through a high resistance, across the two capacitor electrodes in order to maintain a constant charge on the electrodes. This polarization voltage is in the range of 100–200 V.

The application of polymer electrets as charged membranes can replace the bias that is normally needed [32] (see Figure 3.8). It was already shown

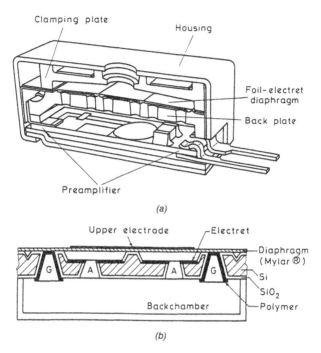

Clamping plate

Housing

Foil-electret diaphragm

Back plate

Preamplifier

(a)

Upper electrode — Electret

Diaphragm (Mylar®)

G A A G

Si

SiO₂

Backchamber — Polymer

(b)

FIGURE 4.24. Practical microphone transducer units using polymer foil electrets: (a) conventional and (b) micromachined version. (Redrawn with permission (a) after the figure of Baker [31] from the *Proc. of the Symp. at the 2. Chemical Congress of the North American Continent,* Las Vegas, 1980, ©1981 American Chemical Society and (b) with permission after Voorthuyzen et al. [34], ©1989 IEEE.)

a few decades ago that many films of acrylics and cellulose esters, polystyrene, and vinyls can be charged, but no successful usage was achieved. Then, in the work of Sessler [33], a new course was taken [31]. The electron beam poling of polytetrafluoroethylene (PTFE) and poly(fluoroethylene propylene) (Teflon-FEP) resulted in electrets retaining an electric charge for an indefinitely long time (see Section 3.1.2).

Excellent devices for telephone speech transducing are now manufactured from these polymers [31]. The typical structure is shown in Figure 24(a). After a rapid development in the 1970s, polymer-electret acoustic sensors today represent a mature technology with only a few innovations. The systems are widely used in practice and polymer-electret microphones now dominate the field of acoustic transduction.

A new development is the use of Teflon-FEP electrets in silicon-based microsensors [34,35]. Figure 4.24(b) shows the structure of an electret microsensor developed by Voorthuyzen et al. [34]. The miniature microphone is composed of nine small unit cells operating parallel, which are fabricated on the same silicon chip. The figure shows the cross section of one unit cell. One-sided metallized Mylar® (DuPont) foils were applied as diaphragms. The Si wafer was shaped by anisotropic etching (see Section 1.1.5). On the front side, the air cavities necessary for condenser microphones are etched, while on the reverse side of the wafer, two types of holes are etched through it. One type (A) ends in the air cavity and connects it to the back chamber behind the back plate. The other type (G) is etched around the air cavity area, in order to be able to attach the diaphragm to the back plate in a later processing step. The V-grooves etched around the structures simplify the dicing of the individual microphones. An electron-beam charged Teflon-FEP layer was used as electret. It was deposited by heat-sealing and patterned by plasma-etch processes. The sensitivity of the sensors is about 20 mV/Pa in the frequency range of 50 Hz to 10 kHz. They were used in hearing aid systems.

It must be mentioned here, however, that SiO_2 was found to be a very good electret as well. Thus, newer acoustic microsensors do not apply polymer as electrets [34].

4.3.2 Piezoelectric Transducers

Piezopolymer transducers contain a poled polymer, such as PVDF or one of its copolymers, P(VDF-TrFE), for example, as a piezoelectric element that operates as a sound-pressure sensor.

One type of microphone consists of films with uniform piezoelectric activity in the thickness. These transducers work as bending devices if they are arranged in a carved geometry with clamped edges. Microphones with plane geometry are possible when one applies monomorphs or bimorphs (see Section 3.1) that vibrate as stiff plates rather than membranes. Monomorphs and bimorphs can be made from a single film by means of electron-beam poling. Such prototypes show sensitivities of 0.5 mV/Pa [32].

Ultrasonic transducers with piezoelectric polymer film are alternatives to conventional piezoelectric transducers operating in air. The construction principle of a cylindrically shaped foil transducer investigated by Harnish et al. is shown in Figure 4.25(a) [35]. The transducer can be used both as transmitter and receiver. By variation of the geometric transducer data (length, curvature, opening angle, number of parallel curves), the acoustic properties of foil transducers can be adopted to different measurement tasks over a wide range. The following properties are of special interest: frequency response characteristic as both transmitter and receiver, pulse response, directivity pattern, and electric impedance.

This type of PVDF transducer is suited for the frequency range of 40–200

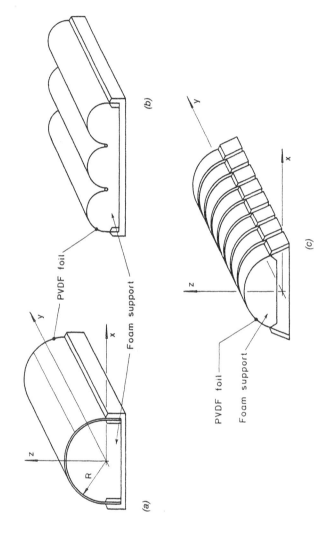

FIGURE 4.25. Construction of piezopolymer ultrasonic transducers: (a) cylindrically shaped, (b) multiple curved, (c) linear multi-element structures. (Redrawn from the figure of Harnisch et al. [36], with the kind permission of Elsevier Sequoia S. A., Lausanne, Switzerland, publisher of *Sensors and Actuators A and Sensors and Actuators B.*)

kHz, and the sensitivity is 0.2–1 mV/Pa in the receiver mode and 20–50 mPa/V in the transmitter mode (at a distance of 1 m to the transducer) [36]. The electric impedance is typically 1–10 kohm, much lower than the one of eletrostatic transducers. The acoustic aperture angle in the xz-area [see Figure 4.25(a)] is about 40°, and it is very small in the yz-area if the length is much larger than the wavelength of sound in propagation medium.

Multielement transducers can also be built easily to control the directivity pattern. Figure 4.25(b) shows a multiple curved transducer, which supplies a sharp directivity pattern in both xz- and yz-areas [36,37]. On the other hand, Figure 4.25(c) shows an example of a linear multielement transducer using a simply curved PVDF foil. By means of time-shifted driving of every single element, an electronically controlled directivity pattern can be obtained [36]. Thus, the parallel arrangement of linear multielement transducers leads to an array structure that is the base for ultrasonic scanning within a limited solid angle.

Silicon-based ultrasonic microsensors are produced by the combination of microfabrication technologies and PVDF films. These subminiature transducers are based on sound-pressure–induced modulation of the drain current in a field effect transistor. A lot of development has been done on this PVDF-MOSFET structure, often called POSFET, in the last decade. Several versions of POSFET were realized by Mo et al. [38]. The best realized structure can be seen in Figure 4.26.

The fabrication of the devices is an interesting example of how the processes described in Chapter 1 can be combined in practice. It starts from a lightly doped p-type $<100>$ silicon substrate. First, deep and shallow boron diffusions are performed to form a p^+ rim and a p^+ silicon etch stop layer as mechanical support for diaphragms formed later. Then the n^+

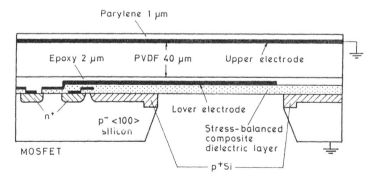

FIGURE 4.26. Micromachined POSMOS transducer structure. (Redrawn from the figure of Mo et al. [38], with the kind permission of ELsevier Sequoia S. A., Lausanne Switzerland, publisher of *Sensors and Actuators A* and *Sensors and Actuators B*.)

source-drain diffusion can be performed. A wet oxidation and then CVD processes are applied to deposit a 1-μm-thick stress-balanced composite $SiO_2/Si_3N_4/SiO_2$ diaphragm layer. Cr and Au are then evaporated on the insulating layer and are patterned into 1-mm squares to serve as lower electrodes. Also, the integrated interconnection system can be shaped. The Si-etch window on the backside of the wafer is also defined with the lower electrode at its centre. The substrate is anisotropically etched from the backside. The Si-etch will stop at the p^+ layers. A reactive ion etching is the final step of the diaphragm formation; it removes the p^+ silicon layer underneath the lower electrode. Next, a 40-μm-thick PVDF sheet, with Au metallization on its top, is bonded onto the substrate using nonconductive epoxy. The upper electrode and the Si substrate are both grounded, and the signal is supplied by the lower electrode. Device performances are about 1 mV/Pa sensitivity and 1.6 kPa lower detection limit.

Possible applications include medical imaging and nondestructive evaluation since large arrays of small size transducers improve image quality. Micromachining processing on a silicon basis enables minimal cross-talk between neighbouring elements [38].

The third direction of developments is concerning piezoelectric ceramic/polymer composites for airborne ultrasonic application. A transversal vibrator-type Pb(Zr,Ti)O_3 ceramic, (PZT)/polymer 2-2 composite sandwich layer transducer (SLT) was investigated by Möckl et al. [39]. The structure of the SLT and the driving and sensing circuitry in transmitter and receiver mode, respectively, are shown in Figure 4.27. Thin rectangular foils of piezoceramics are the active elements with thickness t_c. The ceramic plates are metallized on both sides and poled perpendicularly to the electrodes. Two plates of polymer material, generally polyethylene or polypropylene with thickness t_p applied to both sides of a piezoplate form the elementary cell of the sandwich transducer. Each SLT consists of one or more of these elementary cells. Sound is radiated from all six of the transducers, but only the radiation of one of the front sides of the sandwich is used in practice. Sound emission from the remaining faces is suppressed by casting some foam made of rubber or polyurethane around the element, leaving free the radiating area.

The ratio of ceramic thickness, t_c, to the one of the sandwich $t = t_p + t_c$, the PZT volume fraction, controls acoustic properties like acoustic impedance, bandwidth, vertical radiation pattern, or electroacoustical efficiency. The choice of the material for the passive layers of the sandwich structure offers further variability in the design of transducers. Polymers with different elastic properties are applicable, depending on whether a broadband transducer for short pulse generation or a heavy duty, single-mode power source is required [39].

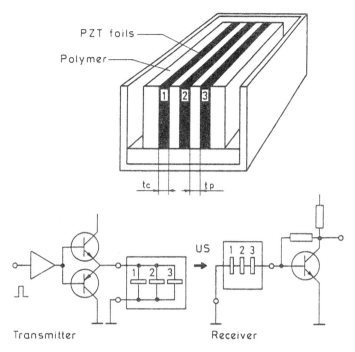

FIGURE 4.27. Structure and circuitry of the sandwich transducer. (Redrawn from the figure of Möckl et al. [39], with the kind permission of Elsevier Sequoia S. A., Lausanne, Switzerland, publisher of *Sensors and Actuators A* and *Sensors and Actuators B*.)

4.3.3 Hydrophones

Microphones designed for underwater use are known as hydrophones. Most hydrophones are piezoelectric. They are often used for sound transmission, as well as sound reception functions. Piezoelectric polymer films are viewed as the next generation material for hydrophones. The trend is for large area hydrophones for ships and submarines [28]. Since PVDF film can be produced in large areas and also in thick sections, it is a good alternative to ceramics in hydrophones. When properly matched to another polymer as the carrier, PVDF has been designed into sonobuoys. Its low density and low acoustic speed give rise to a low acoustic impedance near to that of water; hence, there is good direct acoustic impedance matching when it works in water.

The structure of a composite cylindrical hydrophone composed of a PVDF film and an organic outer shell is shown in Figure 4.28 [40]. A PVDF film with a thickness of 90 μm was glued to the inner wall of the hydrophone's outer shell made of an organic material with high strength and

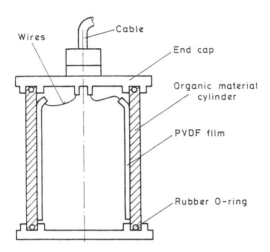

FIGURE 4.28. Schematic diagram of the cylindrical hydrophone. (Redrawn from the figure of Yiquan and Binwen [40], with the kind permission of Elsevier Sequoia S. A., Lausanne, Switzerland, publisher of *Sensors and Actuators A* and *Sensors and Actuators B*.)

low elastic modulus. Both ends of the shell are elastically sealed. The sensitivity of the hydrophone, *S*, can be expressed as

$$S = \frac{E_1 \cdot t \cdot (2 \cdot g_{31} + g_{32})}{E_2 \cdot [1 - (a/b)^2]}$$ (4.12)

where

E_1 = the Young's modulus of the PVDF film
E_2 = the Young's modulus of the shell material
t = the thickness of the PVDF film
a,b = the inner and outer diameters of the shell, respectively
g_{31}, g_{32} = the piezoelectric voltage constants of the PVDF film

Equation (4.12) shows that the thicker the PVDF film is, the higher is the sensitivity, and the lower the elasticity modulus and the thinner the shell is, the higher is the sensitivity. Performance parameters of the cylindrical hydrophone realized by Yiquan and Binwen are summarized in Table 4.5 [40].

4.3.4 Fibre-Optic Microphones

A fibre-optic microphone is a device in which the incident sound waves modulate the light guided in optical fibres without the help of electrical/optical and inverse conversion. It gives the possibility for sensing acoustic signals and transmitting them directly on a fibre-optic communication link. The main advantage is that there is no electrical interference. Moreover, the microphone may be used in surroundings where the danger of explosion exists.

Fibre-optic microphones depend on the fact that sound waves cause a phase or intensity modulation of light waves. Phase modulation is due to the pressure sensitivity of the optical phase constant of an optical waveguide and can be detected in an interferometric set-up. On the other hand, intensity modulation of the transmitted light may be obtained, according to Figure 4.29(a) [41], by directing a light beam emitted from a transmitting fibre after reflection from a vibrating membrane into a receiving fibre.

A microphone, as sketched in Figure 4.29(a), was built by Garthe using multi-mode fibres. With a circular membrane (diameter 25 mm) made of polyester (Mylar®, thickness 2 μm) and the distance of the membrane, x, adjusted to the point at which the steepness of the power coupling ratio function reaches its maximum, the device produced a modulation of 0.04%/Pa. Since optical-waveguide microphones yield an optical output, their sensitivity cannot be compared directly to electro-acoustic transducers. The noise-equivalent pressure level (the sound pressure level that produces the same output as the internal noise of the microphone) is the quantity appropriate for comparison. This is located at 50–60 dB, the values of which are not yet as low as those for the silicon microphones [32].

To enlarge the microphone steepness, an essential improvement can be reached when using single-mode instead of multi-mode fibres. However, the coupling between such fibres is very sensitive to radial displacement and requires an extreme accuracy in the positioning of the related parts. The best way to overcome all positioning problems is to avoid using separate parts. This leads to the application of integrated optics, where optical wave-

TABLE 4.5. *Performance Characteristics of the Cylindrical Hydrophone.*

Frequency (Hz)	10–2000
Sensitivity (10^6V/Pa $= 0$ dB)	> -192 dB
Fluctuation	$< \pm 1$ dB
Phase error	$<2.7°$
Depth (under water)	$\geqslant 300$ m
Diameter	30 mm
Length	40 mm
Acceleration (1 V/1 g $= 0$ dB)	< -64–80 dB
Capacitance	3300 pF
Specific gravity	$<1.8 \times 10^3$ kg/m^3

Reprinted from Yiquan and Binwen [40], with the kind permission of Elsevier Sequoia S. A., Lausanne, Switzerland, publisher of *Sensors and Actuators A* and *Sensors and Actuators B*.

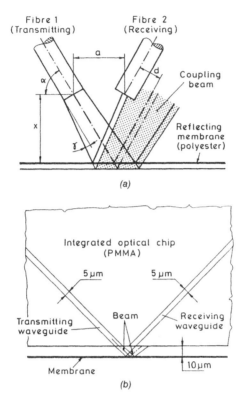

FIGURE 4.29. Intensity modulation-type optical microphones using (a) optical fibres or (b) integrated optics. (Redrawn from the figures of Garthe [41], with the kind permission of Elsevier Sequoia S. A., Lausanne, Switzerland, publisher of *Sensors and Actuators A* and *Sensors and Actuators B*.)

guides are fabricated by particular technologies in or on transparent substrate materials. In the case of the optical microphone, the optical chip contains two waveguides, as shown in Figure 4.29(b) [41], which replace the optical fibres of the original set-up.

One of the possible substrate materials is poly(methyl methacrylate) (PMMA). Its big advantage is the easy fabrication. Integrated optics on a PMMA basis is in the development stage. Two techniques are under investigation: ion implantation and photo-locking [41].

In the case of ion implantation, light ions are shot with high energy into PMMA. The implantation causes an increase in the refractive index, and the substrate is to remain unchanged when masking with an aluminium layer. Thus, waveguides can be implanted in the substrate.

In the photo-locking method, PMMA is doped with UV-sensitive photoinitiator either by in-diffusion or by mixing the dopant to the monomer before polymerization. The refractive index can be enlarged by UV-irradiation; masking can be performed by photolithographically structured silica glass mask [41].

4.4 INFRARED RADIATION SENSORS

Piezoelectric materials also show the pyroelectric (see Section 3.1.3) effect, which is generally applied in infrared sensors. Thus, PVDF and its copolymers also have a large application area in infrared radiation sensing systems. Over the last twenty years, a great deal of research activity has been devoted to infrared detection. There are two main reasons for this: the continuously increasing number of actual or potential applications and the huge progress observed on semiconductor materials and integrated circuit technologies.

Sensors for electromagnetic radiation may be classified as quantum sensors and thermal sensors. Quantum sensors are based on photoemission, photoconduction, or photovoltaic effect. Thermal sensors operate in two steps: first, an incident photon flux is converted into a temperature variation, which can be measured by different physical principles, namely by voltage (thermopiles), a change in resistance (bolometers), a change in pressure (Golay cell), or a change in charge (pyroelectric sensors). Thermal sensors indicate the entire absorbed radiation, regardless of its spectral composition and are, therefore, particularly well suited for the detection of infrared (IR) radiation.

The pyroelectric effect (see Section 3.1.3) is a change in the dielectric spontaneous polarization induced by the temperature change. Concretely, a change in the surface charge and a voltage variation can be measured between the electrodes on the surfaces of the dielectric. Pyroelectric sensors are the fastest of the thermal detectors since temperature changes at the molecular level are directly responsible for the detection process.

Pyroelectric sensors react to temporarily variable IR radiation. Unless a radiation field is nonstationary, as, for example, a laser pulse, it requires modulation in order to enable the transient response of a pyroelectric sensor. Either a chopper is used or heat sources are observed that change naturally, for example, with movement.

PVDF is ideal for the detection of relatively weak radiation sources with temperatures not much above the ambient temperature, such as human bodies or vehicles. This is because the absorption maximum of PVDF provides a good match to the emission maximum of those heat sources. These behaviours stress the application in the field of person detectors and security systems.

Pyroelectric sensors operate at ambient temperature in contrast to quantum detectors, which are often cooled to cyrogenic temperatures to obtain better performances. This is the key point why considerable efforts are applied to the development of solid-state pyroelectric imaging arrays using the most sophisticated silicon processing technologies. However, several drawbacks arise and make high-resolution, large array fabrication difficult.

First, thermal leakage occurs between adjacent pixels. The application of PVDF seems to be a good solution because of the high thermal resistivity of the material. Secondly, a piezoelectric effect also occurs and can degrade the sensitivity. Therefore, relatively low electro-optical performance can be obtained (poor sensitivity, low speed). Integrating the sensor with the readout electronics is the best option to overcome the mentioned problems. The discussion about pyropolymer sensors will be divided according to the following areas: freely suspended pyrosensors and pyroelectric arrays.

4.4.1 Freely Suspended Pyrosensors

Person detectors, which have been developed on the basis of PVDF, are available on the market at low cost and are produced by several companies.

Figure 4.30 shows a typical structure and its circuit diagram (AMP PIR 180–1000) [42]. A thin film metallization is sputtered onto the PVDF foil and formed geometrically to achieve good sensitivity over a wide angle field of view. The transducer element consists of three dual capacitors arranged in the configuration shown in Figure 4.30. This arrangement requires no mirrors for a full (180°) field of view. The sensor also contains the preamplifier FETs and input resistors, which are configured to provide impedance matched outputs.

Dual outputs allow the external balance of the gain and common mode rejection of unwanted signals generated by ambient temperature changes, vibration (due to piezoelectric effect), and other noise sources. For human motion sensing, a conventional Fresnel lens should be applied as a window before the transducer. It has a directivity pattern also shown in the figure and gives an alternating optical signal to the sensor elements when an infrared source is moving in the field of view. The most important characteristics of the sensor are given in Table 4.6 [42]. For a comparison with other type infrared sensors, the noise properties are an important criterion.

The sensors are described by the specific detectivity, D^*, which gives the reciprocal minimum detectable signal power as being equal to the noise power. It is a "figure of merit," the actual detectivity normalized to unit bandwidth and unit area and is a measure of the signal-to-noise ratio. It can be expressed as

$$D^* = (S/N)A\Delta f/P \qquad (4.13)$$

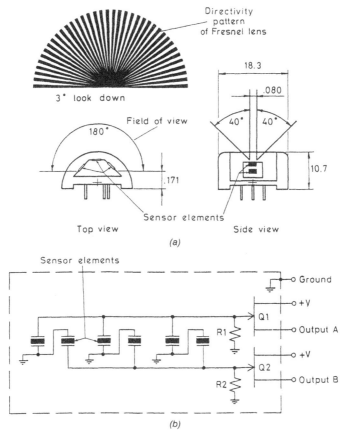

FIGURE 4.30. (a) Configuration and (b) internal schematic of a typical infrared person detector based on PVDF. (Reprinted with permission from AMP PIR 180-100) [42].

where

S/N = the signal-to-noise ratio (ratio of signal voltage to noise voltage)
A = sensitive area of the sensing device
Δf = noise equivalent bandwidth of amplifier in test set-up
P = r.m.s. value of radiant power incident on sensitive area

In the field of security, sensor arrays are sometimes required with a larger angular detection region, furthermore making it possible to determine in which angular region the person to be detected is located. An interesting concept for angle-selective sensor arrays is that of mirror optics, with an axis of rotation perpendicular to the optical axis, e.g., a spherical mirror or spherical parabolic mirror that was published by Mader and Meixner [43].

TABLE 4.6. Performance Characteristics of the Pyrodetector PIR 180-100.

	Characteristics	Units
Detector type	6 Element, dual channel	–
Element size	1.0×2.0	mm
Responsivity	1250	V/W
D^* (specific detectivity)	2.9×10^8	$cmHz^{1/2}W^{-1}$
Operating voltage	3-15	V
Operating current	90 maximum	$10^{-6}A$
Operating temperature	-30 to $+50$	°C
Operating humidity	90% maximum	–
Storage temperature	-40 to $+70$	°C

Reprinted with permission from AMP [42].

It has been proven that PVDF presents a clearly superior spatial resolution to other pyromaterials.

As another application, a laser detector, which was developed by Shu Duo, is shown in Figure 4.31 [44]. Generally, a sensor to measure laser beam intensity or profile is designed as a flat plane. Its active face is usually covered by a black overlayer such as gold-black, carbon black, or black paint. A strict optimization of the structure and materials should be performed for measuring short, high-power laser pulses. Since these overlayers cannot endure a high radiation power density, it is almost impossible to use such flat sensors to measure stronger impulse laser beams, which are important in practice.

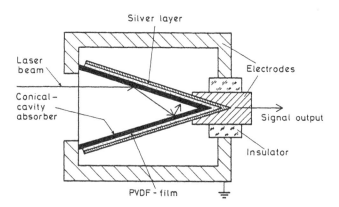

FIGURE 4.31. Sketch of a conical-cavity laser detector. (Redrawn from the figure of Shu-Duo [44], with the kind permission of Elsevier Sequoia S. A., Lausanne, Switzerland, publisher of *Sensors and Actuators A* and *Sensors and Actuators B*.)

To solve this problem, a conical-cavity laser detector, as sketched in Figure 4.31, was designed. A PVDF film, with a two-electrode overlayer on both sides, was set on a base. On the outer face, the film was protected by another insulating layer. The absorbing layer was made by depositing a carbon layer on the cavity, and the PVDF film was closely wrapped over its back surface. As an electrode, a silver layer was deposited on the PVDF surface, while the ground electrode is a copper ring that contacts well with the front edge of the cavity. The limit of the detected radiation power density (100 MW/cm^2) is much higher in comparison with flat sensors and the sensitivity is 5 mV/mJ [44].

4.4.2 Pyroelectric Arrays

During the last few years, pyroelectric IR sensor arrays have found many applications in the field of infrared spectroscopy and thermal imaging. Also, in intruder alarm systems, arrays have several advantages as compared to single detector elements.

Pyroelectric IR sensor arrays have already been realized in a variety of ways. In a pyroelectric vidicon, a freely suspended triglycine sulphate (TGS) crystal or deuterated TGS target is scanned by an electron beam. Lateral heat conduction is reduced by grooves between the sensor elements [45]; however, its big size and the high operating voltage limits the application possibilities.

Considering many sensors integrated into an array, it is useful to connect each sensor to a silicon device that contains the circuit of transistor switches. There are several methods to achieve this.

One is the hybrid technology, where the elementary detectors are thermally isolated from other pixels by reticulation cuts. They are fixed first to an absorber sheet and then face down, flip chip bonded together to the silicon readout circuitry [46]. This bonding technique guarantees the relatively good thermal isolation of the elements from the substrate.

The other possibility is the monolithic technology, where the pyroelectric material is directly deposited or glued to the silicon substrate. This is the area where pyroelectric polymer films have great possibilities. The most important problems that arise in connection with the monolithic technology are as follows:

- Thermal isolation is not very well achieved so that thermal cross-talk may occur, and also the low thermal resistance toward the substrate can destroy electro-optical performances.
- Reliable gluing and poling methods should be developed for the pyropolymer layers.

- Advanced silicon integration and microfabrication processes have to be applied.

Due to the progress of the last few years, these problems seem to be overcome, and arrays have already been achieved.

A lot of work was devoted to thermal and electrical modelling of the system to optimize the type and thickness of materials and element values of readout circuitry, in order to get the best performances [46–49]. When comparing free suspended and monolithic configurations, the conclusion can be drawn that the electrical parameters of the transfer function are essentially the same. The difference lies in the thermal parameters. Due to the worse thermal isolation of the integrated version, the sensitivity of such a sensor is lower for the low-frequency range. The detectivity and the spatial resolution can be improved only by the improvement of the thermal isolation on the chip using different micromachining techniques.

Gluing the PVDF sheet to the silicon substrate proved to be difficult in the past. As a result of technological advancement, it is now a good controlled process. PVDF sheets are glued to the chip by a thin layer of poly(isobutylene) [50], while the copolymer P(VDF-TrFE) sheets can be bonded by their own material [51].

Another problem is that the technology needs unmetallized poled pyropolymer films, which are not available in practice. Removing metallization from poled films may cause losses in the polarization. The best approach would be the direct poling on the chip surface, which means many technological and circuit application difficulties. However, Münch and Tiemann [50] have suggested an interesting solution for that problem.

In Figure 4.32(a), a typical realization of sensor array with PVDF on silicon IC is presented [50]. A p-well CMOS process with aluminium gates (see Section 1.1.7) is used. A new technology for integrating high-resistance gate resistors by p-n junctions operated at zero bias has also been developed. Figure 4.32(b) shows the circuit diagram of the array.

At the first step towards larger sensor arrays, an 8×1 linear array was fabricated. A given sensor element can be selected by an analog multiplexer. (This concept cannot be applied in the case of high pixel number arrays.) The sensors are coupled to an impedance-matching preamplifier with a MOSFET input. A bias resistor for each element is useful in order to eliminate the influence of temperature variations of the whole detector. This resistor is also necessary to adjust the operating point of the MOSFETs.

Special attention has to be paid to these resistors. The basic requirement is that the electrical time constant of each sensor element exceeds the in-

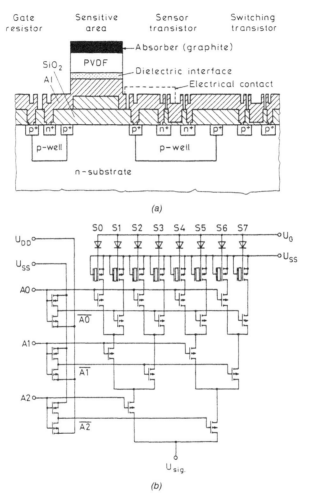

FIGURE 4.32. (a) Cross section and (b) circuit diagram of the integrated IR-sensor array. (Redrawn from the figures of Münch and Thiemann [50], with the kind permission of Elsevier Sequoia S. A., Lausanne, Switzerland, publisher of *Sensors and Actuators A* and *Sensors and Actuators B*.)

verse of the chopper frequency, f_{ch}, while the latter must exceed the thermal time constant of the system:

$$2 \cdot \pi \cdot R_g \cdot C_s \gg 1/f_{ch} \gg 2 \cdot \pi \cdot R_{th} \cdot C_{th} \qquad (4.14)$$

where

R_{th} = the thermal resistance
C_{th} = the heat capacity
R_g = the gate resistance
C_s = the sensor capacitance

This leads to a resistance value in the 10^{10} Ω range for C_s = 10 pF and f_{ch} = 10 Hz. It is evident that this value cannot be achieved by conventional diffused resistors. In compatibility with the CMOS process, the gate resistor is realized by a p-n junction in a separate p-well.

The thick oxide layer serves to reduce parasitic capacitance and heat conduction. A further improvement is achieved by sputtering a thick oxide under the PVDF film. A 25-μm-thick PVDF film was used as pyroelectric material. The upper side is coated with a thin conducting film of graphite, which also serves as the common counter-electrode for all sensors. The sensor electrodes were formed by contact metallization on the silicon chip. The PVDF film was glued to the chip by a thin layer of polyisobutylene.

The use of integrated p-n junctions as gate resistors makes it possible to polarize the pyroelectric layer on the chip without damaging the gate oxides of the sensor MOSFETs. This is done by operating the diode in a forward direction, thus reducing the resistance of the junction by several decades. The transfer function at 10 Hz is about 850 V/W, and the specific detectivity (see Section 4.4.1) is 10^7 cmHz$^{0.5}$/W. The cross-talk between adjacent pixels in the low-frequency range is about 0.3×10^{-3}.

Several similar structures have been realized with similar performances in the past few years, building linear and small (3 $\times$ 3 size) arrays [51,52], which are, in comparison with another type of arrays with 512 $\times$ 512 pixels, only preliminary results. However, the advantages of pyropolymer detectors and the great progress in the last five years indicate that bigger arrays can be expected in the near future. Considerable improvement can be achieved by means of relatively thick (25–50 μm) polyimide thermal insulating and P(VDF-TrFE) copolymer piezoelectric layers with a thin nickel film as absorber.

Another charge coupling and readout possibility is the application of charge coupling devices (CCDs). An infrared image sensor having 64 $\times$ 32 infrared sensitive MOS gates has been developed by Okuyama by

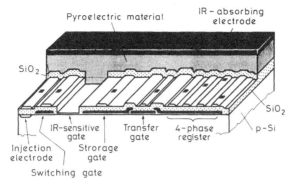

FIGURE 4.33. Structure of one element of the IR-CCD. (Redrawn from the figure of Okuyama et al. [53], with the kind permission of Elsevier Sequoia S. A., Lausanne, Switzerland, publisher of *Sensors and Actuators A* and *Sensors and Actuators B*.)

combining several types of pyroelectric materials and Si-CCD [53]. Figure 4.33 shows an example for a device structure of one element of the proposed pyroelectric IR-CCD. The total active area is about 4×5 mm^2. The charge induced in the pyroelectric material modulates the Si surface potential under the IR-sensitive gate through the series capacitance of the SiO$_2$ and the dielectric connector. Free charge is injected from the injection electrode through the IR-sensitive gate region to the storage gate region and is transferred to the transfer gate region. The charge is finally carried to the gate of an output FET by four-phase driving pulses.

P(VDF-TrFE) was also used as a pyroelectric material without a separate bonding layer because a thin film of this material can easily be obtained. The powder of the raw material was dissolved in methyl ethyl ketone solvent and was put on the wafer to make a thin film of some tens of micrometres in thickness. The film was dried and electrically poled to obtain pyroelectricity. Thin NiCr film was used as an IR absorbing electrode. The device still needs some development to get satisfactory parameters.

4.5 HUMIDITY SENSORS

Recent developments in automated systems have made ever-increasing demands for sensors that can measure the concentration of a chemical compound, also including humidity. As humidity is a very common, continuously changing component of our environment, measurement and/or control of humidity are important not only for human comfort, but also for a broad spectrum of application fields, such as domestic appliances, automotive electronics, medical service, industry, agriculture, and meteorology. The measurement of relative humidity has always been a difficult problem because of the interaction of other variables such as temperature and gas components in the air and because of the lack of stable and reliable low-cost sensors.

There are three major groups of humidity sensors: those that measure mechanical property changes; psychrometric measurements that compare the latent heat of evaporation of a saturated environment to the environment in question; and those that respond to electrical or optical property change such as resistance, capacitance, and colour. The main definitions and equations in connection to humidity measurements are given in Section 3.1.4.1.

Although polymer film humidity sensors still remain modest in practical applications compared with ceramic sensors, their potentiality has been growing, especially in the field of integrated microsensors. Polymer films are well suited to standard IC processing techniques to fabricate small low-cost sensors. Due to the research efforts of the last few years, a lot of different types have been developed, which can be categorized as follows:

- capacitance types
- resistance types, including ionic and electronic conduction versions
- transistor structures (FET; CFT, see Section 2.2)
- resonators (mainly BAW and SAW types, see Section 2.3)
- electrochemical types
- fibre-optic types

The material interaction processes giving the operation possibility on a polymer basis are discussed in Chapter 3. In this chapter, the device structures will be concentrated on.

4.5.1 Capacitance-Type Humidity Sensors

Capacitance-type humidity sensors are based on the phenomenon that the sorption of molecules with a high dipole moment, such as H_2O, leads to a change in the dielectric constant of the otherwise low-permittivity polymer films.

Several types of capacitive polymer-based humidity sensors are available in the market. Figure 4.34(a) shows a typical one (type Philips H 1) [54]. It consists of a perforated plastic case containing a stretched membrane of nonconductive polymeric foil coated on both sides with gold electrodes forming a parallel plate capacitor. Figure 4.34(b) shows its capacitance-humidity characteristics.

A thin-film type humidity sensor was developed by Vaisala [55], which is now widely used in a lot of humidity measuring instruments. Figure 4.35(a) illustrates the structure of the sensor. The lower twin electrodes are de-

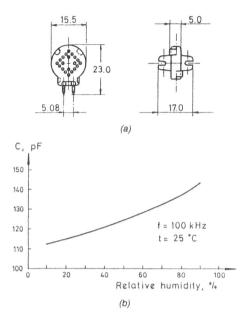

FIGURE 4.34. A polymer-based capacitive humidity sensor [54]: (a) configuration and (b) characteristic. (Reprinted with permission from Philips Components.)

posited onto a glass substrate by iridium evaporation. Cellulose acetate was dissolved in ethylene dichloride and then applied to the surface to form thin humidity-sensitive dielectrics. On top of this, gold is evaporated as the upper electrode, which is thin and porous enough to permit quick lateral moisture transport. The upper acts as a common counter-electrode to the lower twin electrodes. As shown in Figure 4.35(b), the capacitance is roughly proportional to the ambient humidity. The sensor exhibits hysteresis less than 2% RH; the 90% response time is less than 1 sec.

The porosity of the humidity-sensitive polymer is important to obtain quick responses. Most polymers, such as polyimide, exhibit practically no moisture sensitivity without cracks. A stress-induced fracture technique was proposed by Delapierre et al [56]. Their thin-film structure consists of a tantalum electrode, which is oxidized anodically in order to avoid any risk of short-circuiting, even when there are holes in the polymer; a polymer film is then spread with a photoresist spinner; a Cr-Ni-Au electrode connects the porous counter-electrode, which is obtained by chromium evaporation under conditions such that the polymer film is tensible stressed. These stresses generate a very large number of cracks in the polymer, transforming it into many small islets protected by a thick layer of chromium [56].

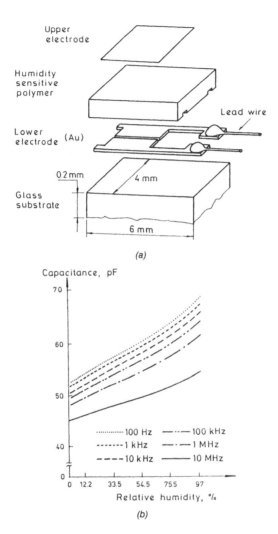

FIGURE 4.35. A polymer thin-film humidity sensor of Vaisala: (a) structure and (b) charac-
teristic. (Redrawn from the figures of Yamazone et al. [55], with the kind of permission of
Elsevier Sequoia S. A., Lausanne, Switzerland, publisher of *Sensors and Actuators A* and
Sensors and Actuators B.)

It was demonstrated that the characteristic of the sensor has a shift depending on the measurement history, which is only really significant when
the humidity exceeds 75 % RH. This behaviour can be caused by the fact
that high moisture content can modify the structure of the polymer (see
Section 3.1.4).

Recent improvements in polymer processing technologies applied in the
microfabrication of integrated circuits enables humidity sensors to be fabri-

cated on Si chips. Thus, it becomes possible to produce small, low-cost interchangeable sensors and to integrate humidity sensors with other types of sensors (e.g., temperature sensors) and/or other signal-handling circuitry on the same chip.

Polyimides, even photosensitive ones, have been successfully applied on Si chips as humidity-sensitive dielectrics [57–61]. The sensitivity and response depend upon the sensor geometry, including the electrodes. Figure 4.36(a) shows the top view and the cross section of a silicon-based polyimide humidity sensor developed by Denton et al. [57]. This is a typi-

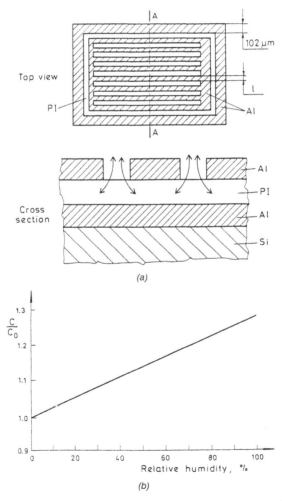

FIGURE 4.36. A silicon-based PI thin-film humidity sensor: (a) structure and (b) characteristic. (Redrawn with permission from the figures of Denton et al. [57], ©1985 IEEE.)

cal Al-PI-Al parallel plate capacitor structure. A typical steady-state capacitance humidity characteristic can be seen in Figure 4.36(b).

Results about several similar structures have been reported over the last few years [59–61]. A sensor with a circular top electrode was realized by Schubert and Navin [59]. The polyimide was patterned in an O_2/CF_4 plasma using dry etch processing (see Section 1.4.5).

A double comb interdigitated electrode system on the top of photolithographically patterned polyimide dielectrics, combined with self-aligned contact CMOS technology, was reported by Boltshauser et al. [60]. The latter structure was also realized as a suspended version with a floating polysilicon counter-electrode using selective under-etching of silicon [61]. The sensors are in the development stage.

4.5.2 Resistance-Type Humidity Sensors

Resistance- (or impedance-) type humidity sensors are made of conductive polymers, which can be categorized into ionically conductive and electroconductive types (see Chapter 3).

The sensing mechanism of humidity sensors using ionically conductive polymer electrolytes is based on the fact that their ionic conductivity increases with an increase in water adsorption due to increases in the ionic mobility and/or charge carrier concentrations (see Section 3.3.3).

A thin-film humidity sensor called Hument was developed and placed on the market by Sicovend AG [62]. Figure 4.37(a) shows a schematic view of the sensor. The thin humidity-sensitive film is prepared by copolymerizing 2-hydroxyl-3-methacryl oxypropyl trimethyl ammonium chloride with methacrylic ester on an alumina ceramic substrate with Au electrodes of interdigitated form. A porous polymer overcoating film covers the top of the humidity-sensitive film to protect it from contamination by oils, dust, etc. The resistance value of the coated electrode varies according to the change in relative humidity as an exponential function, as shown in Figure 4.37(b). The characteristics are stable even when low and high humidities are repeatedly and continuously measured. The hysteresis is extremely low; its value is about 1% RH. The response time is about 2 min in the absorption, while it is a little longer in the desorption process.

A similar sensor was fabricated using a photosensitive polymer by Hijikigawa et al. [63]. The substrate with the pair of interdigitated electrodes is coated by the mixed aqueous solution of styrene-sulphonate monomers, a cross-linking agent (N,N'-methylene-bisacrylamide), and vinyl polymers using spin casting. The coated film is then polymerized and cross-linked by ultraviolet irradiation in a nitrogen atmosphere. The humidity-sensitive film is finally covered with a moisture-permeable cellu-

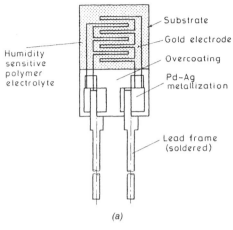

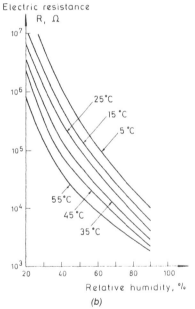

FIGURE 4.37. A typical polymer-based impedance-type humidity sensor [62]: (a) schematic view and (b) characteristics. (Reprinted with permission from Sicovend).

lose ester film for protection. The resistance versus relative humidity characteristics of the sensor are similar to those of the former one.

A thick-film hybrid module with a frequency output signal was fabricated by Tsuchitani et al. The module consists of an integrated humidity sensor, a temperature sensor chip, and measuring circuitry on the same ceramic substrate, as shown in Figure 4.38(a). Fabrication of the module is based on the following procedure [64]:

- screen printing and firing of comb-shaped gold electrodes, wiring, and thick-film resistors, then glass coating
- bonding of chip thermistor, chip capacitors, and operational amplifiers to the circuit using surface mounting technology
- formation of a humidity-sensitive layer by spreading the liquid of a copolymer of ionic monomer NaSS (sodium styrene sulphonate) and nonionic monomer HEMA (hydroxyethyl methacrylate) on the substrate and then drying
- formation of a protective film of silicone rubber on the humidity-sensitive layer by dipping

Impedance changes of the humidity sensor and the temperature sensor are converted into frequency using astable multivibrators. It is possible to feed the output signal of the module directly into a microcomputer. The relative humidity-oscillation frequency characteristics are shown in Figure 4.38(b). The temperature dependence is 0.5–0.6 % RH/°C in the temperature range between 10 and 50°C. The 90 % response time is 3–4 min, since the module has a relatively large heat capacity.

An important problem encountered in adopting polymer materials to humidity sensors is water resistivity. Polymer electrolytes are generally hydrophilic and soluble in water, so that they have a poor durability against water or dew condensation. This problem can be mitigated by cross-linking a hydrophilic polymer with an appropriate cross-linking reagent and/or copolymerizing hydrophilic monomers with hydrophobic ones.

A lot of materials (halogenated or other type polymer salts, cross-linked and quaternized copolymers, organopolysiloxanes, etc.) have been investigated [65–69], which were previously described in Section 3.3.3. These investigations have proven that the response times are determined by the film thickness and porosity and by the hydrophobicity of the constituent ionic monomer while the mole fraction of the constituent ionic monomer to nonionic monomer strongly affects the sensing characteristics.

Electronic conduction–type humidity sensors can be fabricated from conductive polymer composites. The swelling of the polymer due to water absorption counteracts ohmic contacts between dispersed conductive particles, and thus, the electronic resistance of the film will be increased. The switching effect of the polymer composites (see Section 3.2.1) results in a sharp increase in the resistance as the relative humidity approaches 100 %, as shown in Figure 3.19 [55]. This behaviour is especially useful in dew point sensors.

Electroconductive conjugated polymers (see Chapter 3) were also in-

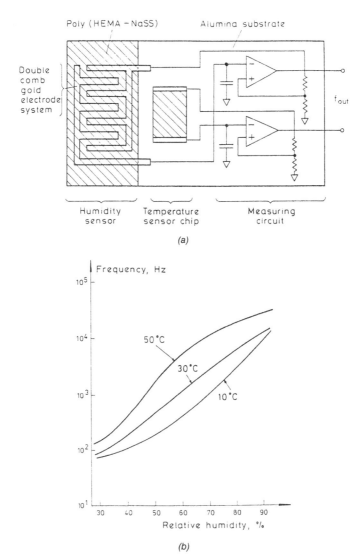

Poly (HEMA – NaSS)   Alumina substrate

Double
comb
gold
electrode
system

f_{out}

Humidity   Temperature   Measuring
sensor   sensor chip   circuit

(a)

Frequency, Hz

10^5

10^4

50°C

30°C

10^3

10°C

10^2

10^1

30   40   50   60   70   80   90

Relative humidity, %

(b)

FIGURE 4.38 (a) Schematic structure and (b) characteristics of a polymer-based thick-film humidity sensor module. (Redrawn from the figures of Tsuchitani et al. [64], with the kind permission of Elsevier Sequoia S. A., Lausanne, Switzerland, publisher of *Sensors and Actuators A* and *Sensors and Actuators B*.)

vestigated as sensitive layers in resistive humidity sensors [70–72]. Conducting polymers with a conjugated backbone are known for their interaction with small gaseous molecules giving rise to changes in the electrical conductivity. Polyaniline, for instance, shows an increase in conductivity when exposed to water vapours, owing to the possibility of proton exchange between the water molecules and the protonated and unprotonated forms of this polymer.

Furlani et al. [70,71] have investigated homogeneously iodine-doped polyphenylacetylene (PPA) and $SnCl_2$ doped polyethynylfluorenol (PEFl) prepared via catalytic oligomerization. The synthesized polymers were doped by dissolving them and the doping agent in common organic solvents. The solutions were aged and then spun on silica substrates where chromium electrodes had been previously evaporated and photolithographically patterned. The current responses to relative humidity at the two different material types are shown in Figure 4.39. The main drawback of this type of material is the relatively high cross-sensitivity to different gas components.

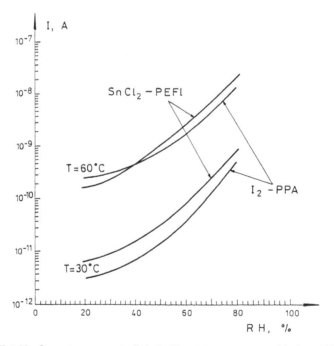

FIGURE 4.39. Current response to RH at different temperatures, of I_2-doped PPA and $SnCl_2$-doped PEFl thin films. (Redrawn from the figures of Furlani et al. [70,71], with the kind permission of Elsevier Sequoia S. A., Lausanne, Switzerland, publisher of *Sensors and Actuators A* and *Sensors and Actuators B*.)

More recently, a relative humidity sensor based on conductivity measurements of the ionically conductive polymer P(DMDAAC) has been fabricated [72]. This sensor consists of a platinum conductivity cell, comprising four parallel platinum "finger" electrodes on a ceramic substrate spin-coated with the polymer film. The latter was cross-linked with a 10 Mrad γ-irradiation dosage. The sensor exhibits a nonlinear conductivity response to RH in the 20–80% range.

4.5.3 Transistor-Type Humidity Sensors

A microchip humidity sensor based on an FET was developed by Hijikigawa et al. [73]. THe humidity sensor was integrated with a temperature-sensing diode, as shown in Figure 4.40(a). The FET humidity sensor is, in fact, a douplex-gate IGFET structure, prepared from an n-channel MISFET with a meandering gate. Cross-linked cellulose acetate 1 μm thick is stacked as a humidity-sensitive membrane between the lower gate electrode and the upper gate electrode. The latter is a porous gold metallization layer. The two gates are electrically connected with a sufficiently large fixed resistance.

An equivalent circuit of this sensor is shown in Figure 4.40(b). A DC voltage, U_0, and a small AC voltage, u_{in}, are applied to the upper gate electrode. It can be easily shown that the output voltage, u_{out}, is related to the capacitance of the humidity-sensitive membrane, C_s, as follows:

$$u_{out} = u_{in} \cdot R_L \cdot g_m / (1 + C_i/C_s) \qquad (4.15)$$

where

R_L = the load resistance
C_i = the gate insulator capacitance
g_m = the transconductance of the FET

u_{out} can be correlated almost linearly with relative humidity, as shown in Figure 4.40(c). The integrated, temperature-compensated sensor shows a good accuracy (hysteresis less than 3% RH), with a response time less than 30 sec.

Charge flow transistors (CFTs) (see Section 2.2.1) with moisture-sensitive polymer coatings were examined several times for humidity measurement purposes by Senturia et al. [57,74]. Delay-time measurements were carried out on a CFT device with a humidity-sensitive polymer [poly(p-aminophenylacetylene) (PAPA)] suggested to observe ambient humidity.

A floating-gate charge flow transistor (see Figure 4.41), a so-called sur-

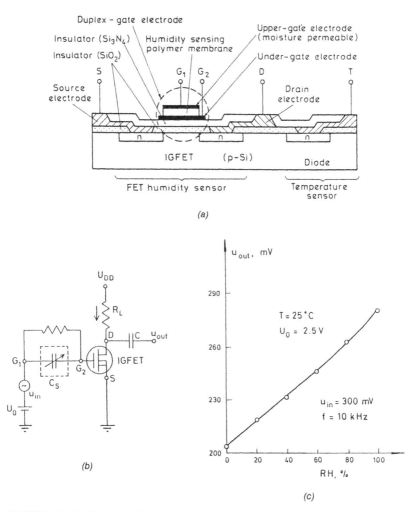

FIGURE 4.40. (a) Structure, (b) equivalent circuit, and (c) characteristic of the duplex gate FET humidity sensor using polymer membrane. (Redrawn from the figures of Yamazone et al. [55], with the kind permission of ELsevier Sequoia S. A., Lausanne, Switzerland, publisher of *Sensors and Actuators A* and *Sensors and Actuators B*.)

324

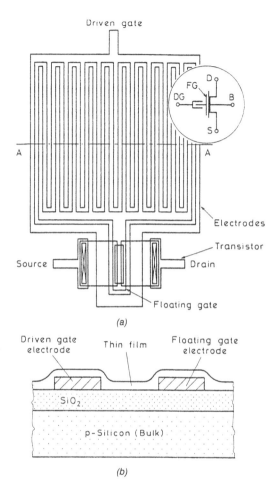

FIGURE 4.41. A floating-gate charge-flow transistor with humidity-sensitive polymer coating (PEO): (a) top view, (b) cross section A-A. (Redrawn with permission from the figures of Garverick and Senturia [74], ©1982 IEEE.)

face impedance measurement (SIM) device, was also fabricated by Garverick and Senturia [74]. The SIM device combines a guarded interdigitated electrode pair with a depletion-mode n-channel MOSFET. One electrode (the floating gate, FG) is connected to the gate of the FET; the other electrode (the driven gate, DG) is available externally. The device has four terminals: source (S), drain (D), driven gate (DG), and substrate (B). A voltage signal applied to the driven gate results in current flow between driven and floating gates. Because the floating-gate-to-substrate capacitance is small, even small interelectrode currents can produce substantial changes in the floating gate voltage. This is evidenced by a change in the channel conductance of the FET, which can be monitored with the aid of an on-chip reference FET. Device model and measurement technique are described elsewhere [74]. Wafers were coated with a thin film of the moisture-sensitive resistor material, poly(ethylene oxide) (PEO), spin cast from aqueous solution. The device enables the measurement of sheet resistances in the range 10^9 to 10^{16} ohm/square in the AC frequency range of 1 Hz to 10 kHz. This structure was also used for the resistivity measurements of polyimide [57].

4.5.4 Resonator-Type Humidity Sensors

Resonator-type (BAW or SAW) devices are attractive for chemical microsensor applications because of their small size, low cost, high sensitivity, and good reliability. The sensors are based on the piezoresonance effect described in Section 2.3: the resonance frequency changes depending on the amount of gases sorbed by a polymer coating (see Section 3.4).

Application of SAW devices to the study of fundamental properties of hygroscopic polymer/water systems was reported by Brace et al. [75]. The basic device used is a SAW dual-delay line oscillator. One propagation path of the delay line was coated with a film of the polymer to be studied, while the other path remained uncoated. The devices were fabricated on each 0.5-mm-thick 128° rotated y-cut, x-propagating $LiNbO_3$ substrate. They used fifteen-finger-pair interdigital transducers (IDTs) of Al or Ni, having a periodicity of 52.5 μm, and centre-to-centre spacing between transducers of 5 mm. Conventional photolithography was used to define the pattern. Hygroscopic polymers were deposited by spin coating from a solution onto the delay lines, followed by thermal treatment. Polyimide (PI) and cellulose acetate butyrate (CAB) films, 1–5 μm thick, were deposited and baked for one hour at 130°C in air and in vacuum, respectively. Figure 4.42 shows a typical variation of an SAW oscillator frequency with relative humidity for a polyimide-coated delay line at 23.5°C [75].

Recently, poly(ethylenimine) (PEI) has been applied as a chemical interface for carbon dioxide and water on a 40-MHz quartz-based SAW device [76]. The strong interference from water vapour indicates the possibility to develop a humidity sensor on that basis.

Glow discharge polymer (see Section 1.4.5) layers deposited alternately on both sides of an AT-cut quartz-type BAW resonator were studied by Radeva at el. [77] with the aim of preparing a sensor for measuring relative humidity. Several polymer types were investigated and poly(hexamethyldisiloxane) (PHMDS) has proven to be the most suitable material for the mentioned purposes.

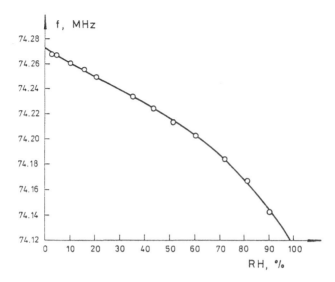

FIGURE 4.42. Variation of SAW oscillator frequency with relative humidity for polyimide-coated delay lines. (Redrawn from the figure of Brace and Sanfelippo [75], with the kind permission of Elsevier Sequoia S. A., Lausanne, Switzerland, publisher of *Sensors and Actuators A* and *Sensors and Actuators B*.)

4.5.5 Electrochemical Humidity Sensors

As was discussed in Section 3.9, conductive polymers may change their electrical behaviour, including the contact potential with an increase of water sorption due to the protonation effect. This can be a basis for electrochemical humidity sensing methods.

A very interesting potentiometric humidity sensor structure was published by Kuwano et al. [78] Figure 4.43(a) shows the schematic structure of the sensor. It uses conducting polythiophene polymer film as a sensing electrode. The Pt film serves the better contacting possibility. The counter-electrode is silver and AgI-Ag_2MoO_4, a high ionic conductivity (HICON) glass, is used as an electrolyte, which contains MoO_4^- ions. The potential-relative humidity characteristics of the Pt electrode are shown in Figure 4.43(b).

4.5.6 Fibre-Optic Humidity Sensors

Fibre-optic sensors are a new attractive field of sensorics. Here, the polymer-based fibre-optic sensors will be concentrated on. The structures and advantages of fibre-optic sensors were described in Section 2.6.

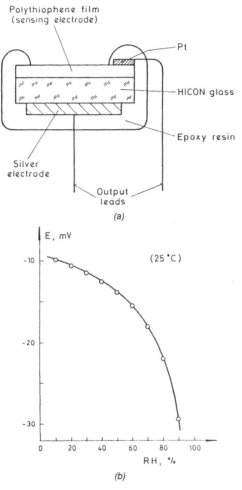

FIGURE 4.43. A potentiometric humidity sensor using polythiophene electrode: (a) structure and (b) characteristics. (Redrawn from the figures of Kuwano et al. [78] with permission from the Electrochemical Society of Japan.)

Optical-fibre systems are partly of interest as a posible technique to detect gas species in the atmosphere. Several approaches have been reported for fibre-optic sensors to detect the humidity of environment. As was described in Sections 2.6 and 3.7, a part of fibre-optic sensors is operating on colourimetric, fluorescence, and/or luminescence effect of indicators that may be embedded into polymer material. Generally, the latter is just a host material.

A fibre-optic humidity sensor was suggested by Ballantine and Wohltjen

[79], which utilizes the colour change of Co^{2+}, Cu^{2+}, or V^{5+} salts when they are hydrated or dehydrated by ambient air. Thus, $CoCl_2$ was embedded in poly(vinyl pyrrolidone) (PVP) and deposited onto the core of a fibre-optic waveguide by immersing it in a solution and then slowly withdrawing it. The film was then dried with a heat gun. The colour of these films (which, in the case of Co^{2+} salts, changes from blue when dry to pink when wet) was measured through the fibre by internal reflection spectroscopy (see Section 2.6) at 660 nm [79].

By using composite films of hydrolyzed Nafion® and dyes having a terminal *N*-phenyl group, an optical humidity sensor type was presented by Sadaoka et al. [80,81]. For thin-film preparation, Nafion® powder was dispersed in a water-alcohol solution. Thin films were fabricated on alumina or glass substrates by dipping and/or spin coating. After the film was dried, dyes were entrapped in it by immersion into the dye-water solution. The films were dried again. A *Y*-type quartz fibre was fixed just in front of the sensor. Light from a D_2/I_2 lamp was launched into the fibre and directed to the sensor. The reflected and modulated light was collected by the same fibre. The collected light was analysed by using a spectro-multichannel photodetector in the range of 400 to 800 nm. The sensing mechanism was discussed in Section 3.8.

Various indicator dyes were used by Sadaoka et al. [80,81], an example of which is shown in Figure 4.44. The spectra in the reflection mode of Nafion®-crystal violet composite were examined as a function of the relative humidity. The signals were measured as the ratio of the reflected intensity to that in dry air. A reflection minimum was observed at 630 nm. It is well known that crystal violet undergoes solvatochromism in which the colour is very sensitive to the acidity of protons, which is modified by the sorption of water molecules.

4.6 GAS SENSORS

At present, in different branches of industry and agriculture, gases and vapours that can exert an injurious and toxic effect on workers are widespread. In connection with this, the problem of the reliable detection of the leakage of such gases into the atmosphere during their storage and use is at the forefront. Moreover, the increasing demand for direct monitoring of gas concentrations, both in the atmosphere and in different liquids, maybe even in blood, in the field of anaesthesia and in vivo monitoring in intensive care, in environmental protection, fire detection, and chemical process control makes the development of new low-cost, solid-state sensors very timely. A lot of devices for the sensing of various gases have been described in the literature.

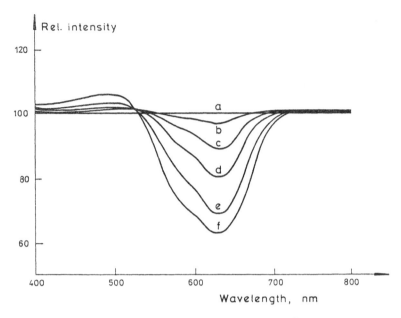

FIGURE 4.44. Humidity dependence of the spectra for Nafion® composite with crystal violet: $a = 0$; $b = 17$; $c = 29$; $d = 38$; $e = 56$; $f = 71\%$ RH. (Redrawn from the figure of Sadaoka et al. [81], with the kind permission of Elsevier Sequoia S. A., Lausanne, Switzerland, publisher of *Sensors and Actuators A* and *Sensors and Actuators B*.)

Many types of conventional electrochemical gas sensors are already based on polymer membranes. In the new generation of sensors, the advantages of both polymers and microsensor processing has been utilized. The great progress in the application of polyelectrolytes brought new possibilities for solid electrolyte sensors, which can also be realized in plate form using film deposition technologies.

It was shown in Chapter 3 that a lot of polymers can show selective behaviour when interacting with different molecules, which can be determined by their fabrication technology, for example,

- Electroconductive conjugated polymers by the doping-dedoping procedure or by the entrapment of specific molecules are capable of substrate recognition.
- Polysiloxanes or other type sorbent polymers can show selective sorption behaviour when introducing functional groups or cage compounds in the polymeric or organic side chains.
- Semipermeable polymeric membranes have different permeabilities for different molecules.

These types of polymers can be applied widely in impedance (resistive or capacitive), mass-sensitive (resonator), calorimetric, and optical sensor devices. However, a few operation principles still exist, for example, the photopyroelectric gas sensing method, which is not involved in the previous list. The description of gas sensors operating with gas-sensing polymer films according to their structure and operation principles can be divided into the following sections:

- conventional electrochemical sensors using permselective polymer membranes
- semiconductor microsensors with polymer membranes
- solid polymer electrolyte-based planar electrochemical microsensors
- impedance-type sensors based on resistivity or permittivity changes
- mass-sensing–based resonators using polymer sorbents
- calorimetric sensors based on catalytic effects on polymer surfaces
- fibre-optic gas sensors
- other gas sensing devices with polymers, for example, photopyroelectric gas sensors

4.6.1 Classical Electrochemical Gas Sensors Using Permselective Polymer Membranes

Electrochemical gas sensors for the concentration measurements of inorganic gas components in gas-phase or dissolved in a liquid-phase medium have been well known for a long time. Both potentiometric and amperometric principles are used either by simple electrodes or by compound electrochemical cells (see Section 2.4).

Polymers are used in these sensors as permselective membranes. The operation of the sensors is determined by the electrochemical reaction on the sensing electrode and by the permeability of membrane for different compounds.

There are two main types of the membranes (see also Section 3.6):

- heterogeneous microporous membranes, through which the gases can permeate in the gas phase by ultrafiltration (for example, capillary membranes: cellulose acetate, PTFE, PVC)
- homogeneous (or dense) film-type membranes, through which gases can permeate by segregation and absorption (silicone rubber, PTFE, polyethylene, Mylar®, etc.).

The operation of the conventional amperometric sensors can be illustrated by the Clark-type O_2 sensor (see Figure 4.45). The measurement of the O_2 partial pressure (p_{o_2}) of liquids is made almost exclusively by the polarographic/amperometric technique (see Section 2.4.2); a constant volt-

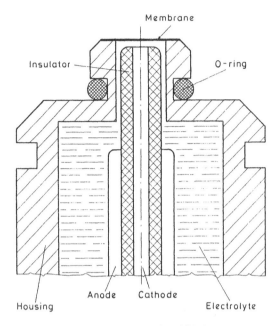

FIGURE 4.45. Structure of the conventional Clark-type oxygen sensor.

age is applied in the polarographic plateau region, and the limiting current is proportional only to the p_{o_2} of the solution. The polarographic technique was modified by Clark by separating the cathode anode and electrolyte from the liquid sample by means of a diffusion-controlling membrane permeable to oxygen but impermeable to water, ions, proteins, or blood cells [82]. The latter two are present in samples at biomedical applications. The most commonly used membranes are PTFE, polypropylene, and polyethylene. All these membranes are also good electrical insulators.

Generally, the cathode of the O_2 sensor is platinum or gold, the reference electrode is silver coated with silver chloride (Ag/AgCl), and the electrolyte solution usually contains potassium chloride with a buffering agent. Oxygen cathodes are usually polarized at about -700 mV with respect to the reference electrode so that the sensor is operating in the diffusion-controlled region of the polarogram.

Various electrochemical gas sensors for different gases are based on the same membrane materials and operation principles. An amperometric H_2 sensor, for example, also uses PTFE membrane. The cathode is Pt, the anode and the reference electrode are made of Cu, and the electrolyte is $H_2SO_4 + CuSO_4$.

Another possibility of application of polymers in electrochemical sensors is to use them as electrode-modifying membranes in potentiometric sensors. The Stow-Severinghaus CO_2 sensor electrode is an example for a potentiometric sensor with the structure: pH electrode/$NaHCO_3$ solution/PTFE membrane [83].

A new type of electrode-sensor tries to avoid the application of an intermediate liquid electrolyte using conductive polymer membranes. Graphite rods coated with poly(dimethyldiallyl-ammonium chloride) [P(DMDAAC)] and subjected to immobilization by γ-irradiation have been shown to respond linearly to dissolved oxygen concentration [84]. The oxygen is detected electrochemically by reduction at a potential of -0.4 V versus Ag/AgCl. The concentration range of linearity is 1.4 to 3.9 ppm. The swollen polymer has the ability to attenuate the signal due to a molecule that adsorbs onto graphite, while at the same time allowing oxygen to permeate through to the electrode surface. It also has the ability to preconcentrate negatively charged species.

Semipermeable membranes can be used not only directly in electrochemical cells, but Figure 4.46 shows another principle. The gas sensor, which can operate without a membrane as well, is covered by the permselective membrane; thus, it can be applied in liquids. A porous PTFE membrane can be used to detect oxygen in water [85]. Membranes with pores of proper measurements can realize waterproofness; it can pass water vapour and oxygen but not liquid water (see pervaporation in Section 3.6).

4.6.2 Semiconductor Sensors Using Polymer Membranes

The increasing application of sensors based on micromachining technologies has also presented a new challenge for gas-sensing devices.

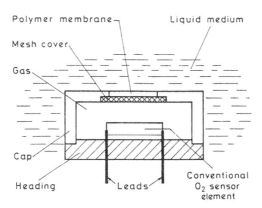

FIGURE 4.46. Cross section of a gas sensor unit with porous polymer membrane.

Lundström et al. have investigated TMOS (thin-film/metal-oxide/semi-conductor) structures (see Section 2.2.1) for a long time [86]. Recently, they have also fabricated devices using polymer membranes [87]. The basic structure in their experiments is a TMOS capacitor, which also operates on the field effect principle; however, it is not a field effect transistor. The basis of operation is similar to the TMOS transistor described in Section 2.2.1, the difference being in the measurement technique. Exposure of a TMOS capacitor to gas can cause a shift of the capacitance-voltage characteristic along the voltage axis. A polymer film on top of the thin-film metallic gate can act as a permselective membrane. Al/Ir and Al/Rh thin-film gates were deposited using electron beam evaporation on Si/SiO₂ structures, respectively. Positive and negative photoresists were applied to the samples by spinning and baking. The resists were used as polymer membranes on the sensors.

The negative and positive photoresists have fundamentally different molecular structures. The negative resist is a cross-linking polymer based on synthetic polyisoprene structurally analogous to natural rubber. A positive resist has polar groups in a polymeric benzene ring structure. Because of the different microstructures, different selectivity properties can be expected when using them as membranes. The capacitors were tested for several gases at 150°C. Ir-gate TMOS structures are initially sensitive to hydrogen, ammonia, and ethanol. The application of membranes increased the response to ammonia, decreased to hydrogen, and almost destroyed the ethanol sensitivity [87]. At Rh-gate TMOS devices, the negative photoresist decreased the sensitivity to water vapour while the positive photoresist increased it to ammonia.

Nehlsen et al. [88] have investigated ZnO thin films deposited onto micromachined Si/SiO₂ substrates. Different catalyst metals have been sputter deposited, and highly gas-selective semipermeable plasma-polymerized membranes (made using an RF PECVD technique, see Section 1.4.5) have been mounted with a SiO₂ membrane-carrier frame above the sensors to enhance the concentration of a single component in the gas mixture closed in the cavity. Hexamethyldisiloxane films with a thickness of 100 nm have shown a SO_2/CO separation rate of thirty-seven.

Lechuga et al. reported results with Schottky barrier devices, Pt/n-GaAs, with discontinuous platinum films that are sensitive detectors of ammonia over a wide temperature range [89]. When a polymer layer of polyetherimide (PEthI), an amorphous polyimide, is combined with the device, the diode sensitivity increases, and also, its selectivity can be improved. The diodes were fabricated on Si-doped GaAs epilayers grown by molecular beam epitaxy (see Section 1.1.2). The schematic cross section of the sensor is shown in Figure 4.47(a). The devices have a dual metallic con-

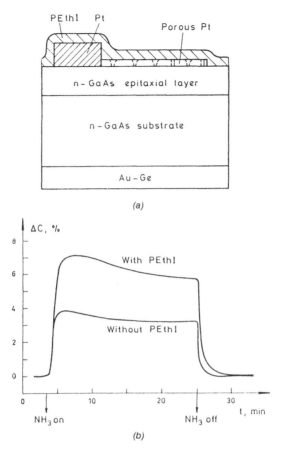

FIGURE 4.47. (a) Schematic cross section and (b) response of the Schottky-barrier type NH₃ sensor. (Redrawn from the figure of Lechuga et al. [89], with the kind permission of Elsevier Sequoia S. A., Lausanne, Switzerland, publisher of *Sensors and Actuators A* and *Sensors and Actuators B*))

figuration: a platinum dot was first deposited for electrical contact. Then, a thin, porous platinum layer was evaporated by electron-beam over and outside the contact dot. Back-ohmic contact had been made previously by Au-Ge thermal evaporation. PEthI films were applied by casting a solution of PEthI in chloroform and subsequent annealing. The response of the devices was tested by the capacitance changes of the Schottky barrier. The effect of the PEthI coating resulted in considerable improvement in the sensitivity for ammonia, as shown in Figure 4.47(b).

A very special device structure, the so-called suspended gate field effect transistor (SGFET) (see Figure 4.48) with sensing conductive polymer

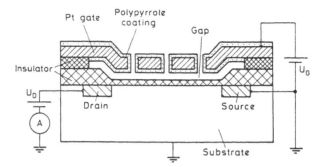

FIGURE 4.48. Schematic cross section of the suspended metal gate field effect transistor. (Redrawn with permission after the figure of Josowicz and Janata [90] from *Analytical Chemistry,* Vol. 58, ©1986, American Chemical Society.)

films for gas sensor applications have been developed by Janata et al. [90]. It uses the change of electron work function of a chemically sensitive polymer layer in response to interaction with a gas or vapour. The changes are detected by the threshold voltage shift of the FET (see Section 2.2.1). Polypyrrole (PPy)-based devices respond to lower aliphatic alcohols at room temperature and with a time response of seconds. It was also demonstrated that, under suitable electropolymerization conditions where aromatic solvents such as nitrotoluenes are incorporated into the PPy matrix, SGFETs exhibit selective sensitivity to vapours of aromatic compounds.

Another direction of development of membrane-coated sensors is the realization of conventional liquid electrolyte-based electrochemical cells made by microfabrication technologies. A micromachined miniature Clark-type oxygen sensor was fabricated by Suzuki et al. [91]. The structure is shown in Figure 4.49. A glass substrate with a silver working electrode, gold counter-electrode, and Ag/AgCl reference electrode is bonded to a silicon substrate by field-assisted bonding at 250°C by applying − 1200 V to the glass substrate against the silicon one in a nitrogen atmosphere. The silicon substrate has anisotropically etched grooves to provide small cavities to accommodate an electrolyte solution.

Because the electrochemical reactions are localized in a very small amount of electrolyte, electrochemical cross-talk between the electrodes must be eliminated. Therefore, the container grooves are etched only over each electrode area, and they are connected by long narrow grooves. The oxygen-permeable membrane, made of poly(fluoroethylene propylene) (PFEP), was affixed thermally to the silicon substrate over a cavity above the working electrode (see Figure 4.49). The electrolyte was incorporated by dipping the whole chip in the electrolyte solution in a centrifuge tube,

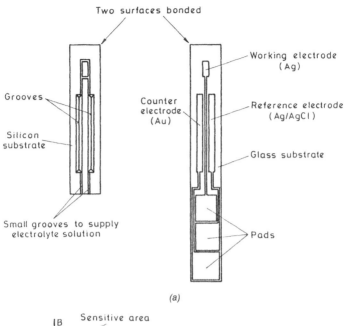

(a)

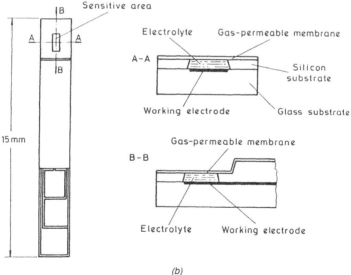

(b)

FIGURE 4.49. Structure of the micromachined Clark-type oxygen sensor: (a) the two parts of the sensor, (b) the completed structure. (Redrawn from the figure of Suzuki et al. [91], with the kind permission of Elsevier Sequoia S. A., Lausanne, Switzerland, publisher of *Sensors and Actuators A* and *Sensors and Actuators B*.)

placing it in a chamber and evacuating. The response of the miniature sensor was about 30 nA/ppm O_2, with an average response time of 30 sec and an almost zero residual current. The stable operation time is 10 h.

4.6.3 Solid Polymer Electrolyte-Based Electrochemical Microsensors

The development of solid polymer electrolytes (SPEs) provides a possibility to fabricate electrochemical sensors without liquid solution, so there is no need for electrolyte refreshing cycles. Moreover, solid-state sensors will be available, which operate at room temperatures.

Nafion® membranes are widely used in SPE-type gas sensors. An amperometric SPE electrochemical oxygen sensor with a compact structure, fast response, and long life was reported by Yan and Lu [92]. The configuration is shown in Figure 4.50(a). A piece of acid-treated Nafion® membrane serves as the electrolyte. On one side of the membrane, two pieces of hydrophobic gas-diffusion electrodes composed of Teflon®-bonded Pt-black are pressed: the bigger one is the sensing electrode, while the smaller one directly faces the surrounding air and serves as the reference electrode. The counter-electrode is a chemically deposited Pt layer. A plastic plate with a capillary is sealed on the surface of the sensing electrode.

A potentiostatic measurement was carried out at -0.6 V, with respect to the reference electrode, resulting in an almost linear amperometric response, as shown in Figure 4.50(b) (applying current-voltage conversion at the output). The temperature coefficient is about $0.3\%/°C$, which is nearly by one order of magnitude lower than that of the Clark-type sensors (see Sections 4.6.1 and 4.6.2). Response times are in the range of a few seconds, and lifetimes can be estimated in years, supposing that the ambient relative humidity remains in the range of 32 to 96%. When it passes this limit, the sensor performance can temporarily deteriorate [93].

A further advantage of the application of SPEs is that they can act both as electrolyte and as permselective and/or diffusion-controlling membrane at the same time. Therefore, planar structure amperometric microsensors can be built up using them.

Recently, several attempts have been made to develop planar-type amperometric microsensors applying micromachining and thin- or thick-film technologies. However, these early prototypes suffer from a sluggish response, short lifetime, and big cross-sensitivities. A few examples without detailed descriptions are as follows:

- Buttner et al. [94] reported wetted Nafion®-based H_2S sensors with electrodes on silicon substrate.

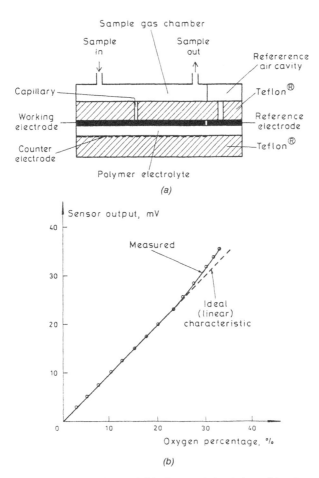

FIGURE 4.50. (a) Configuration and (b) characteristic of the solid polymer electrolyte based oxygen sensor. (Redrawn from the figure of Yan and Lu [92], with the kind permission of Elsevier Sequoia S. A., Lausanne, Switzerland, publisher of *Sensors and Actuators A* and *Sensors and Actuators B*.)

- A CO sensor was developed by Otagawa et al. [95] using acid-treated Nafion® with RF sputtered thin-film Pt electrodes patterned with photolithography and subsequent etching on alumina and glass substrate. The electrode pattern is typical for planar electrode arrangements, as shown in Figure 4.51.
- Tieman et al. [84] reported an oxygen sensor with P(DMDAAC) SPE, which was immobilized on ceramic substrates with screen-printed thick-film Pt electrodes using the technique explained in Section 4.6.1.

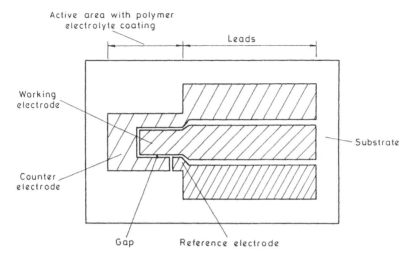

FIGURE 4.51. Basic planar sensor design using solid polyelectrolyte. (Redrawn after Otagawa et al. [95], with the kind permission of Elsevier Sequoia S. A., Lausanne, Switzerland, publisher of *Sensors and Actuators A* and *Sensors and Actuators B*.)

- An NO sensor with acid-treated Nafion® was developed by Maseeh et al. [96] using silicon microfabrication technology. The schematic cross section of the micromachined sensor-structure is shown in Figure 4.52. To decrease the response time, a back-cell sensor configuration was used. The sensed gas reaches the working electrode from the back side through a porous substrate rather than diffusing through the polymer electrolyte. The pores were made by reactive ion etching with a chlorine-based chemistry through a CVD oxide etch mask. The electrodes were deposited by sputtering

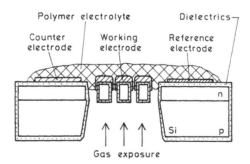

FIGURE 4.52. Schematic structure of the micromachined gas sensor with polyelectrolyte. (Redrawn with permission from the figure of Maseeh et al. [96], ©1991 IEEE.)

thin metallization films. The Nafion® was applied in gel form by mechanical spreading. The sensor showed a 90% response time of 330 msec, which is much better when comparing it with the 6-sec response time of a porous ceramic back-cell design [96].

4.6.4 Impedance-type and Related Sensors

Impedance-type gas sensors are based on the resistivity and/or permittivity changes of conducting, semiconducting, and insulating polymers. They generally use, independently of the electrical behaviour of the sensing film, the interdigital capacitor/resistor structure shown in Figure 2.4. The electrodes are made by thin- or thick-film technology on glass or alumina substrates. Sometimes oxidized silicon substrates are also used. In certain applications a heating- and temperature-sensing resistor must be integrated on the substrate. Meander-type resistors are used on the back side of the substrate, or, in the case of silicon, diffused resistors under the silica film can also be applied.

Research and development work on impedance-type sensors is concerned with material investigation rather than device fabrication. A few recent results about different sensing layers used in impedance-type sensors for various gas components are summarized in Table 4.7 [97–102]. The parameters generally represent a development stage.

Film conductivity or permittivity changes can be detected not only by direct impedance measurements. An interesting example can be mentioned here: workers of the Fraunhofer Institut have built up capacitively controlled field effect transistors using interdigital structures covered by heteropolysiloxanes on silicon surfaces connected to MOSFETs. Their structure is similar to the CFT devices demonstrated in Figure 4.41, except the gate connections [103,104]. These sensors consist of a combination of a conventional MOSFET and a gas-sensitive interdigital capacitor (see photos in Appendix 2).

4.6.5 Mass-Sensitive Gas Sensors

Most molecule/sensor interactions lead to an increase of the sensor's mass, which may be determined with the different types of gravimetric transducers (see Section 2.3), i.e., bulk (BAW), mainly quartz microbalance (QMB), surface acoustic wave (SAW), and flexural plate wave (FPW) devices. The structure and operation principles of the devices were described in Section 2.3 and the sorbent coatings in Sections 3.4 and 3.5. However, a few recent results with different structures and coatings are summarized in Table 4.8 [105–108].

TABLE 4.7. Behaviour of Gas Sensing Polymers in Impedance-Type Sensors.

Sensing Polymer	Sensed Gas or Vapour	Sensed Quantity	Relative Change	Gas Concentration	Comment
Polystyrene [97]	NO_2 in N_2	G*	1–3	10%	at 180°C
Poly(AlPcF) [98]	O_2 in N_2	G	2–10	1–8%	
	NO_2 in N_2		3	200 ppm	
Polyaniline [99]	NH_3	R*	3	1%	
Polypyrrole [100]	NH_3	R	0.5	1%	
Polyphenylacetylene [101]	CO_2	C*	4%		
	CO		3.3%	2.7 kPa	at 100 Hz
	CH_4		2.5%		
Poly(ethylene glycol) [102]	dimethyl-formamide	C	20%	0.15%	
Poly(cyanopropyl-methylsiloxane) [102]	n-hexane, ethanol	C	3%	0.15%	

*G – conductivity, R – resistance, C – capacitance.

TABLE 4.8. Properties of Gravimetric Gas Sensors.

Resonator-Type	Substrate	Sorbent Coating	Sensed Gas or Vapour	Sensitivity Hz/ppm/MHz
BAW-QMB [105]	Quartz	DMPS*	Perchlorethylene C_2Cl_4	0.1 cca. 1 Hz/ppm
SAW [106]	No data	polycarbonate resin polyepichlorohydrin 1.2 polybutadine	acetone dichlorometane acetone toluene	0.34 0.98 0.57 5.83
SAW [107]	$ZnO/Al/Si_xN_y$	DMPS	toluene	0.028
Flexural-plate [107]	$ZnO/Al/Si_xN_y$	DMPS	toluene	0.25
BAW [108] cantilever	PVDF/glass	—	SF_6	0.025

*DMPS—Dimethylpolysiloxane.

343

It is worth mentioning that polymer materials can be applied in resonant gas sensors not only as a sorbent coating, but also as a wave propagation medium. A simple mechanical resonator sensor was fabricated by Block et al. [108] from a glass plate acting as a vibrating cantilever beam and interrogated with aluminized, monaxially stretched piezoelectric PVDF polymer foils glued on both sides of the plate. One of these foils acts as a transmitter, the other as receiver. The sensitivity of the device allows binary gas mixtures to be controlled at an accuracy level of a few percent (see Table 4.8) [108].

Wenzel and White have realized flexural plate wave (FPW) chemical sensors fabricated by silicon micromachining [107]. The structure of the back-cell type device is shown in Figure 4.53. The core of the device is an ultrasonic delay line consisting of a composite plate of low stress silicon nitride, aluminium, and zinc oxide. Interdigital transducers on the ZnO layer launch and receive the plate waves.

A selective and sensitive molecular sensing system based on three QMBs coated with plasma polymer films was described by Sugimoto et al., which is a good example for principal-component analysis of the dynamics of piezoelectric responses for molecular recognition [109]. The AT-cut QMBs were coated by RF sputtered poly(chloro-trifluoroethylene) (PCTFE), by pyrolytic graphite, and by cosputtered polyethylene (PE) and polytetrafluoroethylene (PTFE). Plasma polymer films are likely to contain stable radical sites and conjugated moieties that can interact with target molecules. A dynamic piezoelectric response is the result, which can be classified using the principal-component analysis [110]. Shifts in resonance frequency are caused by mass changes induced by sorption or desorption of the analyte molecules. Using Equation (2.11), the normalized saturation mass values

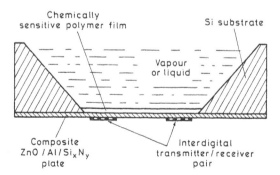

FIGURE 4.53. Schematic cross section of the flexural-plate-wave chemical sensor. (Redrawn from the figure of Wenzel and White [107], with the kind permission of Elsevier Sequoia S. A., Lausanne, Switzerland, publisher of *Sensors and Actuators A* and *Sensors and Actuators B*.)

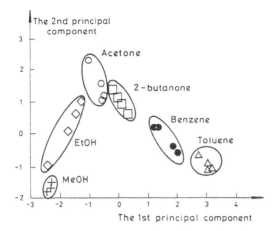

FIGURE 4.54. Molecular classification map using two parameters of the response curve. (Redrawn from the figure of Sugimoto et al. [109], with the kind permission of Elsevier Sequoia, S. A., Lausanne, Switzerland, publisher of *Sensors and Actuators A* and *Sensors and Actuators B.*)

can be calculated from the frequency shift values. The time constants of the transient piezoelectric response curves can also be calculated, which are partly characteristic for the analyte molecules. The combination of two parameters, time constant and saturation mass, from three sensor probes can provide a six-dimensional vector. Molecular classification mapping is a two-dimensional projection of this six-dimensional space derived from molecular sensing measurements. Figure 4.54 shows the most valuable result of the molecular classification mapping. In this figure, neither axis has any explicit meaning, and the four points correspond to four concentration levels [109].

4.6.6 Calorimetric Sensors

Calorimetric gas sensors (see Sections 2.5 and 3.5) are partly based on heat generation of catalytic processes. Generally, they must be heated to achieve sufficient catalytic activity. Pellistors for H_2 and CO_2 detection have been well known sensors for a long time. These sensors suffer from interference caused by water vapour, and changes in humidity can also produce spurious signals. Various methods have been considered to overcome this serious problem.

A bimetallic Pt/Ir catalyst prepared with an inherently hydrophobic support polymer, namely poly(styrene-divinylbenzene) (PSDB), has been fabricated and successfully used by Marcinkowska et al. for an ambient-temperature CO sensor whose response is independent of humidity [111].

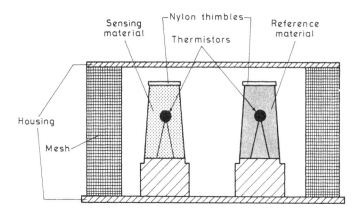

FIGURE 4.55. Schematic structure of the catalytic CO gas sensor. (Redrawn from the figure of Marcinkowska et al. [111], with the kind permission of Elsevier Sequoia, S. A., Lausanne, Switzerland, publisher of *Sensors and Actuators A* and *Sensors and Actuators B*.)

The optimum catalyst for the CO sensor contained a total of 10% by platinum, plus iridium supported on porous PSDB granules. Thermistors were used as temperature-sensitive elements. One was centred in a catalyst held in a nylon mesh thimble. The reference side of the bridge (see Figure 2.23) was identical, except that it had plain granular PSDB instead of catalyst loaded in the formed thimble. The schematic structure of the sensor is illustrated in Figure 4.55. The sensor signal is proportional to the CO concentration up to at least 600 ppm with a sensitivity of 63 nV/ppm, almost independent of the humidity [111].

Reactions between the sensor material and the detected molecules under chopped flow conditions were monitored by Schierbaum et al. [105,112]. Changes in the equilibrium concentrations of absorbed molecules lead to time-dependent nonequilibrium heat generation, which only occurs during pressure variations (see Sections 3.5.2.1 and 3.5.2.3).

A planar thermopile with 64 Cu/Cu-Ni thermocouples in a thin Kapton® (DuPont) foil was used to measure the temperature difference transients [see Figure 3.30(a)] [105]. The active thermocouples were coated by the sensing polymer, PDMS. Typical results obtained in this operation mode are shown in Figure 3.42. Temperature changes are positive during exposure to different partial pressures of C_2Cl_4 in air and negative during subsequent changes to pure air.

4.6.7 Fibre-Optic Gas Sensors

Recently, the development of fibre-optic sensors for the detection of different gas components, both in gases and liquids, has been rapidly increasing. These rather small devices offer the possibility for remote sensing since they are impervious to electrical interference and the capability of operation in both multiple or distributed sensing configuration. Moreover, they have several advantages that can be especially exploited for gas-sensing devices. The most important ones can be summarized as follows:

- Most of the materials are biocompatible, which offers various potential applications for biological sensing, such as measurement of blood gas levels, etc. This area will be discussed in Section 4.8, in connection with the biosensors.
- Fibre-optic sensors generally consist of nonmetallic components; thus, the problems of the conventional corrosion can be overcome by their application. They may have good long-term stability and reliable operation even when sensing corrosive, poisoning, or toxic components.
- These sensors operate without electricity; therefore, the danger of spark-induction is automatically eliminated. That gives great perspective in sensing combustible and/or explosive gas components. There is no need for a special explosion-safe design.

The basic sensor operation principles and the applied materials and material-analyte interaction mechanisms have been described in Sections 2.6 and 3.8, respectively. A few recent results from potential sensor devices are summarized in Table 4.9. The most important parameters are also given in the table. The schematic diagram of the typical experimental arrangement for measuring fibre-optic sensor signals is shown in Figure 4.56 [113]. This arrangement is suitable for absorbance-change measurements, for instance, with core-type or cladding-type sensors. A typical characteristic of a porous fibre ammonia sensor is given in Figure 4.57 [113].

4.6.8 Other Gas-Sensing Principles Using Polymers

There are a few examples of gas-sensing devices that have special

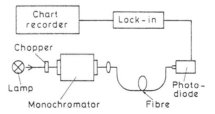

FIGURE 4.56. Typical connection for measuring porous fibre absorbance changes. (Redrawn with permission from the figures of Zhou et al. [113], ©1989 Optical Society of America.)

TABLE 4.9. *Properties of Fibre Optic Gas Sensors.*

Sensor Type	Sensing Polymer	Detected Gas	Measured Quantity	Change/Conc.	Typical Wavelength, nm
Porous fibre [113]	P(MMA + TGDM*) + bromocresol purple	ammonia	intensity (absorbance)	−4.34%/ppm	596
Porous fibre [14]	P(MMA + TGDM) + PdCl$_2$	CO	decay rate	2 ppm/ppm	630
Cladding-based [115]	PTFE	propane	detection limit	(5% low exposive limit)	no data
Cladding-based [116]	COP**	toulene	intensity	−45 ppm/ppm	670
Mach-Zehnder [117] interferometer	DMPS†	perchloro-ethene	intensity	235 ppm/ppm	788
Reflectance [118] interferometer	DMPS	C$_2$Cl$_4$	optical pathlength	0.029 nm/ppm	resolution 2.8

*TGDM—Triethylene glycol dimethacrylate.
**COP—Glycidoxypropylmethyl dimethyl siloxane copolymer.
†DMPS—Dimethylpolysiloxane.

348

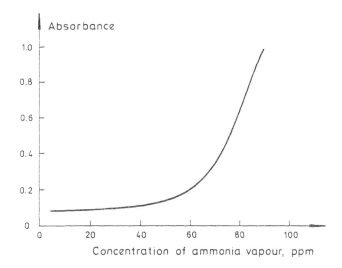

FIGURE 4.57. Response characteristic of the porous plastic optical fibre used for ammonia measurement. (Redrawn with permission from the figures of Zhou et al. [113], ©1989 Optical Society of America.)

features, and it is better to handle them separately from the groups of the previous sections.

A new type of solid-state sensor for the detection of minute concentrations of H_2 gas has been developed by Mandelis and Christofides [119]. The sensor is made of thin PVDF pyroelectric film (see Sections 3.1.2 and 4.4), sputter coated with Pd for the sensor and with Al-Ni double layer for the reference element. The device is called photopyroelectric, P^2E, hydrogen gas sensor. The instrumentation consists of a laser diode, a fibre optic with the divider, and the two pyroelectric devices. The operation of the devices is probably based on the change in pyroelectric coefficient of the film due to electrostatic interactions of absorbed hydrogen ions within the PVDF polymer matrix upon hydrogenation and selective absorption by the metallic (Pd) coating. Hydrogen concentrations as small as 0.075 %, in a flowing $H_2 + N_2$ mixture have been detected using the sensor [119]. The arrangement is illustrated by Figure 4.58.

A microelectrochemical switching transistor was prepared by Thackeray and Wrighton when connecting closely spaced (1-2 μm) Au microelectrodes deposited onto Si/Si_3N_4 surfaces with anodically grown poly(3-methylthiophene) [120]. The latter can be platinized by electrochemical reduction of $PtCl_4^{2-}$. The Pt equilibrates the polymer with the O_2/H_2O or H_2O/H_2 redox couples. Due to the doping-dedoping effect (see Section 1.4.4), the conductivity of the mentioned polymer can be varied by five

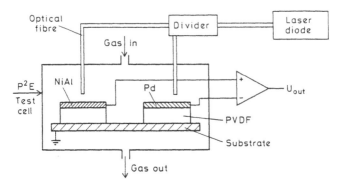

FIGURE 4.58. Structure of the photopyroelectric H_2 sensor [119].

orders of magnitude in acidic solution by changing the atmosphere from H_2 to O_2; thus, O_2 reproducibly turns on the transistor, while H_2 reproducibly turns it off at a fixed pH value in a solution of 0.1 M $HClO_4$.

It can also operate as a pH-sensitive switching device at fixed O_2 concentration; therefore, it is, in fact, a multifunction sensor that can be used either as gas or as a pH-sensitive switch. The structure and operation is illustrated in Figure 4.59(a), and the drain current response versus time upon variation of O_2-H_2 pressure is shown in Figure 4.59(b).

An integrated gas sensor array for odorant sensing, using conducting polymers with molecular sieves, has been investigated by Shurmer et al. [121]. The four-electrode pattern was formed photolithographically in a gold-plated alumina substrate, and they were covered with an electropolymerized polypyrrole conductive-sensing polymer. These steps were identical for all four electrodes. Then their characteristics were modified individually by coating each with a different number of archaidic acid Langmuir-Blodgett films (see Section 1.3.1), which can act as molecular sieves. The sieving behaviour can be influenced by the number of monomolecular layers deposited by the LB technique. The preliminary results indicate that for ethanol, acetone, and ether vapours, the sensitivity falls and then rises with an increase in the number of LB layers, while methanol results in an opposite behaviour [121]. LB films seem to play a complex active role in the sensing mechanism: they act not only as sieves.

4.7 ION-SELECTIVE SENSORS

The monitoring of toxic ionic compounds for controlling freshwater pollution is assuming major importance in environmental protection. Moreover, measurement technologies capable of continuously or semicontinuously monitoring ionic compounds in blood and certain metabolites are desired to get real-time information at the bedside of seriously ill and surgical patients. Therefore, a great demand on reliable and long-life ion sensors has emerged over the last few decades.

A great variety of sensors has been developed using different materials for different ion-selective sensing purposes; however, there are only a few sensing principles that can be used in this field. The basic sensing princi-

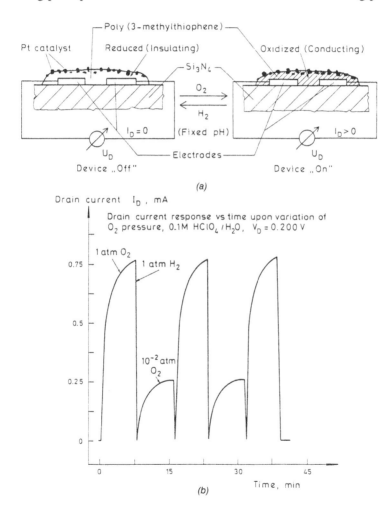

FIGURE 4.59. Chemically actuated switching transistor: (a) structure and (b) response. (Redrawn with permission after the figures of Thackeray and Wrighton [120] from *J. Phys. Chem.* Vol. 90, ©1986 American Chemical Society.)

ples and material interactions resulting in sensor signals are described in Sections 2.2.1, 2.4.1, 2.6, 3.7, 3.8, and 3.9.

Polymer materials have already been widely used in conventional ion-selective electrodes as permselective and/or ion-exchange membranes.

Polymeric ion-selective membranes consist of at least two kinds of components: the polymer, which primarily provides the mechanical support, and the ionophore and other electroactive ingredients, which introduce the desired electrochemical properties. The additional use of suitable plasticizers allows a more subtle adjustment of desired properties of the membranes. Primarily, plasticizers provide the required mobility for sufficient fast kinetics of the ion exchange and/or ion extraction at the phase boundary of the membrane and the sample. To some extent, plasticizers may also take part and/or influence the stability of the ion/ionophore complex, thus influencing the selectivity pattern of the membrane. Plasticized polymers (mainly PVC) with ionophore additives are widely used in almost all types of ion-selective sensors listed later.

Recently, the development of microelectrodes and other type microsensors has required the application of new type polymers, including insulating and/or conducting electropolymerized films, since the latter technology gives a possibility for building up polymers with controlled-size pores when using the doping-dedoping technique.

The ion-selective sensors can be grouped into the following categories:

- liquid electrolyte-based ion-selective electrodes and solid-state microelectrodes
- ISFETs
- multiple ion sensor arrays
- conductimetric sensors
- optrodes

4.7.1 Ion-Selective Electrodes

Ion-selective electrodes (ISEs) have been used for detecting different ion types and especially for pH measurements for a long time. Their general structure and operation principles are described in Section 2.4.1. Originally, they utilized glass membranes. Depending on the glass and inner electrolyte composition, different electrodes were built up for various ion types.

The application of polymers as ion-sensitive membranes has given new possibilities and fabrication techniques for ISEs. Although a limited number of polymer materials are used in the conventional ISEs, the great choice of different ionophores that can be embedded in the polymer matrix gives the possibility to build up many different electrode types. Conventionally, ion exchangers have been used as ionophores, but presently the neutral ion carriers are preferred. Ion carriers also belong to the ionophores and are lipophilic complexing agents with the capability of reversibly binding ions from the analyte solution. The wide choice of ionophores that can interact with various ion types makes it impossible to give a survey in the frame of this book, so we can just refer to the literature [122,123]. A few examples are given in Section 3.7.

The most important polymer materials used as polymer matrix are the following: PVC (high molecular weight), copolymer of vinyl chloride, vinyl acetate and vinyl alcohol PV(C/Ac/A), polyvinyl chloride carboxylated, silicone rubber, and more recently, polyimide (see Section 3.7).

Carrier-based electrodes and microelectrodes are available for the measurement of H^+, Li^+, Na^+, K^+, Cs^+, Mg^{2+}, Ca^{2+}, Sr^{2+}, Ba^{2+}, Tl^{2+}, Cd^{2+}, Pb^{2+}, Ca^{2+}, NH_4^+, and Cl^- activities. With respect to selectivity, these membranes are far superior to the conventional ion-exchange electrodes and show a number of advantages in the technique of preparation as compared to glass microelectrodes. Among the ion carriers, the electrically neutral ones have found a particularly wide field of application as components in ion-selective membrane electrodes.

The preparation of polymeric membranes can easily be performed. At first, a solution must be made either by dissolving the polymer (PVC in tetrahydrofuran, for example) or by mixing the monomer with cross-linking agents. Electrode membrane polymers generally also contain plasticizer. The addition of the ionic carrier determines the electrochemical behaviour of the material. The mixture can be poured into glass or Teflon® (DuPont) forms to get membranes of different shape and finally cured to evaporate the solvent or to perform the cross-linking.

Conventional internal reference liquid electrolyte-based electrodes [see Figure 2.19(a)] have several disadvantages:

- The aging of the electrolyte results in a continuous drift in the output signal; therefore, electrolyte refreshing periods are necessary.
- The small output signal can be measured with sophisticated electronics in an environment with electromagnetic interferences.
- The big dimensions of the electrodes limit their application possibilities.

According to these problems, the recent developments of ISEs have been conducted in the following directions:

- The application of solid inorganic or polymer electrolytes, instead

of liquid electrolytes between the membrane and metallic parts, means one possibility. The other one is to apply only one polymer membrane on the metallic electrode surface. In both cases, stable solid to solid contacts are needed with good adhesion, which means a new requirement for the polymer membrane. In a lot of cases, the problem can be solved by special additives or by the use of new polymer types.
* The application of microsensor technologies can solve the signal transmission and size problems. Compatible polymer materials and deposition technologies are needed in that case.

Plasticized PVC membranes also have great possibilities in solid contact electrodes. Their most important features can be summarized as follows:

* PVC matrix membranes are compatible with a great number of selective compounds: ion exchangers and neutral carriers, which are selective to inorganic and organic ions.
* It is possible to vary in a wide range the concentrations of selective compounds in the membrane and also to vary the chemical nature of its solvent-plasticizer in order to optimize membrane composition.
* It is easy to prepare PVC membranes and electrodes with them. Only a little equipment is necessary; there is no need for high-temperature or high-pressure technologies.

Solid contact ISEs, which are similar to a coated wire electrode, need a stable and reversible transition from electronic conductivity in the metal electrode to the ionic conductivity in the polymeric membrane. Often, in this type of ISEs, the membrane/solid interface is ill-defined, leading to significant potentiometric drift and instability.

According to studies of Harrison et al. [124], the drift may be in connection with the initial adhesion and also the water uptake of the polymer. If the membrane has poor adhesion to the electrode, osmotic pressure (see Section 3.6.2) can peel the membrane from the surface, allowing an inner solution of varying ionic strength to form. It was postulated that the use of $SiCl_4$ as an adhesion promoter not only increases the membrane adhesion to the solid contact, but may also increase membrane self-adhesion, which reduces the rate of plasticizer leaching.

Goldberg et al. [125] have used screen-printed silver epoxy solid contacts and polymeric membranes after silicon-on-insulator CMOS processing. They found that incorporation of $SiCl_4$ adhesion promoter into the polymeric membrane casting solution significantly improves the stability of the membrane/solid contact interface. This improvement can be achieved without adding complexity to the electrode or microsensor fabrication process. The optimum properties of the fabricated potassium electrode were found with a polyurethane-based PU/PV(C/Ac/A) matrix using DOA [bis(2-ethylhexyl)adipate] plasticizer, valinomycin ionophore, and $SiCl_4$ adhesion promoter.

A PV(C/Ac/A) copolymer itself presents good adhesion to a graphite surface. Borracino et al. [126] have fabricated a graphite/copolymer solid ISE for detecting Cd(II) ions. The use of the modified copolymer membrane enhanced the lifetime in comparison to a conventional liquid electrolyte-based ISE.

Cha and Brown have made great efforts to realize chemical-selective membranes using polyimide-matrices [127]. Polyimide compounds have excellent adhesion to integrated circuit surfaces, as it was shown in Sections 3.1.4.1, 4.5.1, and 4.5.3. They are also compatible with microfabrication technologies; however, conventional polyimide curing (imidization) requires temperatures from 300 to 400°C, which would evaporate or destroy the ionophore and plasticizer needed to give the desired electrochemical properties for a membrane. Pre-imidized polyimide solvents are also available; curing of this material means only the removal of the solvent and can be accomplished at low temperatures.

Preliminary experiments have demonstrated the feasibility of using polyimide matrices as ion-selective membranes. Two plasticizers have been identified by Cha and Brown, which are applicable in polyimides. Membrane compositions for ammonium- and calcium-selective membranes are reported, for which potentiometric slope and selectivity are comparable to those of PVC-based membranes at ion concentrations above 10^{-6} mol/l. The expected improvement in adhesion of plasticized PI membranes compared to PVC ones has also been confirmed.

Another approach is the application of double-membrane structures. For example, a layer of PVA containing KCl has been placed between the sensing membrane and the Ag/AgCl electrode contact to mimic the internal reference solution of conventional ISEs [128]. In another case, an external permselective membrane was cast on a conductive membrane made of a PVC matrix over a platinum contact [129].

Rather complicated thin-film ISEs for Na^+, K^+, and H^+ using multilayer structures have been developed by Urban and coworkers [130,131] as shown in Figure 4.60. Figure 4.60(a) demonstrates the structure of the Na^+/K^+ sensors. Metal layers were evaporated on glass substrates and shaped by photolithography. The internal reference electrode was established by the following metal multilayer sequence: Ti/Au/Ag/AgCl. The Ag layer was chemically chlorided. The insulating layer is a PECVD-Si_3N_4 film; the salt layer was evaporated and structured by a lift-off technique. It acts as an

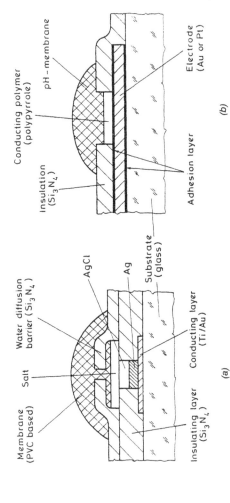

FIGURE 4.60. Cross section of the thin film ion-selective devices with water diffusion barrier: (a) Na⁺/K⁺ sensor; (b) pH sensor. (Redrawn from the figures of Keplinger et al. [130,131], with the kind permission of Elsevier Sequoia, S. A., Lausanne, Switzerland, publisher of *Sensors and Actuators A* and *Sensors and Actuators B*.)

internal solid electrolyte. Finally, the structure was coated with the membrane. For potassium electrodes a composition of PVC-DOS (dioctylsebacate) with valinomycin ionophore and for sodium electrode a PVC-oNPOE (nitrophenyl-octylether) with Na^+-ionophore were chosen. To enhance the long-term stability, an additional silicon nitride layer with a small aperture was introduced to limit the water transport across the membrane in order to obtain a longer lifetime (at least three days). The potential change is about 100 mV when the ion concentration is changed from 1 mmol to 0.1 mol [130].

Figure 4.60(b) shows the schematic cross section of the pH sensor. It consists of an Au or Pt electrode, a conducting polymer film, and neutral carrier membrane. Coatings of the conductive polymer, polyaniline or polypyrrole, were deposited by electropolymerization from aniline/perchloric acid or pyrrole/acetonitrile solutions [131].

Fabry et al. [132] have proposed a combination of polymer electrolyte and inorganic NASICON ion-exchange membrane for Na^+ ion solid-state electrode sensors. The schematic structure of the sensor is shown in Figure 4.61(a), while the reaction model is shown in Figure 4.61(b). Small pellets of sintered NASICON ($Na_3Zr_2Si_2PO_{12}$) were attached to insulating substrates covered with a metal layer for the electrical contact (Ag or Cu).

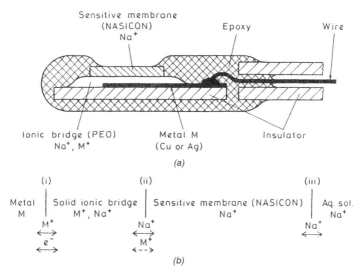

FIGURE 4.61. (a) Schematic view and (b) reaction scheme of the solid-state ionic sensor using solid electrolyte ionic bridge. (Redrawn from the figures of Fabry et al. [132], with the kind permission of Elsevier Sequoia, S. A., Lausanne, Switzerland, publisher of *Sensors and Actuators A* and *Sensors and Actuators B*.)

The connection material between the metal and the membrane is a solid polymer electrolyte. Poly(ethylene oxide) (PEO) with NaI and AgI or CuI dopants was used. The lifetime of these sensors can be greater than 1000 h.

4.7.2 ISFETs

The structure and operation principles of ISFETs, including their measurement techniques (connection, compensation with REFETs, etc.), have been described in Section 2.2.1. Since ISFETs are semiconductor devices, they have several advantages over ISEs, such as lower output impedance, small dimensions, and being well-suited to mass fabrication.

Since their introduction in the early 1970s, much attention has been given to improve the sensitivity, selectivity, and long-term operation of these devices. In order to change the selectivity of the first pH-sensitive ISFETs to other ions, ionophores supported in inert matrices can be deposited on top of the gate insulator. However, four main problems are encountered when so-called solvent polymeric membranes are deposited on top of the ISFET:

- The lifetime is limited, owing to the poor adhesion of the membrane to the SiO_2 or Si_3N_4 surfaces.
- The lifetime depends on the encapsulation of the whole structure.
- The output signal is unstable due to the undefined electrical potential difference across the membrane-insulator interface.
- The lifetime is also limited due to leaching out of membrane constituents.

Strong efforts have been made in recent decades to overcome these problems. Although a great number of articles have been published, only few commercial applications of ISFET devices with polymer-sensing layers exist. The best solution is expected from the combination of different multilayer polymer-matrix structures. The large number of publications and the great variety of materials and ion types to be analysed prevents us from giving a detailed description. Paul Clechet [133] has given a survey about this topic in 1991, and I will quote him (with the kind permission of Elsevier Sequoia S. A., Lausanne, Switzerland, publisher of *Sensors and Actuators A* and *Sensors and Actuators B*) completing his summary with the newest results.

Polymeric membranes have been used much more often than inorganic ones since the early work of Janata et al. [134] on potassium detection using valinomycin embedded in a plasticized polychlorovinyl (PVC) as ionophore. A quadruple function (H^+, K^+, Na^+, Ca^{2+}) ISFET has been constructed by Sibbald et al. [135] by a similar deposition of ionophore-doped polymeric materials, thus emphasizing the ability of these sensitized electronic components to be used for multi-ion detection. A number of ionophores, in particular those already used in classical ion electrodes (e.g., complexing and chelating agents, crown-ether) are available and this method is consequently more versatile than the former.

When ionophore-containing membranes, especially the widely used plasticized hydrophobic PVC ones cited above, are deposited, a number of drawbacks may arise: [136,138]

(1) The poor adherence of these physically bonded membranes to the dielectric can lead to: (a) seepage of water under the membrane, which leads to an eventual short circuit between the insulator and water. Penetration of small molecules, like CO_2, can also disturb the electrical response of the ISFET because of their acido-basic properties which can influence the ionization equilibrium of the superficial $OH(SiO_2)$ or $NH/NH_2(Si_3N_4)$ groups and consequently the signal resulting from the pH response of the dielectric [136]. (b) poor definition of the contact potential, or the thermodynamic equilibrium, between the inorganic and organic materials.

(2) The leaching of the plasticizers from the membrane, which then becomes brittle.

(3) The similar leaching of the ionophores from the membrane, which then becomes insensitive.

(4) Decrease of the channel transductivity, the appearance of hysteresis phenomena and a lengthening of the response time with the thickness change of the membrane.

All these problems can drastically shorten the lifetime and performance of the ion-sensitive membranes and result in a poor reproducibility of these sensors, which actually prevents their commercial development and application. Their improvement is the subject of active research and a number of solutions are now in progress. Most of them are very recent and, as they concern ISFETs, are all compatible with silicon IC technology (essentially photolithography). Only a few are able to overcome all the difficulties enumerated above. For example the suspended mesh of polyimide [139] or platinum with polypyrrole [90] early proposed by Janata's group was essentially conceived to immobilize the polymeric membrane by mechanical means. [First, they fabricated a suspended polyimide mesh over the organic coating membrane using photolithographic techniques. Later, they used platinum mesh covered with electrochemically deposited polypyrrole film (see Section 4.6.2)].

In the same way, the use of photoresists cross-linked by UV irradiation proposed by Kawakami et al. [140] has allowed the suppression of plasticizer and ionophore leaching, while the careful choice of the best plasticizers has, on the contrary, led to improvement of the adherence of PVC membranes [141]. (Clechet [133], pp. 55–56)

Bezegh et al. [142,143] used highly lipophilic plasticizers in the matrix to enhance adhesion of the PVC matrix membrane onto the device. In their approach, a blank membrane containing only the polymer and plasticizer was applied as a continuous coating to the device surface. Then the electrochemical selectivity was introduced by doping this membrane with electroactive ingredients. In the practice, this was performed by dropping ionophores selectively onto the surface of the membrane.

Harrison et al. [124] reported a chemically modified PVC, which contains OH functionality and exhibits enhanced adhesion onto a hydroxyl bearing gate surface via reaction with $SiCl_4$.

An original solution to this problem has been the use of waterproof natural oriental lacquer, the Urushi latex, which seems to exhibit a strong and durable, but not clearly explained, adherence to Si_3N_4 substrates [144,145]. (Clechet [133], p. 56)

Since a hardened Urushi film has a high polish and remarkable durability, Urushi has been used for waterproof lacquer from ancient times. Urushi latex is a water-in-oil type emulsion and consists of Urushiol, which is a mixture of 3-substituted pyrocathecol derivatives with a saturated or unsaturated chain of fifteen carbon atoms, rubber substances that contain the acid form polysaccharides, and several other compounds. Two types of polymerization mechanisms of Urushiol were proposed, as was described by Wakida et al. [145]. The Urushi matrix membranes were fabricated using different ionophores. Almost Nernstian behaviour, good selectivity, and excellent stability were found using these materials.

An interesting recent development is the utilization of Langmuir-Blodgett films whose thickness can easily be controlled to a very low thickness level. Hoffmann and his coworkers have used phthalocyaninato-polysiloxane polymers for amphiphilic films. Adding different ionophores, H^+, Na^+, and, more recently, Ca^{2+}, sensitive films were produced with small response times [146–148].

The chemical anchoring of the membranes to the gate insulation seems to be the most rational approach. The use of silanes or polysiloxanes may solve the problems of membrane adherence. These materials can react both OH and NH/NH_2 superfacial groups on hydrated SiO_2 or Si_3N_4 and polymeric materials. Among others, Battilotti et al. [149] have used this technique. The method is based on the introduction of amino groups onto the surface of the FET gate areas by reacting the surface silanol groups with suitable functionalized organosilanes, followed by the coupling of the hydroxyl moiety of PHEMA with the mentioned amino functions carried out by means of difunctional condensing agents such as toulene 2,4-diisocyanate. Ultraviolet polymerization of the HEMA double bonds and final thermal curing lead to the formation of a copolymer chemically bound to the device. In another work, plasticized PVC membranes with ionophores were bound in the same way [150].

The best solution seems to be the covalent binding, both of the membrane to the substrate and the ionophores and plasticizers to the polymer-matrix. This method has been investigated at the University of Twente for several years [151]. They have developed the following strategy to solve the problems of membrane adhesion and plasticizer/ionophore leaching:

- The membrane should be bound covalently to the chip.
- Intrinsic plastic polymers (elastomers) should be used to avoid the application of external plasticizer.
- The ionophore should be covalently linked to the polymer backbone.

The covalent binding of polymers was also solved by using silanes. A trimethoxysilane, with a methacrylic terminal group, which contains an unsaturated vinyl bond, was used to silanize the dielectric SiO_2 surface partially. The modified surface was then covered by a prepolymer butadiene, which was afterwards polymerized by UV light. During this step, the polymer was also linked through the double bonds of the acrylate groups. (Clechet [133], p. 57)

Figure 4.62 illustrates both the silylation and the polymerization processes. On the basis of this technique, Reinhoudt et al. [152,153] have prepared a multilayered membrane, first by reaction of methacryloxy-propyl-trimethoxy silane and then by coating this layer with a buffered hydrophilic PHEMA. Figure 4.63 shows the thermodynamic equilibrium processes and the architecture of the sensor, called CHEMFET. The PHEMA hydrogel is conditioned with buffered electrolyte solutions, and this eliminates the CO_2 interference and the residual pH sensitivity. However, the effect of the pH-dependent swelling of PHEMA is still not clarified. Covalently attached to the hydrophilic membrane is the hydrophobic sensing membrane that contains the sensing molecule (ionophore/ receptor molecule). The chemical composition of the latter membrane is tuned to the primary ion to be detected. Conventional membranes such as plasticized PVC, acrylic ester, and silicone rubber have been tested; however, the best results can be obtained with copolymers of differently substituted siloxanes [152–154].

Polysiloxane membranes can easily be obtained from commercially available starting materials, such as different RTV silicones and Silopren® (Bayer), by the addition of various ionophores. The elastometric properties of these materials avoid the need for plasticizers. Moreover, polysiloxanes possess OH groups, which can be grafted both to the PHEMA and to many ionophores. Thus, all the problems associated with the use of membranes

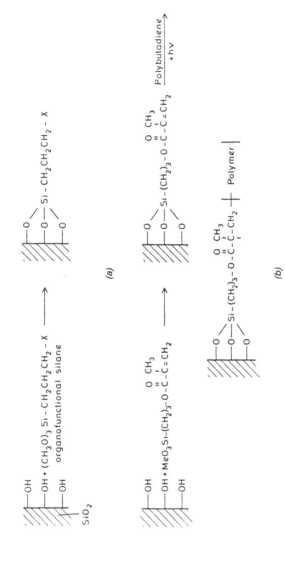

FIGURE 4.62. Reaction scheme for covalent binding of polymers to the ISFET gate insulation surface: (a) silylation, (b) covalent binding and polymerization. (Redrawn from the figures of Sudhölter et al. [151], with the kind permission of Elsevier Sequoia, S. A., Lausanne, Switzerland, publisher of *Sensors and Actuators A* and *Sensors and Actuators B*.)

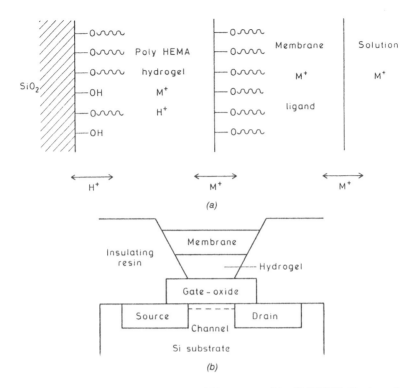

FIGURE 4.63. (a) Equilibrium scheme and (b) structure of the CHEMFET. (Redrawn from the figure of Reinhoudt [153], with the kind permission of Elsevier Sequoia, S. A., Lausanne, Switzerland, publisher of *Sensors and Actuators A* and *Sensors and Actuators B*.)

might be solved by this multi-anchorage method. However, the problems of ionic mobility within the membranes and that of chemical selectivity still remain to be solved.

The theoretical behaviour of ISFETs can be described by the Nernst equation (see Equation (2.9). From that, the variation of the interface potential should be 59 mV per decade of ion activity at 25°C in the case of monovalent ions. The practically realized values at PVC-based or Urushi membranes are generally in the range of 40–55 mV/decade. Because of the theoretical limit of sensitivity, the most important of the ISFET characteristics is rather the selectivity for secondary interfering ions, which can be characterized by the log K_{ij} values [see Equation (2.9)]. At the mentioned sensor types these values are between -1 and -2.5. The operation lifetimes of PVC-based ISFETs are about a few days. The effect of PHEMA hydrogel on the sensitivity of sodium selective sensors with conventional (PVC, Silopren®, etc.) membranes in the presence of 0.1 M interfering Li⁺,

K⁺, Mg²⁺, Ca²⁺ ions was measured by Haak et al. [152]. Their results reveal that the sensitivity for sodium is in the range of 37–60 mV/pNa without PHEMA intermediate layer but near-Nernstian, 58 mV/pNa, with the PHEMA, independently on the type of interfering ion.

Another experimental result of potassium-selective sensors applying PHEMA hydrogel with polysiloxane membranes showed that the cross-sensitivities are smaller than -3.1 with the sensitivity over 55 mV/decade. The long-term stability of this CHEMFET exceeds 200 days [152].

Supramolecular chemistry was also used in the development of ion-selective CHEMFETs [153]. The sensing molecules that have been studied were designed for a high K⁺, Na⁺, Ca²⁺, and Ag⁺ selectivity. Typical examples of such molecules are hemispherands or calix[4]arene crown ethers for K⁺ and tetra-o-alkylated calix[4]arenes in the cone conformation for Na⁺ or Ag⁺. The sensing molecules can be present as mobile hydrophobic species, or they can be covalently attached to the sensing membrane. In both cases a durable sensor can be obtained.

The realization of REFETs (see Section 2.2.1) means the other side of the application problems of ISFETs. Remaining at pH measurements, firmly anchored non-ion-blocking hydrophobic layers on the gates are of importance for the realization of a pH-insensitive FET, which can be used as an all-solid-state reference electrode. It also gives the possibility for temperature compensation [see Figure 2.9(f)].

> One of the first attempts to realize a REFET was that of Matuso [155], who used thin ion-blocking layers of parylene physically attached to Si₃N₄ gates. Probably because of the presence of pinholes, the pH sensitivity was not fully suppressed. Increasing the thickness of the layer was not possible because of the consequence on the REFET response, which must be as close as possible to that of the sensitive ISFET when they are used in a differential measurement set-up [137]. A compromise between a pinhole-free membrane (which requires a thickness of several microns) and the small thickness necessary to maintain the REFET sensitivity at a high level has not yet been found with such ion-blocking materials as parylene, Teflon® or polystyrene [133]. Moreover, such membranes are highly sensitive to the ionic strength of the electrolyte [156]. As the solution which would consist of suppressing the pH sensitivity of the active groups on the dielectric materials by the covalent grafting of silane molecules appears to be impossible (the percentage of remaining free groups must be inferior to 0.01% according to the site binding theory [157], the use of non-ion-blocking membranes described above is very interesting. Such membranes [for example, PHEMA] are insensitive to the ionic strength of the electrolyte and lead to transistors with electrical properties very similar to those of the sensitive ones. (Clechet [133], p. 57)

An extended site-binding model for hydrophobic and blocked organic polymers was proposed by Leimbrock et al. [158]. The model is based on the assumption that two kinds of surface sites exist (COOH and COH groups). These groups can be formed in the presence of oxygen and water in the electrolyte. In this way, surface densities increase during immersion in the electrolyte solution and cause an increasing pH sensitivity. Surface passivation with styrene vapour after the plasma polymerization gives membranes with a better long-term stability than untreated samples. The best properties were achieved by passivated plasma-polymerized tetra-fluoroethylene membranes; however, the measured values are good for application to a REFET system only for a few days [158].

Another approach is to return to the reference electrodes and to develop stable microelectrode structures. A new solution was proposed by Dumschat et al. [159]. A perchlorate sensitive sensor is in contact with a saturated solution of $KClO_4$ and $CaSO_4$. This solution may be stabilized by the photoimageable PHEMA hydrogel. They are located in a cavity, and the solution is contacted by a small opening to the analyte solution. The advantages are that all phase boundary potentials are thermodynamically well defined [159].

4.7.3 Multiple Ion Sensor Arrays

In the last ten years, attempts to integrate several ion sensors on a small chip or substrate have been accelerated by the remarkable progress in the technology of semiconductor and other microcircuit fabrication. One of the great advantages of ISFETs is the possibility for integration in multiple sensor arrays and with integrated multiplexer and other circuitry.

The miniaturized multiple ion sensor arrays are expected to be applied in the biomedical field, such as for ionic species monitoring in blood or urine, since it is possible to use them to measure several ions simultaneously and rapidly with a small amount of sample and to fabricate them at low cost. Until now, the wide ability of such sensor arrays for practical monitoring purposes has been restricted because of the difficulties already described in connection with ISFETs.

Multi-ion sensor ISFET arrays can be built up combining inorganic and organic polymer gate membranes. Igarashi et al. [160], for example, combined a Si_3N_4 H^+-sensitive layer with Na^+- and K^+-sensitive layers, sodium-aluminosilicate glass, and PVC membranes, respectively. Polyester plasticizer was applied in the PVC film.

Tsukada et al. have presented an integrated micro multi-ion sensor using platinum gate ISFETs with several polymeric membranes [161]. It consists of two kinds of ion sensors (K^+- and Na^+-ISFETs) and two CMOS unity gain buffers, as shown in the circuit diagram of Figure 4.64(a). The ISFETs are buffered by high-impedance amplifiers. The configuration gives a linear de-

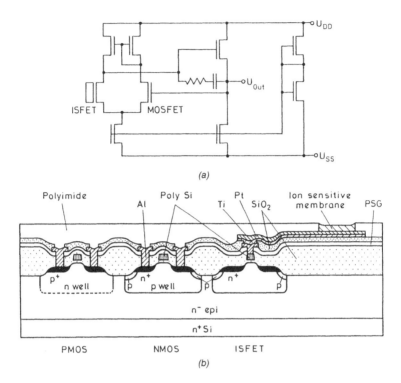

FIGURE 4.64. (a) Circuit diagram and (b) cross-sectional structure of the integrated ISFET (Redrawn with permission from the figure of Tsukada et al. [161], ©1991 IEEE.)

pendence between the ion-selective membrane potential and the output voltage. The schematic structure of the cross section is shown in Figure 4.64(b). In the chemically active area, the ion-selective membranes were formed on Pt/Ti/Al multilayer electrodes that are connected to NMOS gates. The platinum film was used as a protection layer against ion migration and hydration, and Ti was used as an adhesion layer. They were patterned with a lift-off process. The SiO_2 and phospho-silicate-glass were used as passivation films. Finally, a polyimide layer was deposited and patterned photolithographically to form a well. The ion-selective membranes consisted of PVC, ionophore, plasticizer, and additives. Almost Nernstian sensitivities have been detected with selectivity factors less than -2 (with a few exceptions).

Multilayer ceramic technology (see Section 1.2.2) has also been successfully applied to build up multisensors using different ISEs with various polymer membranes [162]. The schematic cross section of one elemental ISE is shown in Figure 4.65. The multilayer interconnection system with the electrode cavity was realized by low-temperature multilayer ceramic technology. After lamination and sintering, integrated thick-film elements were realized by screen-printing and firing cycles, followed by the mounting and soldering of surface-mounted components. The last step is the filling of the cavities with the ion-selective polymers. Each membranized cavity is created as an insulated electrode. The signal caused by the electromotive force is traced by the conductive vias and lines to the backside of the carrier that is hermetically separated from the solution by the ceramic sensor body.

4.7.4 Conductimetric Ion Sensors

Based on a planar interdigitated electrode structure coated with a polymer gel, a conductimetric sensor was reported by Scheppard et al. [163,164] to detect changes in conductivity of a polymer hydrogel layer as the water content of the gel responds to environmental changes, for instance, to pH. A pH-dependent swelling of the proper gel can be observed. The pH-sensitive hydrogel copolymer was prepared from HEMA and DMAEMA (see Section 3.3.2).

The substrate was borosilicate glass, and the gel was prepared by free-radical solution polymerization. The resistance related to the uncoated resistance value as a function of pH and gel thickness is shown in Figure 4.66. The thickness is normalized with respect to the spatial periodicity of the electrode array. In the vicinity of pH 7.7, the resistance change is about of 50% per pH unit.

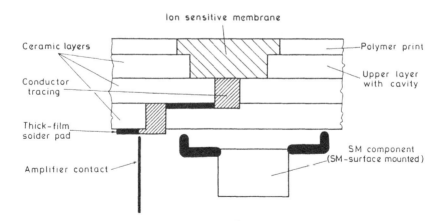

FIGURE 4.65. Structure of the multiple ISE using multilayer ceramic technology. (Redrawn with permission from the figure of Bechtold et al. [162].)

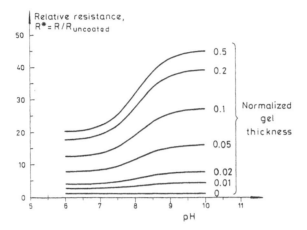

FIGURE 4.66. Characteristics of the hydrogel based conductimetric pH sensor. The resistance is related to the uncoated sensor resistance value, the gel thickness is normalized to the spatial periodicity of the electrode array. (Redrawn from the figure of Sheppard et al. [164], with the kind permission of Elsevier Sequoia, S. A., Lausanne, Switzerland, publisher of *Sensors and Actuators A* and *Sensors and Actuators B*.)

4.7.5 Fibre-Optic Ion Sensors

Optical ion sensors, mostly ion optrodes, are based upon ion concentration–dependent changes of the optical properties of thin ion-permeable films. A few typical optrode structures are shown in Figure 4.67 [165,166]. Most research has been carried out on ion optrodes for using them in the physiological concentration ranges; however, in recent years, the working range has been extended.

The fabrication technologies and applied materials show a lot of similarities to the ones of electrodes and ISFETs; therefore, the problems that are encountered are also similar. Ion-sensitive optrodes are based on colourimetric changes or fluoroescence quenching (see Section 3.8). The sensing materials are plasticized or elastomeric polymers containing ionophores and/or chromo-ionophores or fluorescent dyes. A lot of examples for materials were mentioned in Section 3.8. Commercially available chromo-ionophores with the suggested membrane composition and application fields are given in the literature [123].

The problems that are well known from the fabrication of electric sensing devices are as follows:

- Adhesion of the membrane: covalent binding should be built up between the fibre and membrane. Using silica or glass fibres, the preliminary silylation using organosilanes and/or polysiloxanes can help to solve the problem, as was demonstrated at ISFET gate oxides.

- The leaching of plasticizer, ionophore, and chromo-ionophore must also be prevented by binding them covalently into the polymer matrix.
- The selectivity can be improved using neutral ion carriers and supramolecular compounds in the membrane. The application of a multilayer membrane structure using intermediate sensing films and permselective membranes can also improve the selectivity.

The most important differences between optrode and electrode systems and fabrication problems, which can make the situation easier or more complicated, can be summarized in the following items:

- The adhesion problems are not so critical because the optical behaviour shows less sensitivity to the structure of interfaces; thus, the latter does not need to be very well defined. The mechanical bonding of the sensitive film can also be maintained by using a

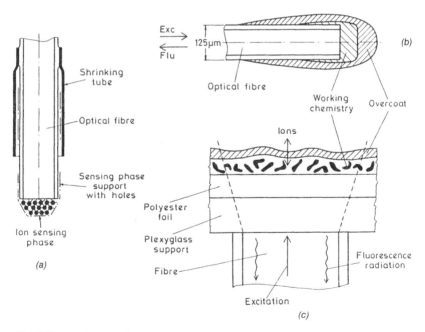

FIGURE 4.67. A few realization types of the optrode sensors. (Redrawn from References [165] and [166], with the kind permission of Elsevier Sequoia, S. A., Lausanne, Switzerland, publisher of *Sensors and Actuators A* and *Sensors and Actuators B*.)

second permselective membrane. Moreover, a lot of practical applications need the use of plastic fibres consisting of PMMA. In this case, the covalent binding of two polymers can be realized much easier than at the silica-polymer interface.

- The fundamental difference between optical techniques and potentiometric measurements is that potentiometric determinations depend on the activity (a) of the ions. Optical measurements are a function of the concentration (c), but not of the activity of the ion. For example, at pH optrodes, the absorption or fluorescence intensity are dependent on the ratio of acid and base forms of the indicator, but not on the activity of the hydrogen ion. The dissociation degree of the indicator dye is not only dependent on the activity, but on the ionic strength and on the temperature. Thus, the unknown ionic strength of the sample whose pH is to be determined affects the equilibrium constant and, hence, the pH calculated from the measured degree of dissociation of the indicator [166]. [pH is defined with the activity of hydrogen ions: $pH = -\lg a_{H^+} = -\lg (k_{H^+}c_{H^+})$, see also Equation (2.17).]

The first examples will be mentioned from the field of pH optrodes. The first absorbance-based fibre-optic pH sensor was specifically developed for the physiological pH range (pH around 7.35) [167]. It is based on monitoring the absorbance of phenol red in polyacrylamide, contained within a proton permeable envelope. Changes in light absorption of phenol red with pH are measured by illuminating the dye through a single optical fibre and sensing the backscattered light through a second fibre that leads to a filter and a light detector. Phenol red was covalently immobilized on the polymer microspheres. Polystyrene microspheres were added to provide more effective light scattering. When phenol red is illuminated by white light, green light is absorbed by the base form of the dye to a degree dependent upon pH, while red light at any wavelength longer than 650 nm is not absorbed in relation to pH and can be used as an optical reference signal [168].

Recently, Hao et al. [169] have also developed a pH sensor based on the indicator phenol red. An interesting new configuration (see Figure 4.68) has been developed for transmission absorbance optrodes. The indicator was immobilized in agarose gel, in a typical polysaccharide material. The absorption spectra of the phenol red is shown in Figure 4.69. The sensor performs reasonably well over the pH range 6.0–8.0 in terms of the sensitivity and reproducibility. The sensor shows a response time of 2–5 min and can operate for up to seven days. The major limiting factor is the leakage of phenol red.

Recent developments of optical pH sensors for medical applications have

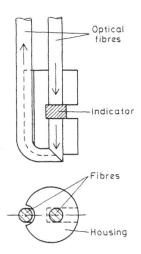

FIGURE 4.68. Structure of the transmission mode two fibre optrode. (Redrawn from the figure of Hao et al. [169], with the kind permission of Elsevier Sequoia, S. A., Lausanne, Switzerland, publisher of *Sensors and Actuators A* and *Sensors and Actuators B*.)

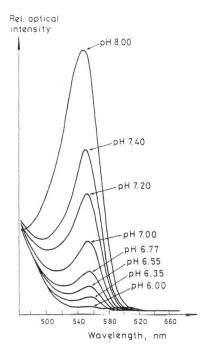

FIGURE 4.69. UV-absorption spectra of the phenol red as a function of pH value. (Redrawn from the figure of Hao et al. [169], with the kind permission of Elsevier Sequoia, S. A., Lausanne, Switzerland, publisher of *Sensors and Actuators A* and *Sensors and Actuators B*.)

been described by Boisde et al. [170]. Two domains were selected: monitoring in blood (pH around 7.35) and determination for gastric pH (1–8). The latter type may also have many applications in fields other than medical areas.

For the first use, the sensor consists of a poly(N-vinylimide-azole) (PVI) grafted onto an optical fibre with immobilized indicators. Measurements near to the transition activity (pK) in a range of 0.3 pH unit can be realized with a low-cost sensor. For the second application, a co-immobilization of several (four) indicator types was applied to get a close linear absorbance pH characteristic in the range from 3 to 10.

Most investigations on fluorescence pH optrodes were performed with HPTS (see Section 3.8) [171]. The excitation-emission spectra are shown in Figure 4.70. The deprotonated form of HPTS bound to cellulose can be excited at 475 nm to give a fluorescence emission at 530 nm. The intensity of the latter is a function of the pH. The acid form of the dye can be excited at 410 nm to give fluorescence from the same band as the base form. When exciting at the isoemissive wavelength of 428 nm, a pH-independent fluorescence signal is obtained [166].

In the last few years great efforts have been exerted to develop various type ion-selective optrodes. They seem to give almost less difficulties than the development of ISFETs.

Table 4.10 gives a summary about the most important parameters of a few recent ion-selective optrode types. The problems that are encountered are the same as at pH optrodes. The most important physical-chemical processes and materials were demonstrated in Section 3.8.

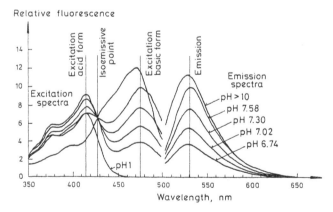

FIGURE 4.70. pH-dependent excitation and corrected emission spectra of the fluorescent optrode with HPTS dye. (Redrawn from the figure of Leiner and Hartmann [166], with the kind permission of Elsevier Sequoia, S. A., Lausanne, Switzerland, publisher of *Sensors and Actuators A* and *Sensors and Actuators B*.)

TABLE 4.10. *Characteristics of Recent Ion Optrodes.*

Detection Type	Polymer	Dye	Ion Type	Detection Limit (mol/l)	Range (mol/l)
Colourimetric optrode	Placticized PVC [165]	Takagi-reagent	K^+	10^{-5}	10^{-2}
	Placticized PVC [172]	Nitrate-ionophore + bromocresol purple	NO_3^-	2×10^{-5}	4×10^{-3}
	Teflon® [173]	Bromophenol-red	NH_4^+	3×10^{-6}	10^{-2}
	Epoxy [174]	TPPS*	$Cd^{2+}, Pb^{2+}, Hg^{2+}$	3×10^{-6}	10^{-3}
Fluorescence optrode [175]	Plasticized PVC	Hexadecyl-acridine orange	K^+	10^{-4}	10^{-1}

*5,10,15,20-tetra (*p*-sulphonatophenyl) porphyrin.

373

The most advanced approach is the application of supramolecular chromo-ionophores in the optrodes [176]. Ionizable chromogenic reagents, such as crowned calix[4]arenes and hemispherands with azophenol dyes incorporated into polystyrene resin, show promise as cation- and amine-selective dyes in optical-fibre sensors. The most selective calix[4]arene shows a 300:1 extraction ratio for K^+ and Na^+ and does not extract Ca^{2+} or Mg^{2+}. An extension of this method for sensing cations suggests that similar azophenol dyes may be used for sensing amines and that selectivity will be a function of lipophilicity and base strength of the amine in addition to its fit within the supramolecular cavity of the reagent. It is possible to develop reagents with high cation selectivity, and rather simple molecular-mechanics calculations may be a rough guide for reagent design.

4.8 SENSORS IN MEDICINE AND BIOLOGY

In this section, a survey will be given about those sensor devices that measure biological parameters of living things. Only one part of them belongs to biosensors. *Biosensor* is generally defined as a sensor that uses a living component or product of a living thing for measurement or indication.

Several sensor types have already been described in the previous sections, which may also have applications in the medical area, including blood pressure sensors; pulse and breathing wave sensors; ultrasonic and infrared sensor arrays; and O_2, pH, and multicomponent ion sensors. However, there are still a few sensor types that can measure inorganic components in living things and therefore have a very special design and cannot be used in other application areas. These sensor types will be discussed in the first part of this section.

The second part deals with the sensors for organic components produced by living things. These sensors mainly contain compounds from living things (enzymes or antibodies); thus, they are real biosensors. There are also sensors that use compounds from living things for an indication of inorganic components. However, they are already out of the scope of this book.

4.8.1 Sensors for Inorganic Compounds

This group of sensors consists mainly of those that are used for blood gas and ion monitoring. Over the past twenty-five years, blood gas, pH, and ion concentration analysis has become one of the most common diagnostic procedures in hospitals. There are a lot of cases when the simple in vitro analysis is not sufficient, so continuous in vivo monitoring is necessary, which requires the application of stable and reliable sensors.

The real-time continuous blood gas monitoring is particularly important for preterm babies who are vulnerable to respiratory illness because of immaturity of the lung or inadequately developed mechanisms for the control of breathing. If the arterial partial pressure of oxygen (p_{O_2}) goes to a low or to a high level for only a short period of time, brain or eye damage, respectively, may follow, and in unfortunate situations, even blindness or death can occur.

The measurement of arterial p_{CO_2} is also important. In the newborn, for example, there is evidence to suggest that rupture of blood vessels in the brain can happen as a consequence of fluctuations in cerebral blood flow caused partly by changes in p_{CO_2}.

Invasive and noninvasive techniques have become available in the last few years for continuous measurement of arterial gas levels in patients. Of these, the transcutaneous technique and the technique of optrodes are now used extensively in medicine [177,178]. Although intravascular monitoring remains an important part of intensive care, catheterisation of arteries is not without risk, especially with newborn babies. This is one of the reasons that have led to the advent and rapid growth of noninvasive blood gas monitoring.

For blood gas, pH, and ion concentration monitoring, the following sensor types can be used:

- miniature membrane-based amperometric sensors
- potentiometric microelectrodes and ISFETs
- fluorescence and colourimetric optrodes

The most important special requirements can be summarized as follows:

- Accurate measurement is needed generally in a narrow detection range.
- Temperature compensation is often necessary.
- Miniaturization, especially at intravascular sensors, is an unavoidable requirement.
- There is a great demand on continuous monitoring of several parameters in the same time; thus, multifunction or multicomponent sensors are needed. In this respect, medical sensors are unthinkable without integration and microsensorics in the future. A few examples will illustrate the present possibilities.

Transcutaneous blood gas sensors are generally membrane-modified, temperature-controlled amperometric sensors: Clark-type operation is used for p_{O_2}, and the Stow-Severinghaus technique for p_{CO_2} measurements (see Section 4.6.1) [82,83].

The p_{O_2} at the surface of human skin at normal temperature is near zero. However, if skin temperature is increased, the p_{O_2} measured at the surface can approach the arterial value quite closely. The estimation of arterial p_{O_2} by this method called transcutaneous technique relies on the correct choice of skin temperature [179]. In newborn infants, a sensor temperature of $43.5 \pm 0.5\,°C$ has been found to be an optimum for reliable monitoring of arterial p_{O_2}. The transcutaneous p_{CO_2} measurement relies on a similar technique.

The recent trend is to use a single combined sensor for both gases. The sensor was reported by Parker et al. [178]. The cross section of such a combined transcutaneous sensor is shown in Figure 4.71. It uses the same electrolyte for both the p_{O_2} and p_{CO_2} measurement and has a common reference electrode and a polymer diffusion membrane. The cathode is a 25-μm diameter platinum wire, and the pH-sensitive glass electrode for CO_2 measurements using the Stow-Severinghaus method ($CO_2 + H_2O \leftrightarrow H^+ + HCO_3^-$) is 2–3 mm in diameter. The schematic cross section also shows the heater coil and the temperature-controlling thermistor. The diffusion membrane is held in position by a Teflon® ring, and the outer annulus is for attachment to the skin by means of a double-sided adhesive disc.

The most important problem of the conventional transcutaneous sensors is that they have very complicated structure: a miniaturized heater coil, a temperature sensor for heating control, enclosed electrolyte, a Pt-cathode, an Ag/AgCl anode, and a polymer membrane must be integrated into a small size device. They are generally expensive because of the difficult mounting and the application of bulk precious metal electrodes.

From a technology point of view, the advantages that can be obtained through microfabrication of electrochemical biosensors include reduced size, reduced sample volume, reduced cost, and fast response. In addition,

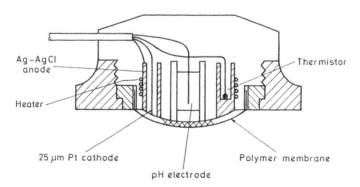

FIGURE 4.71. Schematic diagram of a combined O_2/CO_2 transcutaneous sensor. Redrawn from the figure of Parker [178], with permission from IOP Publishing Limited.)

it should be possible to produce a highly uniform and well-defined micro-structure of the electrode surface area using microelectronic technology. However, a lot of special realization problems have to be solved, and complications often arise as the size of the electrodes and interelectrode distance are reduced (see Section 4.6.2).

A new type transcutaneous p_{O_2} sensor structure has been presented by Harsányi, Peteri, and Deák [180], which is based on the multilayer ceramic technology that is used in the fabrication of high-density electronic interconnection systems. The electrodes are made by thick-film technology using small amounts of precious metals. The heating element is an integrated thick-film resistor and a *p-n* junction is used for temperature sensing.

The structure of the integrated transcutaneous blood oxygen monitoring sensor is shown in Figure 4.72. The disc-shaped ceramic sensor body has a groove at the edge for packaging purposes and a deep cavity in its centre for the temperature sensor chip. The sensor chip should be as close to the skin surface as possible for the necessary accuracy of temperature control. A diode chip is used for temperature-sensing purposes. The horseshoe-shaped heating resistor is screen-printed and fired at 850°C on the top of the ceramic body using common thick-film processing technique. It consists of several segments to give more freedom for heating.

The sensing electrode was fabricated with thick-film gold paste fired also at 850°C for an hour. The Ag film for the Ag/AgCl reference electrode was also formed by a thick-film technique before being electrochemically chlorided in a 0.1 M HCl solution. The total surface area of this Ag/AgCl electrode is 0.7 cm², while the area of the Au electrode is 1.77 mm².

The ceramic body was moulded into a plastic package. The PTFE membrane with a thickness of 10 μm was fastened with a mounting ring to the surface of the sensor, as shown in the cross section in Figure 4.72. A conventional KCl solution with buffering agent was used as an electrolyte. A thin cellophane foil of disc shape was placed between the electrodes and the membrane as a spacer.

The most important problems that arise in connection with the long-term operation of this gas sensor are as follows:

- the limited lifetime of the electrolyte
- the corrosion of the Ag/AgCl electrode

The lifetime of the electrolyte is determined partly by the water evaporation through the membrane and by water consumption of the electrochemical processes. Therefore, the periodical refreshing of the electrolyte is necessary. The length of this period depends on the amount of electrolyte, which is especially critical in the case of miniature sensors. It is, therefore, necessary to minimize the electrode surfaces, and the application of an appropriate spacer may also be important.

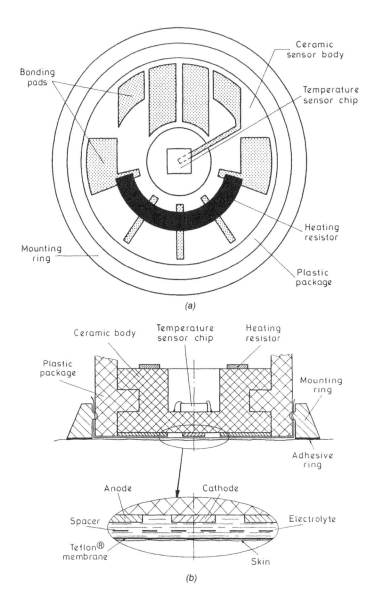

FIGURE 4.72. Structure of the ceramic transcutaneous blood O_2 sensor: (a) top view, (b) cross section.

The possible chemical cross-talk between electrodes in microfabricated amperometric gas sensors has already been studied [181]. It is well known that the reduction of dissolved oxygen molecules at the cathode can also proceed via a two-electron pathway with the formation of hydrogen peroxide. That can distort the sensor signal and lead to electrode consumption at the reference electrode. The situation may be improved when the reference electrode is sufficiently separated from the cathode to reduce the pick-up of the hydrogen peroxide intermediate. The application of multilayer wiring helps to make a better realization.

Intravascular measurement of the biochemicals in blood can often provide the most direct and rapid indication of chemical condition. These sensors must be constructed mostly in the form of flexible catheters or cannulae, with dimensions appropriate for insertion into either a peripheral or central vessel. Each of these designs employ a particular approach to solve the problem of attaching the sensor and, especially, the membrane of an electrode surface safely and securely to the catheter tip. Dip coating seems to be the best candidate to achieve smooth surfaces, which are important to get haemocompatibility. It also allows membrane thickness to be changed simply by varying the number of coats and the solution viscosity. This approach is possible with a number of relevant polymers, such as polyurethanes, polystyrene, and PVC [177].

Recently, two different types of miniaturized carbon dioxide sensors have been developed for in vitro and in vivo applications [182]. The first type is based on the principle of the Severinghaus electrode, with the exception that the glass pH electrode is replaced by a solid-state pH electrode, whereby an electropolymerized conducting polymer (polypyrrole) is employed. The membrane system that is necessary for the realization of this type of sensor is performed with a hydrogel layer and a hydrophobic gas-permeable membrane. The second type of carbon dioxide sensor is an amperometric one, which makes use of a three-electrode configuration. The concentration of carbon dioxide, which is absorbed cathodically, is measured by the oxidation (anodic stripping) of a chemisorbed product on a polarizable metallic electrode (working electrode). The electrode arrangement is also covered by a hydrogel layer and a hydrophobic gas-permeable membrane. Both types are realized by means of thin-film technology. The use of this technology makes it possible to scale down and to integrate this type with other sensors.

Electrochemical techniques have been the most popular to date for invasive monitoring of ions, although at present there is also considerable interest in optical methods. Simple ion sensors are described in Section 4.7.

An integrated chemical sensor for blood analysis with multiple ion and gas sensors, composed of four ISFETs (for pH, Na^+, K^+, and Cl^- ions, respectively) and two gas sensors (for p_{O_2} and p_{CO_2}) on a 4 mm $\times$ 4 mm chip, was realized by Tsukada et al. [183] using semiconductor processing. The cross-sectional structure of the individual elements in the integrated sensor is shown in Figure 4.73. The ISFETs are based on an Si_3N_4-gate structure and use plasticized PVC membranes with various ionophores. The O_2 sensor is a miniaturized Clark-type sensor, consisting of a sputtered Pt cathode and a chemically chlorided Ag/AgCl reference electrode pat-

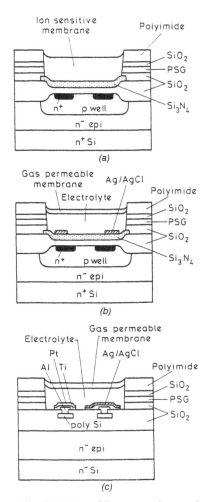

FIGURE 4.73. Cross-sectional structure of the sensor elements in an integrated chemical sensor: (a) ISFET, (b) CO_2 sensor, and (c) O_2 sensor. (Redrawn from the figure of Tsukada et al. [183], with the kind permission of Elsevier Sequoia, S. A., Lausanne, Switzerland, publisher of *Sensors and Actuators A* and *Sensors and Actuators B*.)

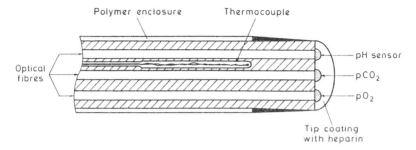

FIGURE 4.74. Triple-fibre sensor for blood monitoring. (Redrawn after the figure of Wolfbeis [185] from *Int. Journal of Optoel.*, 1991, Vol. 6, No. 5, with permission from Taylor & Francis Ltd.)

terned by a lift-off process. An electrolyte gel consisting of PVA [poly(vinyl alcohol)] was placed into the micropool and coated with silicone rubber. The CO_2 sensor is a miniaturized Severinghaus-type sensor using an Si_3N_4-gate pH ISFET instead of the glass electrode. Here, the electrolyte consists of PVA, NaCl, and $NaHCO_3$.

A similar pH/O_2 sensor was developed by Schelter et al. [184], combining a conventional pH-ISFET with a three-electrode Clark-type O_2 sensor. The application of a PHEMA hydrophilic membrane makes it possible to avoid the use of internal electrolyte and hydrophobic gas-permeable membranes. The PHEMA membrane is permeable to oxygen and small ions but protects the working electrode from proteins that adhere to and poison the surface; therefore, better parameters can be achieved.

The sizes of the latter two types are still too large for direct intravascular application; therefore, the application of a sample-taking system with a catheter and a sensor flow-through cell is necessary for continuous monitoring [184].

Blood gas and ion monitoring may also be carried out with optrodes. Oxygen in blood can be measured with an optrode using perylene dibutyrate fluorescent dye (see Section 3.8). The fibre-optic approach is advantageous in intravascular measurements because it offers a way for sensor miniaturization so that even a fibre bundle can be introduced into a radial artery of a patient through a catheter. One example is shown in Figure 4.74 [185]. The working chemistries of the three sensors are placed at the tip of the three fibres, while a thermocouple gives a direct reading of the sensor temperature at the tip. The issue of thrombogenicity has been addressed by designing a smooth tip surface and covalently immobilizing heparin on the surface.

The ambulatory monitoring of pH in the upper gastrointestinal tract is

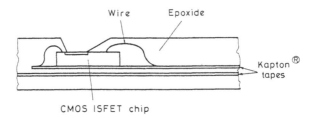

FIGURE 4.75. Structure of the gastric pH catheter. (Redrawn from the figure of Thybaud et al. [186], with the kind permission of Elsevier Sequoia, S. A., Lausanne, Switzerland, publisher of *Sensors and Actuators A* and *Sensors and Actuators B*.)

also increasingly used for diagnostic and research purposes. That field means another application of inorganic component sensing beside the blood monitoring. Such measurements are currently performed with integrated glass electrodes. A severe drawback of this type may be encountered if several sensors must be installed in the gastro-oesophegeal tract.

A new construction was reported by Thybaud et al. [186] using ISFETs, which permits the realization of multiple mounting on a single catheter. The structure is shown in Figure 4.75. It consists of several types of polymers; however, the ISFET itself is not a polymer-based one. The gastric probe is composed of a supple PVC catheter in which the ISFETs are mounted on a floppy substrate of Kapton®. The passivating film is epoxide. The reference electrode is also installed at the catheter tip.

4.8.2 Sensors for Organic Biochemical Compounds

Figure 4.76 shows the general construction of biosensors that can be used for detecting organic biochemical compounds. They consist of the receptor, which is generally a polymer bed with the entrapped or covalently immobilized receptor particles (enzyme molecules, antibodies of even bacterial cells), and the transducer, such as an ion-selective electrode, an FET, an optrode, or a resonator. When exposed to the analyte, the receptor particles respond selectively in the same way, and this response is converted by the

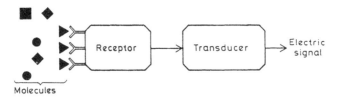

FIGURE 4.76. The general structure of biosensors.

transducer into a useful signal. The respond mechanisms can be categorized into groups of:

- direct interaction mechanism, when the response is a result of the interaction between the analyte and receptors
- indirect interaction when the analyte-receptor reaction leads to the formation of a third chemical component and the sensor responds to the presence of the latter one

As receptors, immobilized enzyme molecules are used in most cases. Enzymes are catalysts, substances that enable biochemical processes to proceed at temperatures compatible with life. Many enzymes offer a high-substrate specificity, meaning they will react with just one substrate out of many; this is very useful in sensors, where selective measurement of a given analyte is to be carried out. Enzyme-based biosensors are transducers with immobilized enzymatic receptors. This type of device has wide applicability and has been used in the determination of urea, glucose, amino acids, cholesterol, penicillin, etc. In principle, any compound can be determined by enzyme-based sensors.

The various sensors differ only

- in the transducer type (electrode, optrode, amperometric cell, ISFET, resonator, etc.)
- in the immobilized enzyme type (urease, glucose oxidase, cholesterol oxidase, penicillinase, etc.)
- in the immobilization technique

Since enzymes are well suited to their catalytic role, they do tend to be unstable structures, which can limit the sensor lifetime to days or even hours. Immobilization techniques can help extend this life, and a variety of methods have been developed. Immobilization means the conversion of an enzyme from a water-soluble, mobile state to a water-insoluble, immobile state. The two most important immobilization techniques are the

- physical bonding, adhesion, or entrapment of the enzyme to or in an inert polymer film or gel [plasticized PVC, polyacrylamide (PAA), etc.]
- chemical binding, covalent immobilization to the sensing surface, and cross-linking of the enzyme

The sensor properties are determined by the enzymes; however, polymers can have an important role in the immobilization and in the charge transfer. Because enzymes can operate at biological temperatures, it makes sense to utilize them in medical applications. The most widespread use of enzyme sensors is probably in the field of urea- or glucose-sensing in humans.

The determination of urea in blood and urine is important in clinical analysis. Urea is catalysed by urease according to the reaction [187]:

$$\text{urea } [(NH_2)_2] + 2H_3O^+ \xrightarrow{\text{urease}} >> CO_2 + 2NH_4^+ + H_2O \tag{4.16}$$

In conventional urea sensors, pH electrodes, ammonium-selective or ammonia gas electrodes, and carbon dioxide sensors have been used to detect the changes indirectly, according to the reaction scheme in Equation (4.16). These principles were also adopted to the so-called urea microsensors in which an ISFET, a microelectrode, or an optrode is used as transducer. Table 4.11 gives a summary about a few recent results.

The response of urea sensors based on the measurement of pH change in an enzymatic membrane depends on the buffer capacity and the initial pH of the sample solution. Van der Schoot et al. [188] have presented an automatic system based on a urea-FET pH-REFET pair to control the compensating current, which is then direct measure for the substrate concentration.

The determination of glucose is important in the chemical field, and the development of bioelectrochemical devices considerably helps routine laboratory work. The development of miniaturized and implantable enzyme sensors employing microtransducers is required in the medical field.

Recently, several amperometric glucose microsensors, which use wire electrodes, have been reported. Although a wire-based glucose microsensor can be made small, its fabrication is not suitable for mass production because each wire must be individually coated by a glucose-oxi-

TABLE 4.11. Summary about Urea Sensors.

Transducer	Polymer	Urease Immobilization Technique	Detection Range (mol/l)
NH_4^+-ISFET [187]	plasticized PVC + ionophore	cross-linking	10^{-6}–10^{-1}
pH-ISFET [188] (+pH-REFET)	PAA (polyacrylamide)	entrapment	10^{-3}
pH-ISE [189]	aminated PVC	cross-linking	10^{-4}–10^{-1}
pH-ISE [190]	carboxylated PVC + H^+-ionophore	cross-linking	10^{-4}–10^{-1}
NH_3 sensitive Pt/n-GaАs Schottky diode [191]	PEI (polyetherimide)	adsorption	4×10^{-4}
pH-optrode [192]	PAA + phenol red	entrapment	4×10^{-5}–25×10^{-5}

dase–immobilized membrane. On the other hand, planar microsensors are suitable for mass production because the membrane can be deposited on all the individual devices simultaneously.

Conventionally, a hydrogen peroxide sensor could be used for the determination of glucose because glucose oxidase (GOD) catalyses glucose oxidation according to the following reaction scheme:

$$Glucose + O_2 + H_2O \xrightarrow{GOD} >> Gluconolactone + H_2O_2$$
(4.17a)

The produced H_2O_2 can be measured electrochemically:

$$H_2O_2 \rightarrow 2e^- + 2H^+ + O_2$$
(4.17b)

It is common to detect the change of concentration of either oxygen or hydrogen peroxide electrochemically for the determination of glucose concentration. There are, however, some disadvantages of these types of glucose sensors; variations of dissolved oxygen may cause fluctuations on the electrode response, and the dynamic range of detection can be decreased by the lack of dissolved oxygen. Human blood glucose concentration is normally about 4–5.5 mM, but it may reach the value of 30 mM in diabetes mellitus. An ordinary dissolved oxygen amount in blood is estimated at only 2.2 mM, which is insufficient to oxidize the total blood glucose in this case.

Since the enzyme itself becomes reduced in the first stage of the enzymic reaction, an alternative approach is to transfer electrons with mediators other than oxygen. Effectively, the mediator acts as an electron acceptor. In the oxidized form, it takes up electrons from the enzyme and then, in the reduced state, transports them to the electrode where it becomes re-oxidized. Many kinds of electron mediators have been reported so far. Ferrocene derivatives were found to be excellent mediators between the reduced form of GOD and the electrode. Immobilization of GOD and a mediator on electrodes is the major subject in the development of this type microglucose sensors. Their disadvantage may be the poor sensitivity at low-metabolite concentrations due to simultaneous electron transfer from the active centre of the enzyme to water-soluble oxygen.

Mediatorless electrodes are based on the direct electron transfer combining oxidase and peroxidase reactions. In that case, oxidase/peroxidase enzymes should be immobilized; thus, bienzyme-based sensors are prepared. Since this book is devoted to polymers, enzymatic reactions will not be discussed in more detail.

As was mentioned, polymers have an important role in the immobilization of enzymes and mediators and also in the charge transfer process. Conventional solutions apply the entrapment of GOD and other components in the polymer matrix.

Xie and Ren have used cellulose acetate for the entrapment of GOD in a thermal calorimetric biosensor in which the reaction heat of the catalysed chemical process is detected by a thick-film resistor bridge [193]. The polymer was bound to the substrate by an intermediate silastic/cellulose acetate film.

Hamppet et al. used a PVA/PE copolymer for the immobilization of GOD in a potentiometric sensor [194]. Haemmerli et al. have applied PVPy dip coating [195]. The enzyme was dissolved in the PVPy precursor.

A thin-film amperometric sensor was presented by Koudelka et al. [196]. GOD was immobilized using glutaraldehyde as a cross-linking agent and bovine serum albumin as a carrier protein. This enzymatic membrane was coated by an outer glucose diffusion limiting polyurethane membrane by dip coating.

Urban et al. [197] have produced Pt-electrode–based glucose sensors covered with a polymer multilayer structure. GOD was entrapped in a photo-patterned PHEMA hydrogel membrane. To prevent electrochemical interferences and fouling of the Pt-electrode, an electropolymerized semi-permeable membrane was deposited, at first, to the electrode surface. A second enzyme membrane layer containing the enzyme catalase was immobilized above the oxidase enzymes. To separate the two enzyme layers and to introduce an additional diffusion barrier, a third PHEMA membrane without enzyme was placed between the oxidase and catalase enzyme membranes.

Another approach is the covalent binding of the enzyme onto the electrode surface. Enzymes can easily be bound to graphite surfaces and even to Pt surfaces after activation with silanes. The covalently coupled enzyme films can be coated by a polymer membrane to stabilize the enzymatic film. Nafion® can be used [198], which acts as a good charge-transferring membrane.

A microporous enzyme membrane attached to a platinum electrode surface was prepared by Lee et al. [199] to build up a glucose sensor. The enzyme electrode was prepared in three steps: casting the block copolymer prepared from isoprene and (4-vinylphenyl) dimethyl-2-propoxysilane into a thin film on a platinum electrode. Cross-linking occurs only in the domains of the latter one, while the former remains intact and can be etched easily out of the film, leaving a microporous polymer film on the Pt surface. GOD was then immobilized in the microporous membrane.

The application of electrochemically deposited electroconductive conjugated polymers (see Sections 1.4.4 and 3.9) seems to be the best solution. First, polyaniline [200], but, recently, mostly polypyrrole [201–203] (PPy), were used for that purpose. Polypyrrole has found various applications as an immobilization matrix for enzymes in the construction of amperometric of potentiometric enzyme electrodes. The enzyme can be immobilized on the electrode surface either by covalent binding to a functionalized polypyrrole layer by entrapment through electrochemical polymerization of pyrrole from aqueous solutions in the presence of the enzyme. The latter method is a simple one-step procedure, and PPy/GOD layers can easily be formed on platinum or other conductive surfaces by this technique. Although a lot of work has been done on PPy/enzyme electrodes, the mechanism by which the enzyme is immobilized still remained uncertain [203].

Two strategies for the functionalization of different electrode materials and the immobilization of glucose oxidase to these surfaces are shown in Figure 4.77 [201]. The electrochemical oxidation of graphite [see Figure 4.77(a)] leads to the formation of carboxylic groups on the surface, which can be activated by water-soluble carbodiimides, and the reaction of the derivatives with the e-amino groups of lysine residues of glucose oxidase results in a covalent binding of the enzyme.

Figure 4.77(b) shows the functionalization of PPy films. After having been nitrated, the nitro derivative can be reduced electrochemically. Immobilization of the enzyme can be carried out with these surfaces by activating the carboxylic side chains of the protein with carbodiimide [201].

Katsube et al. [202] have developed a SOS-ENFET–based (see Section 2.2.1) glucose sensor with a thermophilic enzyme, glucokinase. The structure is shown in Figure 4.78. The substrate used was sapphire, which provided an anticorrosive system in the electrolyte solution. Iridium oxide film was applied as an electrically conductive gate electrode in an extended gate structure. A thermophilic enzyme, glucokinase, was immobilized in a PPy matrix, which was deposited on the FET gate electrode by electrochemical polymerization. The characteristic of the sensor is also shown in the figure.

Theoretically, similar techniques can be applied not only for the preparation of glucose sensors, but for other enzyme sensors as well. A bienzyme Pt-based electrode for the detection of total cholesterol by incorporating cholesterol esterase and cholesterol oxidase in PPy films was reported by Yon Hin and Lowe [204], for example.

Immunosensors can also be built up based on the immobilization of antibodies. The basis of an ImmunoFET is the covalent attachment of antibodies or antigens to an inert hydrophobic layer at the gate of a FET. This measures the change in charge resulting from the formation of antibody/antigen complex. Schasfoort et al. [205] reported an Immunofet for the

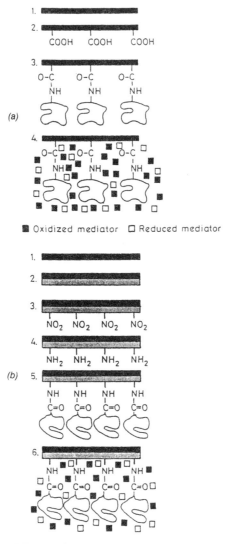

FIGURE 4.77. Immobilization of glucose oxidase to functionalized electrode surfaces: (a) oxidized graphite electrode (1—graphite electrode, 2—electrochemical oxidation, 3—activation and immobilization of the enzyme, 4—adsorption of an appropriate mediator); (b) functionalized PPy electrode (1—Pt or graphite electrode, 2—electropolymerization of pyrrole, 3—nitration of the conductive polypyrrole film, 4—electrochemical reduction, 5—activation and enzyme immobilization, 6—adsorption of an appropriate mediator). (Redrawn from the figure of Schuhmann et al. [201], with the kind permission of Elsevier Sequoia, S. A., Lausanne, Switzerland, publisher of *Sensors and Actuators A* and *Sensors and Actuators B*.)

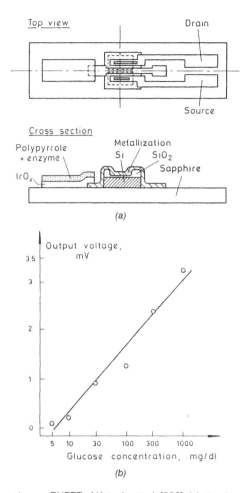

FIGURE 4.78. The glucose ENFET of Katsube et al. [202]: (a) structure and (b) response. (Redrawn with the kind permission of Elsevier Sequoia, S. A., Lausanne, Switzerland, publisher of *Sensors and Actuators A* and *Sensors and Actuators B*.)

competitive immunobiological detection of progesterone. A polystyrene-agarose membrane was deposited on the gate of an ISFET, followed by incubation with the monoclonal antibody antiprogesterone.

Resonator-type gravimetric sensors are also highly sensitive for the adsorption of biological compounds. A Love-plate device, based on surface skimming bulk wave (see Section 2.3.3) propagating in a piezoelectric substrate coated with a polymer (PMMA) layer, was used by Gizeli et al. [206]. Investigation of protein interaction with the PMMA film was performed by following the adsorption of human immunoglobulin-G (IgG). The deposition of a protein multilayer, consisting of IgG/anti-IgG and protein A, was detected by the sensor.

The reverse process was followed with the ImmunoFET produced by Colapicchioni et al. [207]. At first, protein A was chemically immobilized on the gate surface by using a layer of polyaminosiloxane cross-linked with an agent such as glutaraldehyde. The antigen determinations were carried out using an immunoenzymatic assay, as shown in Figure 4.79. After having adsorbed an excess of antibody (Anti-hIgG) on the protein A, the bound antigen was detected by means of a second antibody conjugated with an enzyme having the function of marker. The amount of antibody-enzyme conjugate on the device, and consequently the electrical response obtained in the presence of glucose, was therefore directly proportional to the antigen concentration of the sample. The whole film can be removed in an acidic medium, and the deposition may start again.

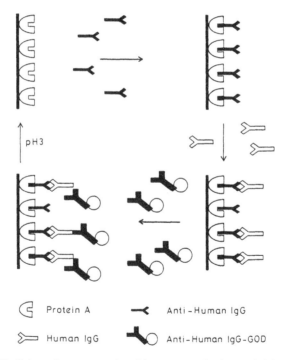

FIGURE 4.79. Schematic representation of the sequence for the sandwich array of Human IgG. (Redrawn from the figure of Colapicchioni et al. [207], with the kind permission of Elsevier Sequoia, S. A., Lausanne, Switzerland, publisher of *Sensors and Actuators A* and *Sensors and Actuators B.*)

Fibre-optic glucose sensors are based on the same enzymatically catalysed chemical process as shown in Equation (4.17). The difference is that they are based not on the measurement of H_2O_2, but on either one of the O_2 or pH. In a typical glucose optrode, an oxygen-sensitive material is covered with a membrane on which glucose oxidase is immobilized. The oxygen optrode acts as the transducer. It was concluded from the experiments that oxygen transducer-based optrodes are superior to pH sensor-type transducers [185].

Papkovsky et al. [208] have developed a luminescent-type glucose sensor. It is based on the oxygen-sensing mechanism of luminescent Pt-octaethylprophine-polystyrene membrane. GOD was immobilized in the latter.

4.9 OTHER APPLICATIONS

There are a few application possibilities of polymers in sensors, which could not be included in the former sections. Sometimes, these are very special and unique devices; however, they demonstrate the great possibilities of polymers in that field. A few examples are

- liquid component sensors
- very special piezoelectric and pyroelectric PVDF sensors
- multichannel chemical taste sensors and electric noses

Actuators are generally discussed together with sensors, although they have a very different function: an electrical signal must be converted into another quantity, which also means an interference in a process flow.

Polymers can be utilized in actuators in the following fields:

- piezoelectric actuators
- optical actuators such as light deflectors
- micromechanical structures to prevent corrosion and for the improvement of their friction properties

4.9.1 Liquid Component Sensors

Sensors for chemical quantities have been discussed in the previous sections categorized in the following groups:

- gas sensors
- sensors for ionic compounds in liquids
- sensors for the detection of gas or biochemical compounds in liquids

Chemical sensors are needed sometimes for detecting inorganic or organic nonionic compounds in liquids, which were not included in the former groups. A few examples will be given in this section.

PHEMA hydrogel membranes were used in amperometric sensors for detecting free chlorine or H_2O_2 in water [209,210]. The membrane has the advantage that it can be photolithographically polymerized and patterned, enabling the fabrication of the complete cell with IC-compatible methods. The basic cell consists of a three-electrode electrochemical cell realized on a silicon wafer. The working and counter-electrodes consist of a Pt layer on top of a Ti adhesion layer, and the reference electrode is a chemically chlorided Ag film. The PHEMA film was applied after a silanization/oxidation process to get an optimum adhesion to both the SiO_2 and Pt surfaces.

In the free-chlorine sensor the following electrode reactions take place [209]:

Cathode:

$$HClO + 2e^- \rightarrow OH^- + Cl^- \qquad (4.18a)$$

Anode:

$$2H_2O \rightarrow O_2 + 4H^+ + 4e^- \qquad (4.18b)$$

The optimum operational polarization voltage for the detection of free chlorine was found to be $+50$ mV versus the reference electrode. Sensors with membrane thicknesses of 10 and 50 μm were found to give approximately linear characteristics between 0.1 and 5 mg/1 free-chlorine concentrations with sensitivities of 2 and 0.4 nA/(mg/1), respectively. The same structure was used to detect H_2O_2 with a 32 μm thick membrane [210]. A linear response with a sensitivity of 12.7 nA/mM was found over the concentration range of 10^{-5}–10^{-3} M.

The potentiometric response of PVC/dicyclohexyl-18-crown-6 membrane-based ion sensors to nonionic alcohols has been measured by Anzai and Liu [211]. The electrochemical cell was composed of Ag/AgCl | 10 mM NaCl or KCl | PVC/crown ether membrane | sample solution | reference Ag/AgCl electrode. The electrode potential was changed by the addition of alcohols in NaCl of KCl solution. The electrodes were found to be sensitive to lipophilic alcohols such as 1-hexanol, 1-heptanol, 1-octanol, and 1-nonanol at the millimolar level or lower with a response of 5–10 mV, whereas low-lipophilic alcohols, including methanol, ethanol, ethylene glycol, etc., have no influence on the sensors' response.

These results suggest the specific adsorption or penetration of the lipophilic alcohols on or in the PVC membrane [210].

Utilizing electrically neutral analyte carriers, it is possible to fabricate optrodes for neutral species such as ethanol [212]. Plasticized PVC membrane (with an uncharged chromo-ionophore of *N*-acetyl-*N*-dodecyl-4-tri-fluor-acetylaniline)-based optrodes can show a decreasing absorbance upon addition of ethanol to the aqueous phase. An absorbance change from 0.35 to 0.1 on the 305-nm wavelength can be measured when changing the ethanol concentration from 0 to 32 vol%.

4.9.2 Special Purpose Sensors

The wide application possibilities of a PVDF polymer (see Sections 3.1.1, 3.1.3, and 4.2–4.4) involve a few very interesting and unique solutions. Piezoelectric PVDF copolymer films can be used as dust detectors in particle trajectory-capture cell systems [213]. The pyroelectric behaviour of PVDF can be used for material identification.

Gao and coworkes have proposed a thermal sensor that is capable of measuring thermal properties of objects grasped by a robot manipulator [214]. It can measure the heat flow transient from a heat source into an unknown object. A schematic diagram of the sensor is shown in Figure 4.80(a). It is composed of three layers. The top layer of thermally conductive ruber forms the outer covering. The bottom layer is an active heat source that produces a surface temperature of about 40°C; the exact value depends on the ambient temperature. The PVDF-sensing film inner layer records the temperature changes at the junction between the sensor and the material being sensed. Figure 4.80(b) shows the thermal response to different materials. A material can be successfully identified against a library of thermal response curves [214].

A similar photopyroelectric (P^{2}E) effect–based material determination method was applied by Miezkowski et al. [215]. The sensor used single-surface electroded PVDF films and a remote (i.e., noncontacting) metal pin positioned close to the unmetallized surface for detecting photothermal wave signal propagation through the sample generated by laser pulses. With the new capacitively coupled P^{2}E technique, the local thermal wave signal can be detected, and the thermal wave field is imaged by recording the signal at different scanning positions.

Chemically sensitive polymers can be used in multisensor systems for taste, flavour, and smell sensors. These are generally arrays of chemical sensors, each with a nonspecific output, coupled to a pattern recognition system that is based on a principal component analysis (see Section 4.6.8) and on the simulation of the neural processing of the human brain.

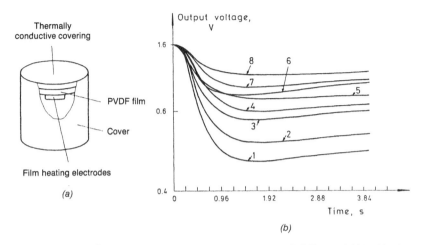

FIGURE 4.80. (a) Schematic diagram and (b) responses of the PVDF material identification thermal sensor (1—copper; 2—steel; 3—aluminium; 4—PCB metal surface; 5—rubber; 6—polyflon; 7—PCB back surface; 8—paper. (Redrawn from the figures of Gao et al. [214], with the kind permission of Elsevier Sequoia, S. A., Lausanne, Switzerland, publisher of *Sensors and Actuators A* and *Sensors and Actuators B*.)

A multichannel taste sensor was reported by Hayashi et al. [216]. It consists of a multichannel electrode with transducers composed of lipid membranes immobilized with a polymer, i.e., eight different kinds of lipid analogs mixed with PVC and plasticizer. This multichannel sensor responds to five basic taste substances in five different ways; the taste patterns composed of eight potential responses could be clearly distinguished from each other for the five basic taste substances.

Gardner et al. [217] have described the results about a multisensor system capable of discriminating the smells of different beers. The system consists of an array of twelve conducting polymer thin films electrodeposited onto a microelectrode structure. The twelve conducting polymers act as the active layers in conductimetric odour sensors and respond differentially to the headspaces of beers and lagers. The output from the polymer array can first be preprocessed using a variety of algorithms (e.g., fractional change in conductance, normalized relative response) and then classified using a statistical or neural predictive classifier.

Conductive polymers have also been used in the initial electronic nose reported by Hatfield et al. [218]. The sensor array, the BiCMOS-based application specified IC (ASIC) signal conditioner unit, and the A/D converter were integrated into a hybrid module on a 90 × 40 mm multilayer ceramic substrate.

4.9.3 Polymers in Actuators

Polymers can have active and passive roles in actuators and micro-mechanical components. Piezopolymers (see Section 3.1.1) can change their mechanical properties (stress, strain) when switching electric fields on them. Therefore, electrically controlled mechanical action can be realized. This is well known from the field of piezoelectric crystals; however, polymers still do not have such a wide range of application.

Ink jet printing, which utilizes piezoelectric polymers, resolves around two principles [215]:

(1) A piezoelectric bender deflects the ink jet when not in use but lets it pass when needed.
(2) A strip of the piezoelectric PVDF film, when suitably energized, acts as a pump in ink jet printers by changing its thickness. In some cases, the film pumps ink on command. In other cases, when powered at a specific frequency, the film pumps continuously.

Some exciting new applications have recently been proposed in the field of optics and light control. A PVDF film wrapped around a glass fibre can alter, by compression, the refraction index. The action modulates the light conducted in the fibre. Light modulation can also be achieved by piezoelectric bending of fibres. In transparent optical switches, piezoelectric benders deflect an optical fibre end within the focal plane of a lens and thereby redirect light beams [220]. This allows channel switching between parallel arrays.

Another application is the use of PVDF in diffraction gratings, enabling a light valve to be constructed. A pair of grated piezoelectric films can create the necessary micromotion to operate a matrix of independently controllable optical switches for flat video displays [221].

Thin films of liquid crystal polymeric materials with large second order optical nonlinearity open the possibility of new devices for laser technologies: deflection of light, light valves, and optical switches [222]. Photochromic liquid crystal polyacrylates can be prepared by the copolymerization of vinyl monomers with spiro-pyran and *p*-substituted phenylbenzoate groups. Poled films of this type show anomalously high optical nonlinearity under electric fields. A second harmonic generation was observed, the intensity of which depends on the electric field. This behaviour can be utilized in light-controlling actuators.

Friction is a major design concern and a key to success in microelectromechanical systems (valves, micromotors, etc.) (MEMS). It is expected that the frictional behaviour of mini- or microcomponents would be different from that of macrodevices, because of the large surface to bulk ratio, as well as reduced size and forces.

Fluorocarbon polymers, such as PTFE, are known to have many exceptional properties that make them very attractive for micromechanical purposes: the extremely high electrical resistivity, chemical stability, low Young's modulus, and very low friction coefficient to almost all other material surfaces (0.04 with respect to steel) [223]. Thin films of PTFE polymer can be deposited and patterned by several methods that are compatible to microelectronic processing techniques (see Section 1.4). Moreover, these films can prevent metallic structures from corrosion. Because the films do not stick to any substrate, special construction methods (e.g., mechanical anchoring) are necessary. Another problem of these films is the thermal runaway.

More recently, conducting electropolymerized films have also been investigated by Bartlett et al. and Gardner et al. [224,225] for covering moving parts in micromechanical devices. The friction coefficient spanned a range of 0.7 down to less than 0.07, and specific values could be achieved through an appropriate choice of the polymer system and the deposition conditions. A few typical values measured at various polymers are as follows [225]:

polypyrrole:	0.07–0.62
polyaniline:	0.2–0.3
poly(N-methylpyrrole):	0.18–0.22
poly(5-carboxyindole):	0.1–0.15

4.10 REFERENCES AND SUPPLEMENTARY READING

1. Doljak, F. A., "Polyswitch PTC Devices—A New Low-Resistance Conductive Polymer-Based PTC Device for Overcurrent Protection," *IEEE, Transactions on Components, Hybrids and Manufact. Techn.*, Vol. CHMT-4, No. 4 (1981), pp. 372–378.

2. Bueche, F., "A New Class of Switching Materials," *J. Appl. Phys.*, Vol. 44, No. 1 (1973).

3. Hu, K. A., Moffatt, D., Runt, J., Safari, A. and Newnham, R., "V_2O_3-Polymer Composite Thermistors," *J. Am. Ceram. Soc.*, 70 (8) (1987), pp. 583–585.

4. Xu, D., Xu, X. and Zou, Sh., "Ion-Beam-Modified Polyimide as a Novel Temperature Sensor: Fundamental Aspects and Applications," *Rev. Sci. Instrum.*, 63 (1) (1992), pp. 202–206.

5. Halbo, L., editor, *Polymer Thick-Film Technology (PTF), Design Guidelines and Applications*, Lobo Grafisk, Oslo (1990).

6. Hicks, W. T., Allington, T. R. and Van Johnson, "Membrane Touch Switches: Thick Film Materials Systems and Processing Options," *Proc. 30th Electronic Comp. Conf.*, San Francisco, CA (1980).

7. Hottinger Baldvin Messtechnik GmbH: Dehnungsmesstreifen mit Zubehör, Katalog G 24.01.5.

8. Vishay Measurements Group Messtechnik GmbH., DMS-Katalog (1990).

9. Rosengren, L., Söderkvist, J. and Smith, L., "Micromachined Sensor Structures with Linear Capacitive Response," *Sensors and Actuators A,* 31 (1992), pp. 200–205.

10. Harsányi, G., "Polymer Thick-Film Technology: A Possibility to Obtain Very Low Cost Pressure Sensors?" *Sensors and Actuators A,* 25–27 (1991), pp. 853–857.

11. Harsányi, G. and Hahn, E., "Thick-Film Pressure Sensors," *Mechatronics,* Vol. 3. No. 2 (1993), pp. 167–171.

12. Cattaneo, A., Dell'Acqua, R., Forlani, F. and Pirozzi, L., "The Industrial Application of the Piezoresistive Effect, A Low Cost Thick-Film Pressure Sensor," *Proc. 30th Electronic Comp. Conf.* (1980), pp. 429–435.

13. Chatigny, J. V., "Piezofilm – Ein Sensorik-Basismaterial," Reprint from *Elektroniker,* Nr. 7 (1988).

14. Wolffenbuttel, M. R. and Regtien, P. P. L., "Polysilicon Bridges for the Realization of Tactile Sensors," *Sensors and Actuators A,* 25–27 (1991), pp. 257–264.

15. Hillis, W. D., "A High-Resolution Imaging Touch Sensor," *Int. J. Robotics Res.,* 1 (2) (1982), pp. 33–44.

16. Rebman, J. and Morris, K. A., "A Tactile Sensor with Electro-Optical Transduction," *Proc. 3rd Int. Conf. Robot Vision and Sensory Controls,* Cambridge, MA (1983), pp. 210–216.

17. Regtien, P. P. L. and Wolffenbuttel, R. F., "Tactile Imaging Sensor," Sensor '85, Conf. Proc., Karlsruhe (1985), pp. 1.5.0–1.5.9

18. Seekircher, J. and Hoffmann, B., "Improved Tactile Sensors," *IFAC-Symposium on Robot Control (SYROCO '88),* Karlsruhe, ed., Rembold, V., Pergamon Press, Oxford, UK (1988), pp. 317–322.

19. Reston, R. R. and Kolesar, E. S., "Robotic Tactile Sensor Array Fabricated from a Piezoelectric Polyvinylidene Fluoride Film," *IEEE 1990 National Aerospace and Electronics Conf., NAECON 1990,* Dayton, pp. 1139–1144.

20. Domenici, C. and De Rossi, D., "A Stress-Component-Selective Tactile Sensor Array," *Sensors and Actuators A,* 13 (1992), pp. 97–100.

21. Howe, R. D., "A Tactile Stress Rate Sensor for Perception of Fine Surface Features," *Proc. of 1991 Int. Conf. on Solid State Sensors and Actuators (Transducers '91),* San Francisco, CA (1991), pp. 864–867.

22. Johansson, R. S. and LaMotte, R. H., "Tactile Detection Thresholds for a Single Asperity on an Otherwise Smooth Surface," *Somatosensory Research,* 1 (1) (1983), pp. 21–31.

23. Omata, S. and Terunuma, Y., "Development of New Type Tactile Sensor for Detecting Hardness and/or Softness of an Object Like the Human Hand," *Proc. of the 1991 Int. Conf. on Solid State Sensors and Actuators (Transducers '91),* San Francisco, CA (1991), pp. 868–871.

24. André, B., Clot, J., Partouche, E., Simonne, J. J., "Thin Film PVDF Sensors Applied to High Acceleration Measurements," *Sensors and Actuators A,* 33 (192), pp. 111–114.

25. AMP Corporation, Piezo Film Sensor Division: Product Data ACH-01.

26. Gutierrez Monreal, F. J., "An Application of Polymer Materials as Sensors: Design and Realization of a Gas Flowmeter," *Conference Capteurs '86,* Paris (1986) pp. 164–171.

27. Chen, Y., Wang, L. and Ko, W., "A Piezopolymer Finger Pulse and Breathing Wave Sensor," *Sensors and Actuators A, 21–23* (1990), pp. 879–882.

28. Halvorsen, D. L., "Piezo Electric Polymers in Sensing Applications," *Sensor '88*, Congress Proc., Nürnberg (1988), pp. 29–44.

29. Brown, R. H., "Applications of Piezo Film Transducers, in Vibration Analysis," *Sensor '88*, Congress Proc., Nürnberg (1988), pp. 161–172.

30. Ribeiro, D. L., Raposo, M., Dias, C. J. and Marat-Mendes, J. N., "Piezoelectric Weight-in-Motion Sensor," *Proc. Int. Conf. Eurosensors VI.*, San Sebastian (Spain) (1992), p. 299.

31. Baker, W. O., "Polymers in the World of Tomorrow," *Proc. of the Symp. at the 2. Chemical Congress of the North American Continent*, Las Vegas (1980), pp. 165–203.

32. Sessler, G. M., "Acoustic Sensors," *Sensors and Actuators A, 25–27* (1991), pp. 323–330.

33. Sessler, G. M., ed. *Topics in Applied Physics, Vol. 33, Electrets*, Springer, Heildelberg (1987).

34. Voorthuyzen, J. A., Bergveld, P. and Sprenkels, A. J., "Semiconductor-Based Electret Sensors for Sound and Pressure," *IEEE Trans. Electr. Insul.*, EI-24 (1989), pp. 267–276.

35. Murphy, P., Hübschi, K., de Rooij, N. and Racine, C., 'Subminiature Silicon Integrated Electret Capacitive Microphone," *IEEE Trans. Electr. Insul.*, EI-24 (1989), pp. 495–498.

36. Harnisch, F., Kroemer, N. and Manthey, W., "Ultrasonic Transducers with Piezoelectric Polymer Foil," *Sensors and Actuators A, 25–27* (1991), pp. 549–552.

37. Manthey, W. and Mágori, V., "Ultrasonic Transducer Arrays for Applications in Air," *Sensor '93*, Congress Proc., Nürnberg, Band 2 (1993), pp. 81–88.

38. Mo, J-H., Andrew, L., Robinson, L., Terry, F. L. Jr., Fitting, D. W. and Carson, P. L., "Improvement of Integrated Ultrasonic Transducer Sensitivity," *Sensors and Actuators A, 21–23* (1990), pp. 679–682.

39. Möckl, T., Mágori, V. and Eccardt, C., "Sandwich-Layer Transducer—A Versatile Design for Ultrasonic Transducers Operating in Air," *Sensors and Actuators A, 21–23* (1990), pp. 687–692.

40. Yiquan, Y. and Binwen, S., "A Cylindrical Hydrophone Made of PVDF Piezoelectric Polymer and Its Performance," *Sensors and Actuators A, 35* (1993), pp. 231–234.

41. Garthe, D., "A Fibre-Optic Microphone," *Sensors and Actuators A, 25–27* (1991), pp. 341–345.

42. AMP Corporation, Piezo Film Sensor Division: Product Data PIR 180-100.

43. Mader, G. and Meixner, H., "Pyroelectric Infrared Sensor Array Based on the Polymer PVDF," *Sensors and Actuators A, 21–23* (1990), pp. 503–507.

44. Shu-Duo, W., "PVDF Film Multifunction Sensors," *Sensors and Actuators A, 21–23* (1990), pp. 883–885.

45. Schöpf, H., Ruppel, W. and Würfel, P., "Integrated Pyroelectric IR Matrix Sensor Arrays," *Sensors and Actuators A, 25–27* (1991), pp. 401–405.

46. Lucas, Ch., "Infrared Detection, Some Recent Developments and Future Trends," *Sensor and Actuators A, 25–27* (1991), pp. 147–154.

47. Ploss, B. and Bauer, S., "Characterization of Materials for Integrated Pyroelectric Sensors," *Sensors and Actuators A, 25–27* (1991), pp. 407–411.

48. Ruppel, W., "Pyroelectric Sensor Arrays on Silicon," *Sensors and Actuators A, 31* (1992), pp. 225–228.

49. Hammes, P. C. A. and Regtien, P. P. L. "An Integrated Infrared Sensor Using the Pyroelectric Polymer PVDF," *Sensors and Actuators A, 32* (1992), pp. 396–402.

50. von Münch, W. and Thiemann, U., "Pyroelectric Detector Array with PVDF on Silicon Integrated Circuit," *Sensors and Actuators A, 25–27* (1991), pp. 167–172.

51. von Münch, W., Nägele, M., Wöhl, G., Ploss, B. and Ruppel, W., "A 3 × 3 Pyroelectric Detector Array with Improved Sensor Technology," *Proc. of Eurosensors VII. Conf.*, Budapest (1993), p. 335.

52. Lienhard, D., Nitschke, S., Ploss, B. and Ruppel, W., von Münch, W., *Proc. of Eurosensors VII. Conf.*, Budapest (1992) p. 274.

53. Okuyama, M., Togami, Y., Taniguchi, H., Hamakawa, Y., Kumita, M. and Denda, M., "Basic Characteristics of an Infrared CCD with Pyroelectric Gate," *Sensors and Actuators A, 21–23* (1990), pp. 465–468.

54. Philips Components: Kapazitiver Feuchtesensor Technische Information Nr. 2322 691 90001, Philips Components Unternehmensbereich der Philips GmbH, Hamburg.

55. Yamazone, N. and Shimizu, Y., "Humidity Sensors: Principles and Applications," *Sensors and Actuators, 10* (1986), pp. 379–398.

56. Delapierre, G., Grange, H., Chambaz, B. and Destannes, L., "Polymer-Based Capacitive Humidity Sensor: Characteristics and Experimental Results," *Sensors and Actuators, 4* (1983), pp. 97–104.

57. Denton, D. D., Senturia, S. D., Analick, E. S. and Schneider, D., "Fundamental Issues in the Design of Polymeric Capacitive Moisture Sensors," Digest of Techn. Papers, *3rd Int. Conf. on Solid-State Sensors and Actuators (Transducers '85),* Philadelphia, PA (1985), pp. 202–205.

58. Glenn, M. C. and Schuetz, I. A., "An IC Compatible Polymer Humidity Sensor," DIgest of Techn. Papers, *3rd Int. Conf. on Solid-State Sensors and Actuators (Transducers '85),* Philadelphia, PA (1985), pp. 217–220.

59. Schubert, P. J. and Nevin, J. H., "A Polyimide-Based Capacitive Humidity Sensor," *IEEE Transactions on Electron Devices,* Vol. Ed-32, No. 7 (1985).

60. Boltshauser, T. and Baltes, H., "Capacitive Humidity Sensors in SACMOS Technology with Moisture Absorbing Photosensitive Polyimide," *Sensors and Actuators A, 25–27* (1991), pp. 509–512.

61. Boltshauser, T., Chandran, L., Baltes, H., Bose, F. and Steiner, D., "Humidity Sensing Properties and Electrical Permittivity of New Photosensitive Polyimides," *Sensors and Actuators B, 8* (1991), pp. 161–164.

62. Sicovend AG., "Elektronische Bauteile, Humidity Sensor," Data Sheet "Hument HPR-MQ," Sicovend AG, Wallisellen, Switzerland.

63. Hijikigawa, M., Miyoshi, S., Sugihara, T. and Jinda, A., "A Thick-Film Resistance Humidity Sensor," *Sensors and Actuators, 4* (1983), pp. 307–315.

64. Tsuchitani, S., Sugawara, T., Kinjo, N., Ohara, S. and Tsunoda, T., "A Humidity Sensor Using Ionic Copolymer and its Application to a Humidity Temperature Sensor Module," *Sensors and Actuators, 15* (1988), pp. 375–386.

65. Sadaoka, Y., Sakai, Y. and Akiyama, H., "A Humidity Sensor Using Alkali Salt-Poly(ethylene oxide) Hybrid Films," *J. Mat. Sci., 21* (1986), pp. 235–240.

66. Sakai, Y., Sadaoka, Y. and Fukumoto, H., "Humidity-Sensitive and Water-Resistive Polymeric Materials," *Sensors and Actuators, 13* (1988), pp. 243–850.

67. Sakai, Y., Sadaoka, Y., Matsuguchi, M., Moriga, N. and Shimada, M., "Humidity Sensors Based on Organopolysiloxanes Having Hydrophilic Groups," *Sensors and Actuators*, 16 (1986), pp. 359–367.

68. Sakai, Y., Sadaoka, Y. and Matsuguchi, M., "A Humidity Sensor Using Cross-linked Quaternized Polyvinylpyridine," *J. Electrochem. Soc.*, Vol. 136, No. 1 (1989).

69. Sakai, Y., Sadaoka, Y., Matsuguchi, M. and Hirayama, K., "Water Resistive Humidity Sensor Composed of Interpenetrating Polymer Networks of Hydrophilic and Hydrophobic Methacrylate," *Proc. of the Int. Conf. on Solid State Sensors and Actuators (Transducers '91)*, San Francisco, CA (1991), pp. 562–565.

70. Furlani, A., Iucci, G., Russo, M. V., Bearzotti, A. and D'Amico, A., "Thick-Films of Iodine-Polyphenylacetylene as Starting Materials for Humidity Sensors," *Sensors and Actuators B*, 7 (1992), pp. 447–450.

71. Bearzotti, A., D'Amico, A., Furlani, A., Iucci, G. and Russo, M. V., "Fast Humidity Response of a Metal Halide-Doped Novel Polymer," *Sensors and Actuators B*, 7 (1992), pp. 451–454.

72. Rauen, K. L., Smith, D. A., Heineman, W. R., Johnson, J., Seguin, R. and Stoughton, P., "Humidity Sensor Based on Conductivity Measurements of a Poly(dimethyl-diallylammonium chloride) Polymer Film," *Sensors and Actuators B*, 17 (1993), pp. 61–68.

73. Hijikigawa, M., Sugihara, T., Tanaka, J. and Watanabe, M., "Micro-Chip FET Humidity Sensor with a Long-Term Stability," Digest of Techn. Papers, *3rd Int. Conf. on Solid State Sensors and Actuators (Transducers '85)* Philadelphia, PA (1985), pp. 221–224.

74. Garverick, S. L. and Senturia, S. D., "An MOS Device for AC Measurements of Surface Impedance with Application to Moisture Monitoring," *IEEE Trans. Electron. Devs.*, ED-29, No. 1 (1982), pp. 90–94.

75. Brace, J. G. and Sanfelippo, T. S., "A Study of Polymer/Water Interactions Using Surface Acoustic Waves," *Sensors and Actuators*, 14 (1988), pp. 47–68.

76. Nieuwenhuizen, M. S. and Nederlof, A. I., "SAW Gas Sensor for Carbon Dioxide and Water. Preliminary Experiments," *Sensors and Actuators B*, 2 (1990), pp. 97–101.

77. Radeva, E., Bobev, K. and Spassov, L., "Study and Application of Glow Discharge Polymer Layers as Humidity Sensors," *Sensors and Actuators B*, 8 (1992), pp. 21–25.

78. Kuwano, J., Maruyama, H., Kato, M. and Ito, M., "A Potentiometric Humidity Sensor Using Polythiophene and High Ionic-Conductivity Glasses," *Denki Kagaku*, 55, No. 3 (1987), pp. 267–267.

79. Ballantine, D. S. and Wohltjen, H., "Optical Waveguide Humidity Sensor," *Anal. Chem.*, 58 (1986), pp. 2883–2885.

80. Sadaoka, Y., Matsuguchi, M. and Sakai, Y., "Optical-Fibre and Quartz Oscillator Type Gas Sensors: Humidity Detection by Nafion® Film with Crystal Violet and Related Compounds," *Sensors and Actuators A*, 25–27 (1991), pp. 489–492.

81. Sadaoka, Y., Matsuguchi, M., Sakai, Y. and Murata, Y., "Optical Humidity Sensing Characteristics of Nafion-Dyes Composite Thin Films," *Sensors and Actuators B*, 7 (1992), pp. 443–446.

82. Clark, L. C., "Monitoring and Control Blood and Tissue Oxygen," *Trans. Am. Soc. Artif. Intern. Organs*, Vol. 2 (1958), pp. 41–48.

83. Severinghaus, I. W. and Bradley, A. F., "Electrodes for Blood PO_2 and PCO_2 Determination," *J. Appl. Physiol.*, Vol. 13 (1958), pp. 515–520.

84. Tieman, R. S., Heineman, W. R., Johnson, J. and Seguin, R., "Oxygen Sensors Based on the Ionically Conductive Polymer Poly(dimethyldiallylammonium chloride)," *Sensors and Actuators B*, 8 (1992), pp. 199–204.

85. Asada, A., Yamamoto, H., "Limiting Current Type of Oxygen Sensor with High Performance," *Sensors and Actuators B*, 1 (1990), pp. 312–318.

86. Lundström, I., Armgarth, M., Spetz, A. and Winquist, F., "Gas Sensors Based on Catalytic Metal-Gate Field-Effect Devices," *Sensors and Actuators*, 10 (1986), pp. 399–421.

87. Hedborg, E., Spetz, A., Winquist, F. and Lundström, I., "Polymer Membranes for Modification of the Selectivity of Field-Effect Gas Sensors," *Sensors and Actuators B*, 7 (1992), pp. 661–664.

88. Nehlsen, St., Weissenrieder, K.-S., Zernack, M. and Müller, J., "Selectivity Enhancement of ZnO Thin Film Gas Sensors Using Catalysts and Gasselective Membranes," *Sensor '93*, Congress Proc., Nürnberg, Band 1 (1993), pp. 143–148.

89. Lechuga, L. M., Calle, A., Golmayo, D., Briones, F., De-Abajo, J. and De La Campa, J. G., "Ammonia Sensitivity of Pt/GaAs Schottky Barrier Diodes. Improvement of the Sensor with an Organic Layer," *Sensors and Actuators B*, 8 (1992), pp. 249–252.

90. Josowicz, M. and Janata, J., "Suspended Gate Field Effect Transistor Modified with Polypyrrole as Alcohol Sensor," *Anal. Chem.*, 58 (1986), pp. 514–517.

91. Suzuki, H., Sugama, A. and Kokima, N., "Micromachined Clark Oxygen Electrode," *Sensors and Actuators B*, 10 (1993), pp. 91–98.

92. Yan, H.-Q. and Lu, J.-T., "Solid Polymer Electrolyte-Based Electrochemical Oxygen Sensor," *Sensors and Actuators*, 19 (1989), pp. 33–40.

93. Yan, H.-Q. and Liu, Ch.-Ch., "Humidity Effects on the Stability of a Solid Polymer Electrolyte Oxygen Sensor," *Sensors and Actuators B*, 10 (1993), pp. 133–136.

94. Buttner, W. J., Maclay, G. J. and Stetter, J. R., "An Integrated Amperometric Microsensor," *Sensors and Actuators B*, 1 (1990), pp. 303–307.

95. Otagawa, T., Madou, M., Wing, Sh., Rich-Alexander, J., Kusanagi, Sh., Fujioka, T. and Yasuda, A., "Planar Microelectrochemical Carbon Monoxide Sensors," *Sensors and Actuators B*, 1 (1990), pp. 319–325.

96. Maseeh, F., Tierney, M. J., Chu, W. S., Joseph, J., Kim, H. L. and Otagawa, T., "A Novel Silicon Micro Amperometric Gas Sensor," *Proc. Conf. Solid State Sensors and Actuators (Transducers '91)*, San Francisco, CA (1991), pp. 359–362.

97. Christensen, W. H., Sinha, D. N. and Agnew, S. F., "Conductivity of Polystyrene Film upon Exposure to Nitrogen Dioxide: A Novel NO_2 Sensor," *Sensors and Actuators B*, 10 (1993), pp. 149–153.

98. Blanc, J. P., Blasquez, G., Germain, I. P., Larbi, A., Maleysson, C. and Robert, H., "Behaviour of Electroactive Polymers in Gaseous Oxidizing Atmospheres," *Sensors and Actuators*, 14 (1988), pp. 143–148.

99. Krutovertsev, S. A., Sorokin, S. I., Zorin, A. V., Letuchy, Ya. A. and Anatova, D.Yu., "Polymer Film-Based Sensors for Ammonia Detection," *Sensors and Actuators B*, 7 (1992), pp. 492–494.

100. Bidan, G., "Electroconducting Conjugated Polymers: New Sensitive Matrices to Build up Chemical or Electrochemical Sensors. A Review," *Sensors and Actuators B*, 6 (1992), pp. 45–56.

101. Haug, M., Schierbaum, K.-D., Endres, H.-E., Drost, S. and Göpel, W., "Controlled

Selectivity of Polysiloxane Coatings: Their Use in Capacitance Sensors," *Sensors and Actuators A,* 32 (1992), pp. 326–332.

102. Hermans, E. C. M., "CO, CO_2, CH_4 and H_2O Sensing by Polymer Covered Inter-digitated Electrode Structures," *Sensors and Actuators,* 5 (1984), pp. 181–186.

103. Fraunhofer-Gesellschaft, München, *Fraunhofer-Verbund Mikroelektronik* (1988), p. 7.

104. Drost, S. and Endres, H.-E., "Gas-Sensitive MOSFETs – An Overview," *Proc. Int. Symp. MICROSYSTEM Technologies,* ICC, Berlin (1991).

105. Haug, M., Schierbaum, K.-D., Gauglitz, G. and Göpel, W., "Chemical Sensors Based upon Polysiloxanes: Comparison between Optical, Quartz Microbalance, Calorimetric, and Capacitance Sensors," *Sensors and Actuators B,* 11 (1993), pp. 383–391.

106. Amati, D., Arn, D., Blom, N., Ehrat, M., Saunois, J. and Widmer, H. M., "Sensitivity and Selectivity of Surface Acoustic Wave Sensors for Organic Solvent Vapour Detection," *Sensors and Actuators B* (1992), pp. 587–591.

107. Wenzel, S. W. and White, R. M., "Flexural Plate-Wave Gravimetric Chemical Sensor," *Sensors and Actuators A,* 21–23 (1990), pp. 700–703.

108. Block, R., Fickler, G., Lindner, G., Müller, H. and Wohnhas, M., "Mechanical Resonance Gas Sensors with Piezoelectric Excitation and Detection Using PVDF Foils," *Sensors and Actuators B,* 7 (1992), pp. 596–601.

109. Sugimoto, I., Nakamura, M. and Kuwano, H., "Molecular Sensing Using Plasma Polymer Thin Film Probes," *Sensors and Actuators B,* 10 (1993), pp. 117–122.

110. Grate, J. W. and Abraham, M. H., "Solubility Interactions and the Design of Chemically Selective Sorbent Coatings for Chemical Sensors and Arrays," *Sensors and Actuators B,* 3 (1991), pp. 85–111.

111. Marcinkowska, K., McGauley, M. P. and Symons, E. A., "A New Carbon Monoxide Sensor Based on a Hydrophobic CO Oxidation Catalyst," *Sensors and Actuators B,* 5 (1991), pp. 91–96.

112. Schierbaum, K. D., Gerlach, A., Haug, M. and Göpel, W., "Selective Detection of Organic Molecules with Polymers and Supramolecular Compounds: Application of Capacitance Quartz Microbalance and Colorimetric Transducers," *Sensors and Actuators A,* 31 (1992), pp. 130–137.

113. Zhou, Q., Kritz, D., Bonell, L. and Siegel, G. H., Jr., "Porous Plastic Optical Fiber Sensor for Ammonia Measurement," *Applied Optics,* Vol. 28 No. 11 (1989), pp. 2022–2025.

114. Zhou, Q. and Siegel, G. H., Jr., "Detection of Carbon Monoxide with a Porous Polymer Optical Fibre," *Int. J. Optoel.,* Vol. 4, No. 5 (1989), pp. 451–423.

115. Ruddy, V. and McCabe, S., "Detection of Propane by IR-ATR in a Teflon®-Clad Fluoride Glass Optical Fiber," *Appl. Spectrosc.,* 44 (9) (1990), pp. 1461–1463.

116. Ronot, C., Archenault, M., Gagnaire, H., Goure, J. P., Jaffrezic-Renault, N. and Pichery, T., "Detection of Chemical Vapours with a Specifically Coated Optical-Fibre Sensor," *Sensors and Actuators B,* 11 (1993), pp. 375–381.

117. Gauglitz, G. and Ingenhoff, J., "Integrated Optical Sensors for Halogenated and Non-Halogenated Hydrocarbons," *Sensors and Actuators B,* 11 (1993), pp. 207–212.

118. Gauglitz, G., Brecht, A., Kraus, G. and Nahm, W., "Chemical and Biochemical Sensors Based on Interferometry at Thin (multi-) Layers," *Sensors and Actuators B,* 11 (1993), pp. 21–27.

119. Mandelis, A. and Christofides, C., "Photopyroelectric (P^2E) Sensor for Trace Hydrogen Gas Detection," *Sensors and Actuators B,* 2 (1990), pp. 79–87.

120. Thackeray, J. W. and Wrighton, M. S., "Chemically Responsive Microelectrochemical Devices Based on Platinized Poly(3-methylthiophene): Variation in Conductivity with Variation in Hydrogen, Oxygen, or pH in Aqueous Solution," *J. Phys. Chem.*, Vol. 90 (1986), pp. 6674–6679.

121. Schurmer, H. V., Corcoran, P. and Gardner, J. W., "Integrated Arrays of Gas Sensors Using Conducting Polymers with Molecular Sieves," *Sensors and Actuators B*, 4 (1991), pp. 29–33.

122. Armstrong, R. D. and Horvai, Gy., "Review Article: Properties of PVC Based Membranes Used in Ion-Selective Electrodes," *Electrochimica Acta*, Vol. 35, No. 1 (1990), pp. 1–7.

123. Fluka Chemica-Biochemica, Chemical Corp., "Selectophore®: Ionophores for Ion-Selective Electrodes and Optodes" (1991).

124. Harrison, D. J., Li, X. and Petrovic, S., "Experimental Determination of Ionophore and H_2O Distribution in Ion-Selective Membranes," *Transducers '89*, Abstracts (1989), pp. 50–51.

125. Goldberg, H. D., Cha, G. S., Liu, D., Meyerhoff, M. E. and Brown, R. B., "Improved Stability at the Polymeric Membrane/Solid-Contact Interface of Solid-State Potentiometric Ion Sensors," *Proc. Conf. Solid State Sensors and Actuators (Transducers '91)*, San Francisco, CA (1991), pp. 781–784.

126. Borraccino, A., Camponella, L., Sammartino, M. P. and Tomassetti, M., "Suitable Ion-Selective Sensors for Lead and Cadmium Analysis," *Sensors and Actuators B*, 7 (1992), pp. 535–539.

127. Cha, G. S. and Brown, R. B., "Polyimide-Matrix Chemical-Selective Membranes," *Sensors and Actuators B*, 1 (1990), pp. 281–285.

128. Smith, M. D., Genshaw, M. A. and Greyson, J., "Miniature Solid State Potassium Electrode for Serum Analysis," *Anal. Chem.*, Vol. 45 (1973), pp. 1782–1784.

129. Diaz, C., Vidal, J. C., Galban, J., Urarte, M. L. and Lanaja, J., "A Double-Membrane Ion-Selective Electrode for the Potentiometric Determination of Potassium," *Microchemical J.*, Vol. 39 (1989), pp. 289–297.

130. Keplinger, F., Glatz, R., Jachimowicz, A., Urban, G., Kohl, F., Olcaytug, F. and Prohaska, O. J., "Thin-Film Ion-Selective Sensors Based on Neutral Carrier Membranes," *Sensors and Actuators B*, 1 (1990), pp. 272–274.

131. Keplinger, F., Urban, G., Jobst, G., Kohl, F. and Jachimowicz, A., "Solid Contact Neutral Carrier pH-ISE with Conducting Polymers," *Eurosensors VII*, Budapest, Abstracts (1993), p. 24.

132. Fabry, P., Montero-Ocampo, C. and Armand, M., "Polymer Electrolyte as Internal Ionic Bridge for Ion Solid-State Sensors," *Sensors and Actuators*, 15 (1988), pp. 1–9.

133. Clechet, P., "Membranes for Chemical Sensors," *Sensors and Actuators B*, (1991), pp. 53–63.

134. Moss, S. D., Janata, J. and Johnson, C. C., "Potassium Ion-Sensitive Field Effect Transistor," *Anal. Chem.*, 47 (1975), pp. 2238–2243.

135. Sibbald, A., Whalley, P. D., and Covington, A. K., "A Miniature Flow-through Cell with a Four-Function CHEMFET Integrated Circuit for Simultaneous Measurements of Potassium, Hydrogen, Calcium and Sodium Ions," *Anal. Chim. Acta*, 159 (1984), pp. 47–62.

136. Van den Vlekkert, H., Francis, C., Grisel, A. and de Rooij, N., "Solvent Polymeric Membranes Combined with Chemical Solid-State Sensors," *Analyst*, 113 (1988), pp. 1029–1033.

137. Bergveld, P., Van den Berg, A., Van der Wal, P. D., Skowronska-Ptasinska, M., Sudhölter, E. J. R. and Reinhoudt, D. N., "How Electrical and Chemical Requirements for REFETS May Coincide," *Sensors and Actuators,* 18 (1989), pp. 309–327.

138. Reinhoudt, D. N. and Sudhölter, E. J. R., "The Transduction of Host-Guest Interactions into Electronic Signals by Molecular Systems," *Adv. Mater.,* 2 (1990), pp. 23–32.

139. Blackburn, G. and Janata, J., "The Suspended Mesh Ion Selective Field Effect Transistor," *J. Electrochem. Soc.,* 129 (1982), pp. 2580–2584.

140. Kawakami, S., Akiyama, T. and Ujihira, Y., "Potassium Ion-Sensitive Field Effect Transistors Using Valinomycin Doped Photoresist Membrane," *Fresenius Z. Anal. Chem.,* 318 (1984), pp. 349–351.

141. Tsukada, K., Sebada, M., Miyahara, Y. and Miyagi, H., "Long-Life Multiple-ISFETs with Polymeric Gates," *Sensors and Actuators,* 18 (1989), pp. 329–336.

142. Bezegh, K., Bezegh, A., Janata, J., Oesch, V., Xu, A. and Simon, W., "Multisensing Ion-Selective Field-Effect Transistors Prepared by Ionophore Doping Technique," *Anal. Chem.,* 59 (1987), pp. 2846–2848.

143. Bezegh, K., Petelenz, D., Bezegh, A. and Janata, J., "Integrated Solid State Probe for Determination of Activity of Sodium Chloride," *Anal. Chem.,* 59 (1987), p. 1423.

144. Wakida, S. I., Yamane, M. and Hiiro, K., "A Novel Urushi Matrix Sodium Ion-Selective Field-Effect Transistor," *Sensors and Actuators,* 18 (1989), pp. 285–290.

145. Wakida, S. I., Yamane, M., Higashi, K., Hiiro, K. and Ujihira, Y., "Urushi Matrix Sodium, Potassium, Calcium and Chloride-Selective Field-Effect Transistors," *Sensors and Actuators B,* 1 (1990), pp. 412–415.

146. Vogel, A., Hoffmann, B., Sauer, TH. and Wegner, G., "Langmuir-Blodgett Films of Phthalocyaninato-Polysiloxane Polymers as a Novel Type of CHEMFET Membrane," *Sensors and Actuators B,* 1 (1990), pp. 408–411.

147. Erbach, R., Hoffmann, B., Schaub, M. and Wegner, G., "Application of Rod-Like Polymers with Ionophores as Langmuir-Blodgett Membranes for Si-Based Ion Sensors," *Sensors and Actuators B,* 6 (1992), pp. 211–216.

148. Kauffmann, F., Hoffmann, B., Erbach, R., "Ca^{2+}-Sensor with Amphiphilic Langmuir-Blodgett Membranes," *Eurosensors VII,* Abstracts, Budapest (1993) p. 30.

149. Battilotti, M., Mercuri, R., Mazzamurro, G., Gainnini, J. and Giongo, M., "Lead Ion-Sensitive Membrane for ISFETs," *Sensors and Actuators B,* 1 (1990), pp. 438–440.

150. Davini, E., Mazzamurro, G. and Piotto, A. P., "Lead-Sensitive FET: Complexation Selectivity of Ionophores Embedded in the Membrane," *Sensors and Actuators B,* 7 (1992), pp. 580–583.

151. Sudhölter, E. J. R., Van der Wal, P. D., Skowronska-Ptasinska, M., Van den Berg, A. and Reinhoudt, D. N., "Ion-Sensing Using Chemically-Modified ISFETs," *Sensors and Actuators,* 17 (1989), pp. 189–194.

152. Haak, J. R., Van der Wal, P. D. and Reinhoudt, D. N., "Molecular Materials for the Transduction of Chemical Information by CHEMFETs," *Sensors and Actuators B,* 8 (1992), pp. 211–219.

153. Reinhoudt, D. N., "Application of Supramolecular Chemistry in the Development of Ion-Selective CHEMFETs," *Sensors and Actuators B,* 6 (1992), pp. 179–185.

154. Van den Berg, A., Grisel, A. and Verney-Norberg, E., "An ISFET-Based Calcium Sensor Using a Photopolymerized Polysiloxane Membrane," *Sensors and Actuators B,* 4 (1991), pp. 235–238.

155. Matsuo, T. and Nakajima, H., "Characteristics of Reference Electrodes Using a Polymer Gate ISFET," *Sensors and Actuators*, 5 (1984), pp. 293–305.

156. Nakajima, H., Esashi, M. and Matsuo, T., "The Cation Concentration Response of Polymer Gate ISFET," *J. Electrochem. Soc.*, 129 (1982), pp. 141–143.

157. Van den Berg, A., Bergveld, P. Reinhoudt, D. N. and Sudhölter, E. J. R., "Sensitivity Control of ISFETs by Chemical Surface Modification," *Sensors and Actuators*, 8 (1985), pp. 129–148.

158. Leimbrock, W., Landgraf, V. and Kampfrath, G., "An Extended Sitebinding Model and Experimental Results of Organic Membranes for Reference ISFETs," *Sensors and Actuators B*, 2 (1990), pp. 1–6.

159. Dumschat, C., Pötter, W. and Cammann, C., "Perchloratreferenz-elektrode für miniatürisierte potentiometrische Sensoren – Erste Grundlagenuntersuchungen an Makroelektroden," *Sensor '93*, Congress Proc., Nürnberg, Band 1 (1993), pp. 135–142.

160. Igarashi, J., Ito, T., Taguchi, T., Tabata, O. and Inagaki, H., "Multiple Ion Sensor Array," *Sensors and Actuators B*, 1 (1990), pp. 8–11.

161. Tsukada, K., Miyahara, Y., Shibata, Y. and Miyagi, H., "An Integrated Micro Multi-Ion Sensor Using Platinum-Gate FIeld-Effect Transistors," *Proc. Conf. Solid State Sensors and Actuators (Transducers '91)*, San Francisco, CA (1991), pp. 218–221.

162. Bechtold, F., Fenzlein, P. G. and Abendroth, D., "New Possibilities of Developing Chemical Sensors in Medicine and Biotechnology Using Low Temperature Cofireable Ceramics (LTMLC)," *Proc. 7th European Hybrid Microel. Conf.*, Hamburg (1989), p. 8.5.

163. Sheppard, N. F., Jr., "Design of a Conductimetric Microsensor Based on Reversibly Swelling Polymer Hydrogels," *Proc. Conf. Solid State Sensors and Actuators (Transducers '91)*, San Francisco, CA (1991), pp. 773–776.

164. Sheppard, N. F., Jr., Tucker, R. C. and Salehi-Had, S., "Design of a Conductimetric pH Microsensor Based on Reversibly Swelling Hydrogels," *Sensors and Actuators B*, 10 (1993), pp. 73–77.

165. Alava-Moreno, F., Pereiro-Garcia, R., Diaz-Garcia, M. E. and Sanz-Mendel, A., "A Comparative Study of Two Different Approaches for Active Optical Sensing of Potassium with a Chromoionophore," *Sensors and Actuators B*, 11 (1993), pp. 413–419.

166. Leiner, M. J. P. and Hartmann, P., "Theory and Practice in Optical pH Sensing," *Sensors and Actuators B*, 11 (1993), pp. 281–289.

167. Peterson, J. I., Goldstein, S. R., Fitzgerald, R. V. and Buckhold, D. K., "A Fiber Optic pH Probe for Physiological Use," *Anal. Chem.*, 53 (1980), pp. 864–869.

168. Bacci, M., Baldini, F. and Scheggi, A. M., "Spectrophotometric Investigations on Immobilized Acid-Base Indicators," *Anal. Chim. Acta*, 207 (1988), pp. 343–348.

169. Hao, T., Xing, X. and Liu, Ch-Ch., "A pH Sensor Constructed with Two Types of Optical Fibers: The Configuration and the Initial Results," *Sensors and Actuators B*, 10 (1993), pp. 155–159.

170. Boisde, G., Blanc, F. and Machuron-Mandard, X., "pH Measurements with Dyes Co-immobilization on Optrodes: Principles and Associated Instrumentation," *Int. J. Optoel.*, Vol. 6 (1991), pp. 407–413.

171. Wolfbeis, O. S., Fürlinger, E., Kroneis, H. and Marsoner, A., "A Study on Fluorescent Indicators for Measuring near Neutral ('physiological') pH-Values," *Fresenius Z. Anal. Chem.*, 314 (1983), pp. 119–124.

172. Lumpp, R., Reichert, J. and Ache, H. J., "An Optical Sensor for the Detection of Nitrate," *Sensors and Actuators B, 7* (1992), pp. 473–465.

173. Reichert, J., Sellien, W. and Ache, H. J., "Development of a Fiber-Optic Sensor for the Detection of Ammonium in Environmental Waters," *Sensors and Actuators A, 25-27* (1991), pp. 421–4182.

174. Czolk, R., Reichert, J. and Ache, H. J., "An Optical Sensor for the Detection of Heavy Metal Ions," *Sensors and Actuators B, 7* (1992), pp. 540–543.

175. Kawabata, Y., Yamamoto, T. and Imasaka, T., "Theoretical Evolution of Optical Response to Cations and Cationic Surfactant for Optrode Using Hexadecyl-acridicine Orange Attached on Plasticized Poly(vinyl chloride) Membrane," *Sensors and Actuators B,* 11 (1993), pp. 341–346.

176. Sandanayake, K. R. A. S. and Sutherland, I. O., "Organic Dyes for Optical Sensors," *Sensors and Actuators B,* 11 (1993), pp. 331–340.

177. Buerk, D. G., *Biosensors. Theory and Applications,* Technomic Publishing Co., Inc. (1993).

178. Parker, D., "Sensors for Monitoring Blood Gases in Intensive Care," *J. Phys. E: Sci. Instrum.,* 20 (1987), pp. 1103–1112.

179. Huch, A., Huch, R. and Lübbers, D. W., "Quantitative Continuous Measurement of Partial Pressure (PO_2 Measurement) on the Skin of Adults and Newborn Babies," *Pflügers Arch. Ges. Physiol.,* 337 (1972), pp. 185–198.

180. Harsányi, G., Péteri, I. and Deák, I., "Low Cost Ceramic Sensors for Biomedical Use: A Revolution in Blood Oxygen Monitoring," *Eurosensors VII,* Abstracts, Budapest (1993), p. 292.

181. Cha, C. S., Shao, M. J. and Liu, Ch.-Ch., "Problem Associated with the Miniaturization of a Voltammetric Oxygen Sensor: Chemical Crosstalk among Electrodes," *Sensors and Actuators B,* 2 (1990), pp. 239–242.

182. Fasching, R., Jobst, G., Keplinger, F., Aschauer, E. and Urban, G., "Electrochemical Thin-Film Carbondioxide Sensors on Novel Polymer Membrane System for in vitro and in vivo Applications," *Eurosensors VII,* Abstracts, Budapest (1993), p. 33.

183. Tsukada, K., Miyahara, Y., Shibata, Y. and Miyagi, H., "An Integrated Chemical Sensor with Multiple Ion and Gas Sensors," *Sensors and Actuators B,* 2 (1990), pp. 291–295.

184. Schelter, W., Gumbrecht, W., Montag, B., Sykora, V. and Erhardt, W., "Combination of Amperometric and Potentiometric Sensor Principles for on Line Blood Monitoring," *Sensors and Actuators B,* 6 (1992), pp. 91–92.

185. Wolfbeis, O. S., "Biomedical Application of Fibre Optic Chemical Sensors," *Int. Journal of Optoel.,* Vol. 6, No. 5 (1991), pp. 425–441.

186. Thybaud, L., Depeursinge, C. and Rouiller, D., "Use of ISFETs for 24h pH Monitoring in the Gastrooesophageal Tract," *Sensors and Actuators B,* 1 (1990), pp. 485–482.

187. Miyahara, Y., Tsukada, K., Miyagi, H. and Simon, W., "Urea Sensor Based on an Ammonium-Ion-Selective Field-Effect Transistor," *Sensors and Actuators B,* 3 (1991), pp. 281–293.

188. Van der Schoot, B. H., Voorthuyzen, and Bergveld, P., "The pH Static Enzyme Sensor: Design of the pH Control System," *Sensors and Actuators B,* 1 (1990), pp. 546–549.

189. Glab, S., Holona, I., Koncki, R. and Hulanicki, A., "Aminated PVC as Material for the Construction of pH-Based Urea Sensors," *Eurosensors VII,* Abstracts, Budapest (1993), p. 44.

190. Glab, S., Koncki, R., Holona, I. and Kopczewska, E., "pH-Membrane Electrodes with Chemically Immobilized Urease," *Eurosensors VII*, Abstracts, Budapest (1993), p. 45.

191. Lechuga, L. M., Mier, G., Calle, A., Golmayo, D. and Briones, F., "Urea Biosensor Using a NH_3 Gas Sensor Based on GaAs Schottky Diode," *Eurosensors VI*, Abstracts, San Sebastian (1992).

192. Gauglitz, G. and Reichert, M., "Spectral Investigation and Optimization of pH and Urea Sensors," *Sensors and Actuators B*, 6 (1992), pp. 83–86.

193. Xie, B and Ren, Sh., "A Versatile Thermal Biosensor," *Sensors and Actuators*, 19 (1989), pp. 53–59.

194. Hampp, N., Eppelsheim, C., Popp, J., Bisenberger, M. and Bräuchle, C., "Design and Application of Thick Film Multisensors," *Sensors and Actuators A*, 31 (1992), pp. 144–148.

195. Haemmerli, S., Schaeffler, A., Manz, A. and Widmer, H. M., "An Improved Micro Enzyme Sensor for Bioprocess Monitoring by Flow Injection Analysis," *Sensors and Actuators B*, 7 (1992), pp. 404–407.

196. Koudelka, M., Gernet, S. and de Rooij, N. F., "Planar Amperometric Enzyme-Based Glucose Microelectrode," *Sensors and Actuators*, 18 (1989), pp. 157–165.

197. Urban, G., Jobst, G., Aschauer, E., Tilado, O., Svasek, P., Varahram, M., "Performance of Integrated Glucose and Lactate Thin Film Microsensors for Clinical Analyzers," *Proc. of Int. Conf. Eurosensors VII*, Budapest (1993), pp. 306–315.

198. Kulys, J., Bilitewski, V. and Schmid, R. D., "Robust Graphite-Based Bienzyme Sensors," *Sensors and Actuators B*, 3 (1993), pp. 227–234.

199. Lee, J.-S., Nakahama, S. and Hirao, A., "A New Glucose Sensor Using Microporous Enzyme Membrane," *Sensors and Actuators B*, 3 (1991), pp. 215–219.

200. Shinohara, H., Chiba-T. and Aizawa, M., "Enzyme Microsensor for Glucose with an Electrochemically Synthesized Enzyme-Polyaniline Film," *Sensors and Actuators*, 9 (13) (1988), pp. 79–86.

201. Schuhmann, W., Lammert, R., Uhe, B. and Schmidt, H.-L., "Polypyrrole, a New Possibility for Covalent Binding of Oxireductases to Electrode Surfaces as a Base for Stable Biosensors," *Sensors and Actuators B*, 1 (1990), pp. 537–541.

202. Katsube, T., Katoh, M., Maekawa, H., Hara, M., Yamaguchi, S., Uchida, N. and Shimomura, T., "Stabilization of an FET Glucose Sensor with a Thermophilic Enzyme Glucokinase," *Sensors and Actuators B*, 1 (1990), pp. 504–507.

203. Hämmerle, M., Schuhmann, W., Schmidt, H.-L., "Amperometric Polypyrrole Enzyme Electrodes: Effect of Permeability and Enzyme Location," *Sensors and Actuators B*, 6 (1992), pp. 106–112.

204. Yon Hin, B. F. Y. and Lowe, C. R., "Amperometric Response of Polypyrrole Entrapped Bienzyme Films," *Sensors and Actuators B*, 7 (1992), pp. 339–342.

205. Schasfoort, R. B. M., Keldermans, C. E. J. M., Kooyman, R. P. H., Bergveld, P. and Greve, J., "Competitive Immunological Detection of Progesterone by Means of the Ion Step Induced Response of an ImmonoFET," *Sensors and Actuators B*, 1 (1990), pp. 368–372.

206. Gizeli, E., Goddard, N. J. and Lowe, C. R., "A Love Plate Biosensor Utilising a Polymer Layer," *Sensors and Actuators B*, 6 (1992), pp. 131–137.

207. Colapicchioni, C., Barbaro, A., Porcelli, F. and Giannini, I., "Immonoenzymatic Array Using CHEMFET Devices," *Sensors and Actuators B*, 4 (1991), pp. 245–250.

208. Papkovsky, D. B. "Luminescent Porphyrins as Probes for Optical (Bio)sensors," *Sensors and Actuators B*, 11 (1993), pp. 293–300.

209. Van den Berg, A., Koudelka, M., Van den Schoot, B. H., Verney-Norberg, E., Krebs, Ph., Grisel, A. and de Rooij, N. F., "An On-Wafer Fabricated Free-Chlorine Sensor," *Proc. Conf. Solid State Sensors and Actuators (Transducers '91)*, San Francisco, CA (1991), pp. 233–236.

210. Van den Berg, A., Grisel, A., Koudelka, M. and Van den Schoot, B. H., "A Universal On-Wafer Fabrication Technique for Diffusion Limiting Membranes for Use in Micro-electrochemical Amperometric Sensors," *Sensors and Actuators B*, 6 (1991), pp. 71–74.

211. Anzai, J. and Liu, Ch-Ch., "Potentiometric Response of PVC/Crown Ether Membrane Electrodes to Nonionic Organic Compounds," *Sensors and Actuators B*, 5 (1991), pp. 171–172.

212. Spichiger, U., Simon, W., Bakker, E., Lerchi, M., Bühlmann, Ph., Haug, J.-P., Kuratli, M., Ozawa, S. and West, S., "Optical Sensors Based on Neutral Carriers," *Sensors and Actuators B*, 11 (1993), pp. 1–8.

213. Tuzzolino, A. J., "PVDF Copolymer Dust Detectors: Particle Response and Penetration Characteristics," *Nucl. Instrum. Methods Phys. Res. A. Accel. Spectrom. Detect. Assoc. Equip.*, Vol. A 316, No. 2-3, pp. 223–237.

214. Gao, C., Wang, Z. and Gao, R., "A PVDF Film Sensor for Material Identification," *Sensors and Actuators A*, 21–23 (1990), pp. 886–889.

215. Mieszkowski, M., Leung, K. F. and Mandelis, A., "Photopyroelectric Thermal Wave Detection via Contactless Capacitive Polyvinylidene Fluoride (PVDF)-Metal Probe-Tip Coupling," *Rev. Sci. Instrum.*, 60 (3) (1989), pp. 306–316.

216. Hayashi, K., Yamanaka, M., Toko, K. and Yamafuji, K., "Multichannel Taste Sensor Using Lipid Membranes," *Sensors and Actuators B*, 2 (1990), pp. 205–213.

217. Pearce, T. C., Gardner, J. W., Friel, S., Bartlett, Ph. N. and Blair, N., "An Electronic Nose for Monitoring the Flavour of Beers," *Analyst* 118 (1993), pp. 371–377.

218. Hatfield, J. V., Hicks, P. J., James-Roxby, P., Neaves, P., Persaud, K. and Travers, P., "Towards an Integrated Electronic Nose Using Conducting Polymer Sensors," *Eurosensors VII*, Abstracts, Budapest (1993), p. 371.

219. Gerliczy, G. and Betz, R., "SOLEF® PVDF Biaxially Oriented Piezo- and Pyroelectric Films for Transducers," *Sensors and Actuators*, 112 (1987), pp. 207–223.

220. Bainerman, J., "Piezoelectrics Switch Optical Signals," *Laser Appl.*, 3 (1984), p. 49.

221. Gale, M. T. and Knop, K., "Surface Release Images for Color Reproduction," Progress Report in *Imaging Science 2*, Focal Press (1980).

222. Krongauz, V., "Photochromic Liquid Crystal Polymers," *Proc. 32 Midwest Symp. on Circ. and Syst*, Champaign, IL (1989), pp. 529–540.

223. Jansen, H. V., Elwenspoek, M., Gardeniers, J. G. E. and Tilmans, H. A. C., "Applications of Fluorocarbon Polymers in Micromechanics and Micromachining," *Eurosensors VII*, Abstracts, Budapest (1993), p. 223.

224. Bartlett, Ph. N., Chetwynd, D. G., Eastwick, V., Gardner, J. W., Harb, S. M., Smith, S. T. and Yao, Z. Q., "Electropolymerized Films for Low Friction Actuator Bearings," *Eurosensors VII*, Abstracts, Budapest (1993), p. 487.

225. Gardner, J. W., Chetwynd, D. G., Smith, S. T., Harb, S. M., Yao, Z. Q., Bartlett, Ph. N., Eastwick-Field, V., "Electropolymerized Films for Low Friction Actuator Bearings" (to be published in *Sensors and Actuators*).

TABLE S.1. Summary about the Sensing Effects, Most Important Polymers, Additives and Sensor Types in Which They Are Applied.

Sensing Effect	Polymer Examples	Typical Additives	Sensor Types
Flexibility, elasticity	PI (Kapton®), PE, Mylar®, Epoxies	—	Mechanical
Piezoresistive	PI, PVAc, PIB, PTFE, PMMA, Polyesters, Epoxies, PE, PU, PVA	Metal powder, C black, V_2O_3, PPy	Mechanical
Percolation			Temperature
Percolation + swelling			Chemical
Piezoelectric	PVDF, P(VDF-TrFE)	—	Mechanical, Acoustic
Pyroelectric			IR
Photopyroelectric			Chemical, Material ident.
Electret	PTFE, Teflon-FEP	—	Acoustic
Permittivity, Thickness and Refraction index changes	CA, PI, PEU, PS, PEG Polysiloxanes (e.g. PDMS)	Functional groups	Chemical (RH, Ions, Molecules in Gases and Liquids
Conductimetric	SPEs, ECPs, their copolymers	Salts, ionic compounds	
Potentiometric	SPEs, ECPs, their copolymers	Salts, ionic compounds	

410

TABLE S.1. (continued).

Sensing Effect	Polymer Examples	Typical Additives	Sensor Types
Potentiometric	PVC, PV(C/A/Ac), Silopren®	Plasticizers and ionophores	Chemical (RH, Ions, Molecules in Gases and Liquids
Gravimetric	CAB, PHMDS, PDMS, PE, PTFE, PCTFE, PIB, PEI, PCMS, PAPMS, PPMS, Fluoropolyol	Functional groups, supramolecular receptors	
Calorimetric	PDMS, PSDB	Catalysts	
Molecular separation	CA, PE, PTFE, PVC, PP, FEP, PCTFE, PDMS, PS, PHEMA, Mylar®, PU, PVA, Silicone rubber, PEthI	—	
Colourimetric, Fluorescence	PVP, PAA, PVC, PVI, PTFE, PS, PHEMA, PMMA, Cellulose, Epoxies, Nafion®	Dyes	
Enzyme- and immunreactions	PVC, PAA, PVA, PE, PEI, PVPy, PU, PMMA, PHEMA, Nafion®, ECPs, (e.g. PPy)	Enzymes, Antibodies	Biosensors
Piezoelectric	PVDF		Actuators
Low friction	PTFE, PPy	—	

Selected Polymers[6]

Polyethylene	$[\,-CH_2-CH_2-\,]_n$

Polyester
$$[\,-R-\overset{\overset{\displaystyle O}{\|}}{C}-O-R'-\,]_n$$

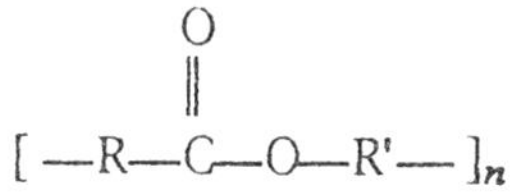

Polypropylene
$$[\,-CH_2-\underset{\underset{\displaystyle CH_3}{|}}{CH}-\,]_n$$

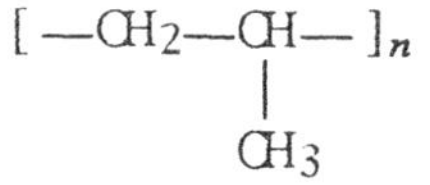

Polysiloxane
$$-\overset{|}{\underset{|}{Si}}-O-\overset{|}{\underset{|}{Si}}-$$

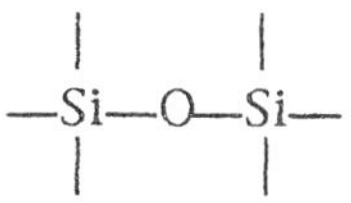

Polytetrafluoroethylene $\quad [\,-CF_2-CF_2-\,]_n$

Poly(vinylidene fluoride) $\quad [\,-CH_2-CF_2-\,]_n$

Poly(vinylidene fluoride-trifluoroethylene)
$$[\,-CH_2-CF_2-CHF-CF_2-\,]_n$$

Poly(vinyl chloride) $\quad [\,-CH_2-CHCl-\,]_n$

Poly(chloro-trifluoroethylene)
$$[\,-CF_2-CFCl-\,]_n$$

---

[6]Special sensing polymers are given in the text.

413

Poly(ethylene oxide) $[-CH_2-CH_2-O-]_n$

Poly(propylene oxide) $[-CH_2-CH-O-]_n$
$|$
CH_3

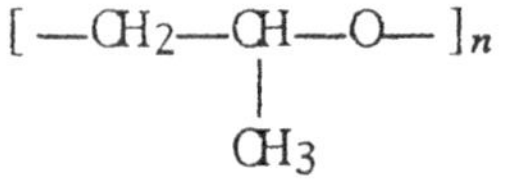

Polyisoprene (natural rubber)

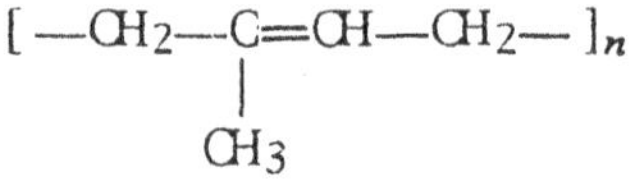

$[-CH_2-C=CH-CH_2-]_n$
$|$
CH_3

Polyacrylamide $[-CH_2-CH-]_n$
$|$
$CONH_2$

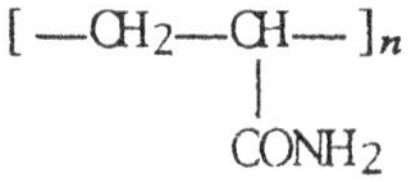

Nylon 6 $[-NH(CH_2)_5CO-]_n$

Polyurethane

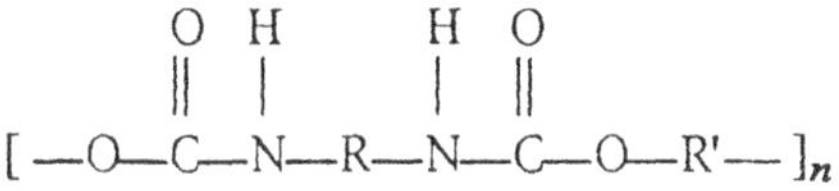

$[-O-\overset{\overset{O}{\|}}{C}-\overset{\overset{H}{|}}{N}-R-\overset{\overset{H}{|}}{N}-\overset{\overset{O}{\|}}{C}-O-R'-]_n$

Poly(vinyl alcohol) $[-CH_2-CH-]_n$
$|$
OH

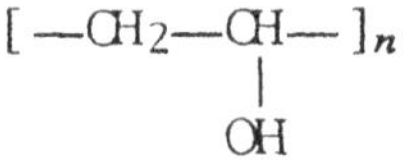

Poly(vinyl acetate) $[-CH_2-CH-]_n$
$|$
O
$|$
$C-CH_3$
$\|$
O

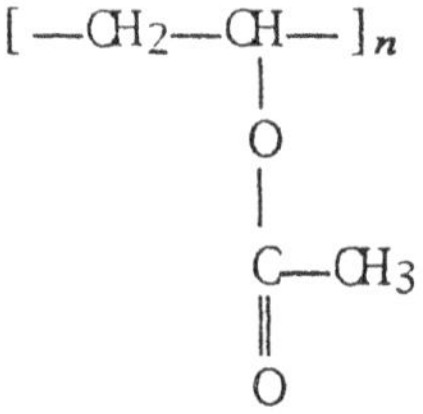

Poly(methyl methacrylate)

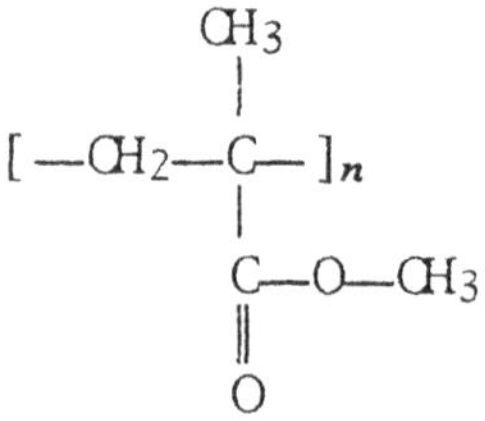

CH_3
$|$
$[-CH_2-C-]_n$
$|$
$C-O-CH_3$
$\|$
O

Silicone rubber

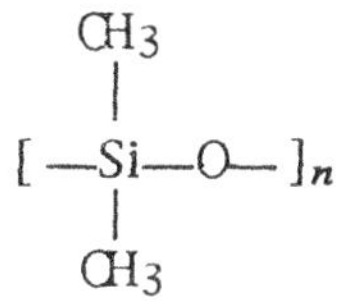

$$[\, -\!\!\underset{\underset{\text{CH}_3}{|}}{\overset{\overset{\text{CH}_3}{|}}{\text{Si}}}\!\!-\!\!\text{O}-\,]_n$$

Polystyrene

$$[\, -\text{CH}_2-\text{CH}-\,]_n$$

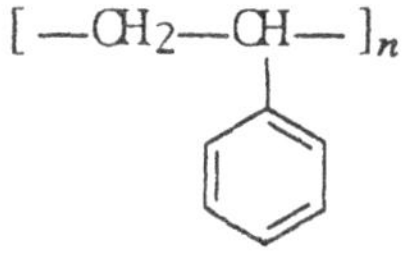

Poly(vinyl pyrrolidone)

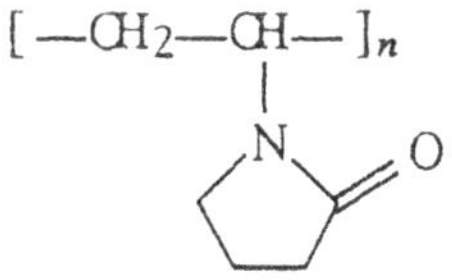

$$[\, -\text{CH}_2-\text{CH}-\,]_n$$

Cellulose acetate

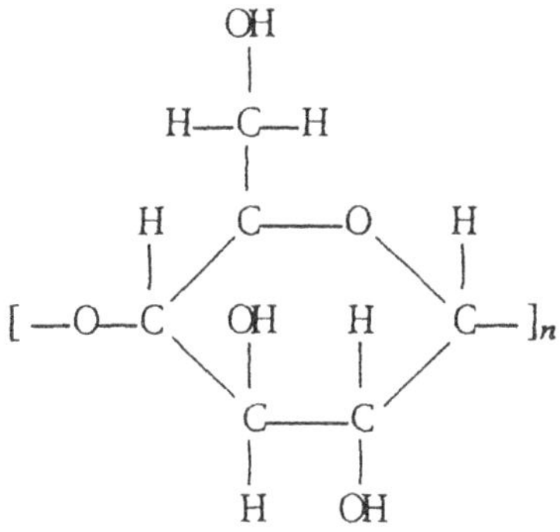

Poly(diphenyl-oxide-pyromellitimide), (Kapton®)

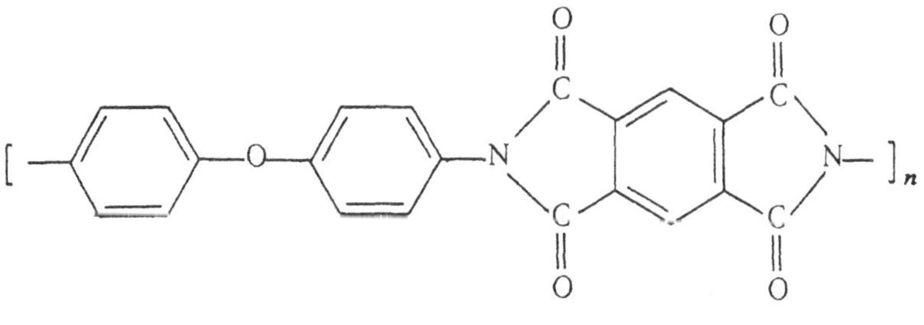

Gas Sensors Based on Heteropolysiloxane Films[7]

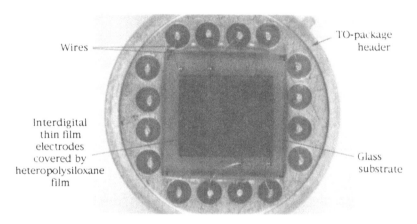

a) Interdigital capacitor gas sensor

[7]Fabricated and photos provided by the Fraunhofer Institut IFT, München.

Interdigital capacitors

TO-package header

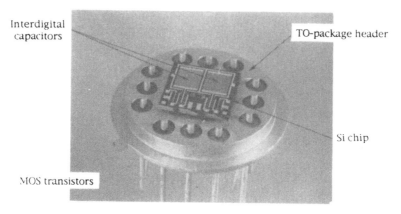

MOS transistors

Si chip

b) MOS transistors controlled by interdigital capacitors.

Humidity Sensor Based on Polymer Film[8]

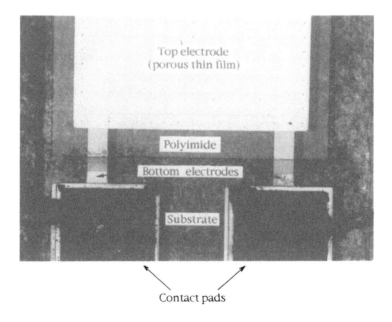

Contact pads

[8]Type HTH-01 Interbip Invest Microel Co. Ltd., Hybrid Devices Plant, 1325 Budapest, P.O.B. 183, Hungary (invented by Ms. Márta Koltai in the research project of Ms. Vera Kolonits, photo made by István Ruff).

419

AC	Alternating Current
APM	Acoustic Plate Mode
ASIC	Application Specified IC
ATP	Adenosine-5′-triphosphate
BAW	Bulk Acoustic Wave
BiCMOS	Bipolar CMOS
bpy	2,2′-Bipyridine
CA	Cellulose Acetate
CAB	Cellulose Acetate Butyrate
CCD	Charge Coupling Device
CED	Cohesive Energy Density
CERMET	Ceramic-Metal
CFT	Charge Flow Transistor
CHEMFET	Chemically Sensitive Field Effect Transistor
CMOS	Complementary MOSFETs
CVD	Chemical Vapour Deposition
DC	Direct Current
DDS	Dodecylbenzene sulphonate
DMAEMA	N,N-dimethylamino-ethyl methacrylate
DMPS	Dimethyl-polysiloxane (also PDMS)
DNA	Deoxyribonucleic acid
DOA	Dioctyl-adipate, bis(2-ethylhexyl)adipate
DOS	Dioctyl-sebacate
ECP	Electroconducting Conjugated Polymer
EDP	Ethylenediamine-pyrocathechol
ELISA	Enzyme-Linked Immunoassay
ENFET	Enzyme-Modified Field Effect Transistor

FET	Field Effect Transistor
FMN	Flavin Mononucleotide
FPW	Flexural Plate Wave
FSO	Full-Scale Output
GOD	Glucose Oxidase
HEMA	Hydroxyethyl Methacrylate
HICON	High Conductivity
HPTS	1-Hydroxy-pyrene-3,6,8-trisulphonate
HQS	Hydroquinone-sulphonate
HRP	Horseradish Peroxidase
HSA	Human Serum Albumin
HTCC	High-Temperature Cofired Ceramic
IC	Integrated Circuit
IDT	Interdigital Transducer
IGFET	Insulating Gate Field Effect Transistor
IgG	Immunoglobulin G
ImmunoFET	Immunosensitive Field Effect Transistor
IR	Infrared
ISE	Ion-Selective Electrode
ISFET	Ion-Sensitive Field Effect Transistor
LPCVD	Low-Pressure Chemical Vapour Deposition
LPE	Liquid Phase Epitaxy
LSER	Linear Solvation Energy Relationship
LTCC	Low-Temperature Cofired Ceramic
LW	Lamb Wave
MBE	Molecular Beam Epitaxy
MEMFET	Membrane-Modified Field Effect Transistor
MEMS	Microelectromechanical System
MISFET	Metal Insulator Semiconductor Field Effect Transistor
MLC	Multilayer Ceramic
MMA	Methyl-methacrylate
MOSFET	Metal Oxide Semiconductor Field Effect Transistor
NAD	Nicotinamide-adenine Dinucleotide
NADH	Reduced form of NAD
NADP	NAD-Phosphate
NaSS	Na Styrene Sulphonate
NMOS	N-channel MOSFET
NMP	N-methylphenothiazine
NPOE	2-Nitrophenyl-octyl-ether
NTC	Negative Temperature Coefficient
P^2E	Photopyroelectric
PAA	Polyacrylamide
PANi	Polyaniline

PAPA	Poly(*p*-aminophenylacetylene)
PAPMS	Poly(aminopropylmethylsiloxane)
PAPPS	(γ-Aminopropylethoxysilane/propylmethoxysilane)-co-polymer
PCB	Printed Circuit Board
PCMS	Poly(cyanopropylmethylsiloxane)
PCR	Pressure Coefficient of Resistance
PCTFE	Poly(chloro-trifluoroethylene)
P(DMDAAC)	Poly(dimethyldiallyl-ammonium chloride)
PDMS	Poly(dimethylsiloxane) (also DMPS)
PDPMPS	Poly(diphenyl/phenylmethylsiloxane)
PE	Polyethylene
PECVD	Plasma-Enhanced Chemical Vapour Deposition
PEFl	Polyethynylfluorenol
PEG	Poly(ethylene glycol)
PEI	Poly(ethylenimine)
PEO	Poly(ethylene oxide)
PET	Poly(ethylene terephthalate)
PEthI	Polyetherimide
PEU	Polyetherurethane
PFEP	Poly(fluoroethylene propylene)
PHEMA	Poly(hydroxyethyl methacrylate)
PHMDS	Poly(hexamethyldisiloxane)
PHMPTAC	Poly(2-hydoxy-3-methacryloxypropyl trimethylammonium chloride)
PI	Polyimide
PIB	Poly(isobutylene)
PIPAMS	Poly(*iso*-propylanoic-acid methylsiloxane)
PMePy	Poly(*N*-methylpyrrole)
PMeTh	Poly(methylthiophene)
PMMA	Poly(methyl methacrylate)
PMOS	P-channel MOSFET
POSFET	PVDF Oxide Semiconductor FET
PP	Polypropylene
PPA	Polyphenylacetylene
PPCN	Poly(paraphenylene azomethine)
PPD	Poly(*o*-phenylenediamine)
PPMS	Poly(phenylmethylsiloxane)
PPO	Poly(propylene oxide)
PPP	Poly(*p*-phenylene)
PPy	Polypyrrole
PPV	Poly(*p*-phenylenevinylene)
PS	Polystyrene

PSDB	Poly(styrene-divinylbenzene)
PSG	Phosphosilicate Glass
PTC	Positive Temperature Coefficient
PTF	Polymer Thick Film
PTFE	Polytetrafluoroethylene
PTh	Polythiophene
PU	Polyurethane
PVA	Poly(vinyl alcohol)
PVAc	Poly(vinyl acetate)
PVC	Poly(vinyl chloride)
PVDF	Poly(vinylidene fluoride)
P(VDF-TrFE)	Poly(vinylidene fluoride-trifluoroethylene)
PVI	Poly(*N*-vinylimide-azole)
PVP	Poly(vinyl pyrrolidone)
PVPy	Poly(vinyl pyridine)
PVS	Poly(vinyl sulphate)
PZT	Lead-zirconate-titanate
QMB	Quartz Microbalance
REFET	Reference Field Effect Transistor
RF	Radio Frequency
RH	Relative Humidity
RTV	Room Temperature Vulcanizing
SAW	Surface Acoustic Wave
SGFET	Suspended Gate FET
SH	Shear Horizontal
SIM	Surface Impedance Measurement
SLT	Sandwich Layer Transducer
SOS	Silicon on Sapphire
SPE	Solid Polymer Electrolyte
SSBW	Surface Skimming Bulk Wave
SURFET	Surface-Modified Field Effect Transistor
STP	Standard Temperature and Pressure
TCNQ	Tetracyanoquinodimethane
TCR	Temperature Coefficient of Resistance
TGDM	Triethylene Glycol Dimethacrylate
TGS	Triglicine Sulphate
TMOSFET	Thin-Film Metal Oxide Semiconductor Field Effect Transistor
TSM	Thickness Shear Mode
TTF	Tetrathiafulvalene
UV	Ultraviolet
VLSI	Very Large-Scale Integration
VPE	Vapour Phase Epitaxy

Gábor Harsányi, Ph.D., graduated in electrical engineering from the Technical University of Budapest in 1981 and received a doctorate in electronics technology from the same university in 1984. He then joined the Microelectronics Company. Since 1984, he has been a member of the research and teaching staff of the Department of Electronics Technology, Technical University of Budapest, in the position of associate professor. His main research interests are the construction and technology of integrated circuits, electronic assemblies, and microelectronic sensors. Recent key topics are reliability physics of high-density electronic interconnection systems, polymers in sensorics, and sensing film structures in fibre-optic sensors. At present, he is head of the Sensors' Laboratory, TU Budapest. In 1992, he received a Ph.D. from the Hungarian Academy of Sciences. Dr. Harsányi is 1993/94 President of the Hungarian Chapter of ISHM (International Society for Hybrid Microelectronics). He received the Best Paper of Session Award at the 1991 International Symposium on Microelectronics in Orlando, Florida, USA. He is also a member of the Technical Program Committee of ISHM – Europe. He has an international reputation in the field of microelectronics and sensorics. His publishing activity is mainly in connection with IEEE, ISHM, and Eurosensors conferences. He is a member of the editorial board of the journal *Hybrid Circuits*. He has conducted several successful research projects both for industrial companies and for government funds. His scientific research, publications, and related coordinating activities are supported by several Hungarian institutions, such as the National Committee for Technological Development (OMFB), the National Scientific Research Fund (OTKA), and the Foundation for the Hungarian Higher Education and Research. He is involved in several international projects, e.g., PHARE-ACCORD and NATO-ARW.